Advances in Fisheries Economics

Advances in Fisheries Economics

Festschrift in Honour of
Professor Gordon R. Munro

Edited by

Trond Bjørndal, Daniel V. Gordon,
Ragnar Arnason and U. Rashid Sumaila

Blackwell
Publishing

© 2007 by Blackwell Publishing Ltd

Blackwell Publishing editorial offices:
Blackwell Publishing Ltd, 9600 Garsington Road, Oxford OX4 2DQ, UK
Tel: +44 (0)1865 776868
Blackwell Publishing Professional, 2121 State Avenue, Ames, Iowa 50014-8300, USA
Tel: +1 515 292 0140
Blackwell Publishing Asia, Pty Ltd, 550 Swanston Street, Carlton, Victoria 3053, Australia
Tel: +61 (0)3 8359 1011

The right of the Author to be identified as the Author of this Work has been asserted in accordance with the Copyright, Designs and Patents Act 1988.

All rights reserved. No part of this publication may be reproduced, stored in a retrieval system, or transmitted, in any form or by any means, electronic, mechanical, photocopying, recording or otherwise, except as permitted by the UK Copyright, Designs and Patents Act 1988, without the prior permission of the publisher.

First published 2007 by Blackwell Publishing Ltd

ISBN-13: 978-1-4051-4161-1

Library of Congress Cataloguing-in-Publication Data
Advances in fisheries economics : festschrift in honour of Professor Gordon R. Munro /
edited by Trond Bjørndal ... [et al.].
p. cm.
Includes bibliographical references and index.
ISBN-13: 978-1-4051-4161-1 (hardback : alk. paper)
1. Fisheries—Economic aspects.   2. Fishery management.   I. Bjørndal, Trond.
II. Munro, Gordon Ross, 1934–
SH334.A38 2007
338.3'727—dc22

A catalogue record for this title is available from the British Library

Typeset by Bjørndal, Gordon, Arnasan and Sumaila
Printed and bound in Singapore by Markono Print Media Pte Ltd

The publisher's policy is to use permanent paper from mills that operate a sustainable forestry policy, and which has been manufactured from pulp processed using acid-free and elementary chlorine-free practices. Furthermore, the publisher ensures that the text paper and cover board used have met acceptable environmental accreditation standards.

For further information on Blackwell Publishing, visit our website:
www.blackwellpublishing.com

Contents

Preface vii

Acknowledgements viii

Tabula Gratulatoria x

Chapter 1   On the Contributions of Professor G.R. Munro to Economics   1
T. Bjørndal and D.V. Gordon

Section 1   Property Rights and Fisheries Management

Chapter 2   Phases in the Evolution of Property in Sea Fisheries   17
A. Scott

Chapter 3   Property Rights Quality and the Economic Efficiency of Fisheries Management Regimes: Some Basic Results   32
R. Arnason

Chapter 4   Resolving the Class II Common Property Problem: The Case of the BC Groundfish Trawl Fishery   59
R.Q. Grafton, H.W. Nelson and B. Turris

Chapter 5   Auctions of IFQs as a Means to Share the Rent   74
D.D. Huppert

Chapter 6   Shadow Prices for Fishing Quotas: Fishing with Econometrics   87
D. Dupont and D.V. Gordon

Section 2   Capital Theory and Natural Resources

Chapter 7   Rational Expectations and Fisheries Management   109
C.W. Clark

Chapter 8   Linking Natural Capital and Physical Capital: A Review of Natural Resource Investment Models   119
A. Charles

Chapter 9   Fisheries Management with Stock Uncertainty and Costly Capital Adjustment   137
M. Doyle, R. Singh and Q. Weninger

Section 3 Game Theory and International Fisheries

Chapter 10	The Incomplete Information Stochastic Split-Stream Model: An Overview R. McKelvey, P.V. Golubtsov, G. Cripe and K.A. Miller	159
Chapter 11	Coalition Games in Fisheries Economics M. Lindroos, V. Kaitala and L.G. Kronbak	184
Chapter 12	Incentive Compatibility of Fish-Sharing Agreements R. Hannesson	196
Chapter 13	Fish Stew: Uncertainty, Conflicting Interests and Climate Regime Shifts K.A. Miller	207
Chapter 14	A Dynamic Game on Renewable Natural Resource Exploitation and Markov Perfect Equilibrium S. Kobayashi	222

Section 4 Applied Fisheries Economics and Management

Chapter 15	The Role of the Fishing Industry in the Icelandic Economy S. Agnarsson and R. Arnason	239
Chapter 16	Factor Use and Productivity Change in a Rights-Based Fishery B.M.H. Sharp and C. Batstone	257
Chapter 17	Scientific Uncertainty and Fisheries Management W.E. Schrank and G. Pontecorvo	270
Chapter 18	Spatial-Temporal Assessment Analysis with Application to the Scotia-Fundy Herring Fishery D.E. Lane	283

Authors	304
Index	305

Preface

This *festschrift* is in honour of one of Canada's foremost economists, Professor Gordon R. Munro.

In recognition of Professor Munro's contribution to the advancement of fisheries economics and management, a group of his former students and colleagues organised a conference in his honour at the University of British Columbia, Vancouver, Canada in August of 2004. The conference was attended by fisheries economists, officials and managers from all over the world. Conference topics were ones to which Professor Munro has made major contributions over his career.

A number of the papers presented at this conference, as well as invited papers, were considered for inclusion in this *festschrift*. After a thorough referee process, the final set of papers was selected and is now presented in this book as a tribute to Professor Munro.

The editors accepted the responsibility for bringing this volume to publication because of the importance of Professor Munro's contribution to the advancement of the economics of the fishery and because of our long-standing relationship with Professor Munro as a mentor, a colleague and a friend.

We thank all those who contributed to the success of the UBC conference, in particular, Mrs. Maria Smith. We thank the authors for their contributions to this *festschrift*. The editors have benefited greatly from the assistance of numerous anonymous referees, to whom we are indebted. We thank Miss Katherine Viner and Miss Venetia Hargreaves-Allen, who served as editorial assistants. We are grateful to Laura Price and Nigel Balmford of Blackwell Publishing for editorial support.

Trond Bjørndal
Daniel V. Gordon
Ragnar Arnason
U. Rashid Sumaila

Acknowledgements

Generous financial support from the Department of Economics, the University of British Columbia; the Fisheries Economics Research Unit, Fisheries Centre; the University of British Columbia; the Department of Fisheries and Oceans, Canada; and the Social Science and Humanities Research Council of Canada, is gratefully acknowledged.

Professor Gordon R. Munro

Tabula Gratulatoria

Individuals

Elvira C. Ablaza
Sveinn Agnarsson
Peder Andersen
Claire Armstrong
Ragnar Arnason
Chris Batstone
John Beddington
Elizabeth Beravale
Trond Bjørndal
Ana Brasão
Anthony T. Charles
Villy Christensen
Colin W. Clark
Parzival Copes
Anthony Cox
Ásgeir Daníelsson
Diane Dupont
Ágúst Einarsson
Ola Flaaten
Daniel V. Gordon
Quentin Grafton
Atle G. Guttormsen
Rögnvaldur Hannesson
Knut Heen
Per Heum
Daniel Huppert
Veijo T. Kaitala
Gunnar Knapp
Shinji Kobayashi
Mª Dolores Garza Gil
Manuel Varela Lafuente

Daniel E. Lane
Audun Lem
Marko Lindroos
Ted L. McDorman
Stephanie F. McWhinnie
Kathleen A. Miller
Edward G.E. Munro
James W.R. Munro
Virginia C. Munro
Joanna and Shawn Newman
Jon Olaf Olaussen
Daniel Pauly
Peter H. Pearse
Pedro Pintassilgo
Tony J. Pitcher
Gorazd Ruseski
Per Sandberg
William E. Schrank
Anthony Scott
Basil Sharp
Massimo Spagnolo
Stein Ivar Steinshamn
U. Rashid Sumaila
Jon G. Sutinen
Louisa and Hideo Takahashi
Peter Tyedmers
Niels Vestergaard
Rolf Willmann
James A. Young

Organisations

Centre for Fisheries and Aquaculture
Management and Economics (FAME)

Centre for Fisheries Economics,
SNF

Department of Economics,
the University of British Columbia

Fisheries Department, Food and
Agriculture Organization of the
United Nations

Institute for Research in Economics
and Business Administration, Bergen

Ministry of Fisheries,
Iceland

Project Seahorse

Social Sciences and Humanities
Research Council of Canada

1

On the Contributions of Professor G.R. Munro to Economics

Trond Bjørndal
Daniel V. Gordon

Professor Gordon R. Munro was born in Vancouver, BC in 1934 and received his BA degree from the University of British Columbia (UBC) in 1956. The years 1956 to 1962 were spent at Harvard University where he received both the AM and PhD degrees. Dr. Munro joined the Department of Economics at the University of British Columbia in 1962 as an Assistant Professor, being promoted to Associate Professor in 1968 and Professor in 1979. He also became an associate of the UBC Fisheries Centre. Professor Munro remained at UBC for 37 years, before retiring in 1999. Professor Munro maintains the rank of Professor Emeritus at UBC.

CAPITAL THEORY AND FISHERIES

Professor Munro, by his academic research, his teaching of a generation of fisheries economists and his policy advisory work has been instrumental in identifying, promoting and introducing improved fisheries management around the world and in Canada in particular. His work in fisheries began in the early 1970s, when he agreed to participate in a multi-disciplinary fisheries research project at the Universiti Sains Malaysia, in Penang, where he had been posted under a Canadian International Development Agency project. A complex policy issue, which the project raised, led him to seek out the assistance of Professor Colin W. Clark, Department of Mathematics, UBC, who was to become a long time colleague and friend. Their joint work on this policy problem led ultimately to the 1975 *Journal of Environmental Economics and Management* (JEEM) article: 'The economics of fishing and modern capital theory: A simplified approach'. This was an important article, both for Professor Munro, in that it was to provide the foundation for everything that was to follow in his work in fisheries economics, and for the fisheries economics discipline, in that this was the first application of modern (i.e. post-1950s) capital theory to the problem of the fishery.

   Professor Anthony D. Scott, in his famous 1955 *Journal of Political Economy* article: 'The fishery: The objectives of sole ownership', pointed to the importance of attempting to set fisheries economics within a capital theoretic framework.

Nonetheless, fisheries economics remained firmly embedded within a static framework for the next 20 years. The Clark and Munro 1975 *JEEM* article, which was a very effective fusion of Professor Clark's mathematical expertise and Professor Munro's economics expertise, did in fact achieve the goal that Professor Scott had set forth 20 years earlier.

The key reason that the article succeeded in achieving this goal is because it brought forth a resource investment rule, which proved to be readily comprehensible, and intuitively appealing, to economists. Some natural resource/environmental economics textbooks, e.g. David Pearce and R. Kerry Turner, in their 1990 book, *Economics of Natural Resources and the Environment*, refer to the equation as the Fundamental Rule (Equation) of Renewable Resource Exploitation. Professor Clark was to use the rule many times in the presentation of his book, *Mathematical Bioeconomics*.

The rule (equation), which can be expressed in several different forms, most commonly appears as:

$$F'(x^*) + \frac{\left(\partial \pi / \partial x^*\right)}{\left(\partial \pi / \partial h\right)|_{h=F(x^*)}} = \delta \quad (1.1)$$

where x denotes the biomass, $F(x)$ the growth of the resource with $F'(x)$ marginal growth, $\pi = \pi(x, h)$ the flow of net economic benefits from the fishery (resource rent) as a function of biomass and harvest (h), and δ the social rate of discount. The rule can be seen as a Modified Golden Rule of Capital Accumulation and represents an implicit expression for x^*, the optimal stock level.

As some comfort to junior academics struggling to publish their work, Professor Munro tells us that he and Professor Clark had great difficulty in publishing their article. He credits Professor Scott, who was quick to see the significance of the resource investment rule, for valuable assistance and support in getting the paper published.

The capital theoretic analysis of fisheries economics was advanced in a major way in the Clark, Clarke and Munro 1979 *Econometrica* article 'The optimal management of renewable resource stocks: Problems of irreversible investment'. This article deals with problems arising from the fact that much non-resource capital used in the fishery (fleet, processing, human) is not readily shiftable out of the fishery, i.e. non-malleable. The existence of such non-malleable capital is of substantial significance in real world fisheries management.

There were numerous other extensions to the theory by a number of authors, and Professor Munro brought all of this together in a mid-1980s 'state of the art' paper, co-authored with Professor A.D. Scott: 'The economics of fisheries management', which appeared in the North-Holland Handbook series, *Handbook of Natural Resource and Energy Economics*, Vol. II. The survey paper was further updated (co-authored with Professor Trond Bjørndal) and published in *The International Yearbook of Environmental and Resource Economics 1998/1999*.

GAME THEORY AND SHARED FISH STOCKS

One of Professor Munro's major fields of interest, before moving into fisheries in the early 1970s, was international economics. His ongoing interest in international affairs carried over into fisheries. Shortly after Professor Munro became involved in fisheries, the UN Third Conference on the Law of the Sea, commenced. The Conference was to have a revolutionary impact upon the management of international fisheries, by leading to the regime of 200 mile Exclusive Economic Zones (EEZs). Given Professor Munro's interests in international affairs, he soon was drawn to fisheries management issues arising from the new Law of the Sea.

Professor Munro is not a man to enter a new area without preparation, and in 1980 he enrolled in a short, but very intensive, course on law for economics professors at the University of Miami Law and Economics Centre. Professor Munro argues that, while the course certainly did not turn him into a lawyer, it did enable him to talk sensibly with legal experts, which was to prove invaluable when he devoted his attentions to EEZ-related fisheries management issues.

Professor Munro first published on EEZ-related issues in 1977, and proceeded to publish a series of papers on these issues through the Law of the Sea Institute conference volumes. He served on the Board of the Institute from 1980 to 1986.

In 1976-77, he came, almost by chance, upon an EEZ management issue, which was genuinely international, namely that of the management of fish stocks shared between and among neighbouring coastal states. In trying to deal with this issue, he realised that, in order to make any sort of progress, it would be necessary to apply game theory. He was given guidance and inspiration in this direction by a then recently published article co-authored by Robert Pindyck of MIT, which applied the theory of cooperative games to the question of OPEC's optimal pricing policy.

Professor Munro's work on this issue led to his 1979 *Canadian Journal of Economics* article, 'The optimal management of transboundary renewable resources', which was later to be awarded the Harry Johnson Prize by the Canadian Economics Association. The article represents, to the best of our knowledge, the first application of game theory to fisheries issues.

While this article provided the basis for all of the subsequent work that Professor Munro was to do on this topic, and may be the article for which he is best known, of his articles on shared fish stock management, the one that is, in fact, most commonly read, is the 1987 *Marine Resource Economics* article, 'The management of shared fishery resources under extended jurisdiction'. The 1987 article is technically considerably less demanding than the 1979 article.

Professor Munro's Finnish colleague, Professor Veijo Kaitala, led off from Professor Munro's 1979 article and further developed the game theoretic analysis of shared fish stock management. Professor Kaitala is really an applied mathematician, who developed an interest in economics. Professors Kaitala and Munro began collaborating seriously in the early 1990s. Together, they addressed the complex and demanding issue of the management of shared fish stocks that are to be found both within the coastal state EEZ and the adjacent high seas - straddling and highly migratory stocks. The best example of this collaborative work is their 1997 *Natural Resource Modeling* article, 'The conservation and management of high seas fishery resources'.

Professor Munro has continued to work on shared fish stock management issues right up until the present day. He has co-authored an FAO Fisheries Technical Paper, with Rolf Willmann and Annick van Houtte (both from the FAO), on *The Conservation and Management of Shared Fish Stocks*: *Legal and Economic Aspects*, published in 2004. The paper is an application of game theory to real world issues. In 2000, he authored a *Marine Resource Economics* article, 'The United Nations fish stock agreement of 1995', and co-authored an *Annals of Operation Research* article on the topic. In 2004 he prepared with Trond Bjørndal a survey article on high seas fisheries management and UN Fish Stocks Agreement, which appeared in *The International Yearbook of Environmental and Resource Economics 2003/2004*.

In examining the management of shared fish stocks, Professor Munro, not surprisingly, devoted quite a bit of attention to the Canada-US Pacific Salmon Treaty. He first published a paper on this topic in 1989, with Robert Stokes, an American economist. He then began working with two more Americans, Kathleen Miller, an economist, and Robert McKelvey, an applied mathematician with a strong interest in game theory. The three of them, with two more Canadians, Ted McDorman (a professor of law) and Peter Tydemers (an ecologist), published a piece on the Treaty in the *Canadian-American Public Policy Occasional Papers Series*, in 2001. Kathleen Miller and Professor Munro then published an article, which arises from this work, 'Climate and cooperation: A new perspective on the management of shared fish stocks', *Marine Resource Economics,* 2004.

INTERNATIONAL FISHERIES CO-OPERATION

Although much of Professor Munro's work on co-operative fisheries management is theoretical, for a decade, from the mid-1980s to the mid-1990s, he was associated with the tripartite (government, universities, private sector) Pacific Economic Co-operation Council (PECC). Within the PECC, he led a Fisheries Task Force, the main work of which consisted of fostering co-operation between and among regional groups of Pacific developing coastal states, under the new Law of the Sea. The groups were the coastal states of south-east Asia, the South Pacific and Pacific Latin America. This represented a genuinely practical application of Professor Munro's knowledge of international fisheries cooperation.

The fishery resource, which these three regions did have, and do have, in common is tropical tuna. Professor Munro published an article in 1990, arising from his work with the PECC, in *Ocean Development and International Law*, a journal, which, although being basically a law journal, accepts articles from economists. The article, 'Extended fisheries jurisdiction and the management of highly migratory species', was well received by both economists and lawyers.

A secondary EEZ-regime issue, which Professor Munro examined, was that of relations, between and among, coastal states and distant water fishing nations. This work led to two articles co-authored with Francis Clarke, 'Coastal states, distant water fishing nations and extended jurisdiction: A principal-agent analysis', and 'Coastal states and distant water fishing nations: Conflicting views of the future' which appeared in *Natural Resource Modeling*, in 1987 and 1991. The articles involve the application of principal-agent analysis, along with international trade theory.

OTHER CONTRIBUTIONS

Over the years Professor Munro has been supervisor or co-supervisor for close to 20 PhD students at UBC, who have done their dissertations in various areas of fisheries management. In addition, he has been external PhD examiner for several students at many universities. Through his own students and, in turn, through their students, he has had and does have a substantial impact on the development of the discipline. In addition, he has had a substantial influence through the work he has done with colleagues. As an example of that, much research in the field of game theoretic applications to fisheries at centres such as Helsingfors and Lisbon is directly or indirectly attributable to Professor Munro. In recognition of his contributions, the International Institute of Fisheries Economics and Trade (IIFET) presented Professor Munro with its Distinguished Service Award in 2004.

Brief references were made above to the international work undertaken by Professor Munro in different countries. To recognise his efforts, the Government of Peru made him a Grand Officer in the Order of Merit for Distinguished Service, in October 1988. He was also decorated by the Government of Chile, where he was made a Commander in the Order of Bernardo O'Higgins the Liberator, in May 1990.

Retirement for Professor Munro has meant a reprieve from university administrative duties and an increase in research time. He is currently working in the areas of subsidies and excess capacity. Professor Munro co-authored several papers in this area with Colin Clark and Rashid Sumaila, appearing in the journals *Fish and Fisheries, Bulletin of Marine Science,* both in 2002, and in an FAO volume in 2003. The three have recently published an article entitled 'Subsidies, buybacks and sustainable fisheries', in the *Journal of Environmental Economics and Management,* 2005. Finally, they have a paper on this topic that is forthcoming in the *Proceedings of the 4th World Fisheries Congress.*

Professor Munro has held numerous positions as research professor at research institutions, universities and government agencies. It is worth noting a few of these positions: Visiting Professor, School of Comparative Social Sciences, Universiti Sains Malaysia, Penang, Malaysia, May 1970-August 1972; Distinguished Research Fellow, Centre for Fisheries Economics, SNF, Bergen, Norway, 1997-2000, 2003-5, and Visiting Expert with the Food and Agriculture Organization of the UN (FAO), under the FAO Programme of Cooperation with Academic and Research Institutions, 1997, 2001, 2002 and 2003. In addition, he has undertaken consultation work for the Department of Fisheries and Oceans (Canada), the Organisation for Economic Co-operation and Development (OECD), and has served on panels established by the National Academy of Sciences (USA), and the Royal Society of Canada.

Gordon R. Munro Publications

Books

Trade Liberalization and a Regional Economy: Studies of the Impact of Free Trade. University of Toronto Press, 1971 (with R.A. Shearer and J.H. Young).

The Economics of Fishing and the Developing World: A Malaysian Case Study. University Sains Malaysia Press, 1978 (with K.L. Chee).

A Promise of Abundance: Extended Fisheries Jurisdiction and the Newfoundland Economy. Economic Council of Canada, Ottawa. 1980.

The Northern Cod Fishery of Newfoundland. Economic Council of Canada, Technical Report No. 18, The Public Regulation of Commercial Fisheries in Canada, Ottawa 1981 (with S. McCorquodale).

Canadian Oceans Policy: National Strategies and the New Law of the Sea. University of British Columbia Press, 1989 (co-editor, D. McRae).

Fisheries and Uncertainty: A Precautionary Approach to Resource Management. University of Calgary Press, 1996 (co-editor, D.V. Gordon).

Canadian Marine Fisheries in a Changing and Uncertain World: Report Prepared for the Canadian Global Change Program of the Royal Society of Canada. National Research Council Press, Ottawa, 1999 (with B. de Young, R. Peterman, A. Dobell, E. Pinkerton, Y. Breton, A. Charles, M. Fogarty and C. Taggart).

Articles and Chapters

British Columbia and the Quest for Free Trade. In J.E. Currie (ed.) *World Trade and the Citizen.* University of British Columbia Press, 1967.

British Columbia's Stake in Free Trade. In R.A. Shearer (ed.) *Exploiting Our Economic Potential: Public Policy and the British Columbia Economy.* Holt, Reinhart and Winston, 1968.

The Malaysian Balance of Payments to 1980: Part I. Reprinted in D. Lim (ed.) Readings *in Malaysian Economic Development.* Oxford University Press, 1975.

The Malaysian Balance of Payments to 1980: Part II. Reprinted in D. Lim (ed.) Readings *in Malaysian Economic Development.* Oxford University Press, 1975.

The Economics of Fishing and Modern Capital Theory: A Simplified Approach. *Journal of Environmental Economics and Management*, 2: 92-106, 1975 (with C.W. Clark). Reprinted in L.J. Mirman and D.F. Spulber (eds) *Essays in the Economics of Renewable Resources.* Elsevier North Holland, Amsterdam, 1982.

Canada and Fisheries Management with Extended Jurisdiction. In L.G. Anderson (ed.) *The Economic Impacts of Extended Jurisdiction*. Ann Arbor Science Publishers, Ann Arbor, MI, 1977.

North America, Extended Jurisdiction and the Northwest Atlantic. In E. Miles and J.K. Gamble, Jr. (ed.) *Law of the Sea: Conference Outcomes and Problems of Implementations*. Ballinger, Cambridge, MA, 1977.

Canada and Extended Fisheries Jurisdiction in the Northeast Pacific: Some Issues in Optimal Resource Management. In J.R. Albers (ed.) *Proceedings of the Fifth Pacific Regional Science Conference*. W. Washington University Press, 1978.

Extended Fisheries Jurisdiction in a Regional Setting: Problems of Conflicting Goals and Interests. In D. Johnson (ed.) *Regionalization of the Law of the Sea*, Ballinger, Cambridge, MA. 1978.

Extended Fisheries Jurisdiction and International Cooperation. *International Perspectives* March/April: 12-8, 1978.

Renewable Resources and Extinction. *Journal of Environmental Economics and Management* 5: 198-205, 1978 (with C.W. Clark).

The Optimal Management of Renewable Resource Stocks: Problems of Irreversible Investment. *Econometrica* 47: 25-47, 1979 (with C.W. Clark and F.H. Clarke).

The Optimal Management of Transboundary Renewable Resources. *Canadian Journal of Economics* 3: 271-96, 1979.

Fisheries and the Processing Sector: Some Implications for Management Policy. *The Bell Journal of Economics* 11: 603-16, 1980 (with C.W. Clark).

Extended Fisheries Jurisdiction and the Distant Water Nations: The Coastal State Perspective. In P. N. Nemetz (ed.) *Resource Policy: International Perspectives*. The Institute for Research on Public Policy, Montreal, 1980.

The Economics of Fishing: An Introduction. In J.A. Butlin (ed.) *Economics and Resource Policy*. Longman, London, 1981.

Fisheries, Extended Jurisdiction and the Economics of Common Property Resources. *Canadian Journal of Economics* 15: 405-25 1982.

Bilateral Monopoly in Fisheries and Optimal Management Policy. In L.J. Mirman and D. F. Spulber (eds) *Essays in the Economics of Renewable Resources*. Elsevier North Holland, Amsterdam, 1982.

Co-operative Fisheries Arrangements between Pacific Coastal States and Distant Water Nations. In H.E. English and A. Scott (eds) *Renewable Resources in the Pacific*. The International Development Research Centre, Ottawa, 1982.

Foreign Access to EEZs and the Derivation of Coastal State Benefits: Methods and Techniques. In *Report of the Expert Consultation on the Conditions of Access to the Fish Resources of the Exclusive Economic Zone*, Food and Agricultural Organization, Rome, 1983.

The Taking of Living Resources: Expectations and Reality. In A.W. Koers and B.H. Oxman (eds) *The 1982 Convention on the Law of the Sea*. Law of the Sea Institute, Honolulu, 1984 (with G. Pontecorvo).

Fisheries, Dynamics and Uncertainty. In A.D. Scott (ed.) *Progress in Natural Resource Economics*. Oxford Press, Oxford, 1985 (with A.T. Charles and C.W. Clark).

Canada, International Trade and the Economics of the Co-operative Fisheries Arrangements. In *Canada and International Trade* 2: 811-51 Institute for Research on Public Policy, Ottawa, 1985.

Irreversible Investment and Optimal Fisheries Management: A Stochastic Analysis. *Marine Resource Economics* 1: 247-64, 1985 (with A.T. Charles).

The Economics of Fisheries Management. In A.V. Kneese and J.L. Sweney (eds) *Handbook of Natural Resource and Energy Economics*, Vol. II. North Holland, Amsterdam, 1985 (with A.D. Scott).

Coastal States, Distant Water Fleets and Extended Fisheries Jurisdiction: Some Long Run Considerations. *Marine Policy* 9: 2-15, 1985.

The Economics of Coastal State-Distant Water Nation Co-operative Arrangements. In *Proceedings of the Second Conference of the International Institute of Fisheries Economics and Trade,* Vol. I. International Institute of Fisheries Economics and Trade, Corvallis, 1985.

Coastal States, Distant Water Fishing Nations and Extended Jurisdiction: A Principal-Agent Analysis. *Natural Resource Modeling* 2: 81-107 1987 (with F.H. Clarke).

The Management of Shared Fishery Resources Under Extended Jurisdiction. *Marine Resource Economics* 3: 271-96 1987.

Developing Coastal States, Extended Fisheries Jurisdiction, and Pacific Economic Cooperation. *Proceedings of the Third Biennial Conference of the International Institute of Fisheries Economics and Trade on Fisheries Trade, Development and Policies*. GERMA, Rimouski, 1987.

International Cooperation for Resource Management: Fisheries. *Foreign Relations*, IV: 38-84, 1989.

Coastal State Rights within the 200 Mile Exclusive Economic Zone. In P. Neher, R. Arnason and N. Mollett (eds) *Rights Based Fishing*. Kluwer Academic, Dordrecht, 1989 (with D.M. McRae).

The Pacific Islands, the Law of the Sea and Pacific Tropical Tuna. In H. Campbell, K. Menz and G. Waugh (eds) *The Economics of Fisheries Management in the Pacific Islands Region*. Australian Centre for International Agricultural Research, Canberra, 1989.

The Canada-United States Pacific Salmon Treaty. In D. McRae and G. Munro (eds) *Canadian Oceans Policy: National Strategies and the New Law of the Sea*. University of British Columbia Press, 1989 (with R.L. Stokes).

Coastal State-Distant Water Fishing Nation Relations. *Marine Fisheries Review* 5: 3-10, 1989.

Extended Jurisdiction and the Management of Pacific Highly Migratory Species. *Ocean Development and International Law* 21: 289-307, 1990.

The Optimal Management of Transboundary Fisheries: Game Theoretic Considerations. *Natural Resource Modeling* 4: 403-26, 1990.

Differential Games and the Optimal Management of Transboundary Fisheries. In R.P. Hamalainen and H.K. Ehtamo (eds) *Dynamic Games in Economic Analysis*. Springer-Verlag, Berlin, 1991.

The Management of Transboundary Fishery Resources: A Theoretical Overview. In R. Arnason and T. Bjørndal (eds) *Essays on the Economics of Migratory Fish Stocks*, Springer-Verlag, Berlin, 1991.

The Management of Migratory Fish Resources in the Pacific: Tropical Tuna and Pacific Salmon. In R. Arnason and T. Bjørndal (eds) *Essays on the Economics of Migratory Fish Stocks*. Springer-Verlag, Berlin, 1991.

Coastal States and Distant Water Fishing Nations: Conflicting Views of the Future. *Natural Resource Modeling* 5: 345-69, 1991 (with F.H. Clarke).

Evolution of Canadian Fisheries Management Policy Under the New Law of the Sea: International Dimensions. In C. Cutler and M. Zacher (eds) *Canadian Foreign Policy and International Economic Regimes*. University of British Columbia Press, 1992.

Mathematical Bioeconomics and the Evolution of Modern Fisheries Economics. *Bulletin of Mathematical Biology* 54: 163-84, 1992.

The Economic Management of Canada's Pacific Resources and the Prospects for Decentralized Control. In *Japan Expert Consultation on the Development of Community-Based Coastal Fishery Management Systems for Asia and the Pacific.* Food and Agriculture Organization, Rome, 1993.

Fishery Diplomacy in the 1990s: The Challenges and Constraints. In K.J. Matris and T.L. McDorman (eds) *SEAPOL International Workshop on Challenges to Fishery Policy and Diplomacy in South-East Asia.* South-East Asian Programme in Ocean Law, Policy and Management, Bangkok, 1993.

Environmental Cooperation Among Pacific Developing States: A Fisheries Case Study. *University of British Columbia Law Review* 27: 201-12, 1993.

Coastal States and Distant Water Fleets Under Extended Jurisdiction: The Search for Optimal Incentive Schemes. *Annals of the International Society of Dynamic Games* 1: 301-17, 1994.

Renewable Resources as Natural Capital: The Fishery. In A. Jansson, C. Folke, M. Hammer and R. Costanza (eds) *Investing in Natural Capital.* Island Press, Washington, 1994 (with C.W. Clark).

The Management of High Seas Fisheries. *Marine Resource Economics* 8: 313-29, 1993 (with V. Kaitala).

The Economic Management of High Seas Fishery Resources: Some Game Theoretic Aspects. In C. Carraro and J. Filar (eds) *Game Theoretic Models of the Environment.* Birkhauser, Boston, 1995 (with V. Kaitala).

The Management of Tropical Tuna Resources in the Western Pacific: Trans-Regional Cooperation and Second Tier Diplomacy. In G. Blake, W. Hidesley, M. Pratt, R. Ridely and C. Schofield (eds) *The Peaceful Management of Transboundary Resources.* International Boundary Research Unit, Graham & Trotman/Martinus Nyhoff, London, 1995.

The Management of High Seas Fishery Resources. In D. S. Liao (ed.) *International Cooperation for Fisheries and Aquaculture Development: Proceedings of the 7th Biennial Conference of the International Institute of Fisheries Economics and Trade*, Vol. 1. Keelung, National Taiwan Ocean University, 1995.

The Management of Transboundary Resources and Property Rights Systems: The Case of Fisheries. In S. Hanna and M. Munasinghe (eds) *Property Rights and the Environment.* International Bank for Reconstruction and Development, Washington, 1995 (with V. Kaitala).

Non-Tariff Barriers to Fisheries Trade: A Survey. *INFOFISH International* 3: 14-8, 1995.

Institutional Change and the Management of British Columbia Fishery Resources. In J.R. Robinson, D. Cohen and A.D. Scott (eds) *Institutions for Sustainable Development of Natural Resources in British Columbia*. University of British Columbia Press, 1995 (with P. Neher).

Approaches to the Economics of the Management of High Seas Fishery Resources. *Canadian Journal of Economics* XXIX: S157-64, 1996.

Approaches to the Economics of the Management of High Seas Fishery Resources. In D.V. Gordon and G.R. Munro (eds) *Fisheries and Uncertainty: A Precautionary Approach to Resource Management*. University of Calgary Press, 1996.

Fishery Resources of the Pacific and International Cooperation. In H. E. English and D. Runnalls (eds) *Environmental Development in the Pacific: Problems and Policy Options*. Addison Wesley Longman, Melbourne, 1997.

The Management of Transboundary Fishery Resources and Property Rights. In B. L. Crowley (ed.) *Taking Ownership: Property Rights and Fishery Management on the Atlantic Coast*. Atlantic Institute for Market Studies, Halifax, 1997.

The Conservation and Management of High Seas Fishery Resources. *Natural Resource Modeling* 10: 87-108, 1997. (with V. Kaitala).

Implementing the Precautionary Approach in Fisheries Management Through Marine Reserves. *Ecological Applications* 8: 572-78, 1998 (with T. Lauck, C.W. Clark and M. Mangel).

Individual Transferable Quotas: Community Based Fisheries Management Systems and 'Virtual' Communities. *Fisheries* 23(3): 12-5, 1998 (with N. Bingham and E. Pikitch).

Transboundary Fishery Resources and the Canada-United States Pacific Salmon Treaty. *Canadian-American Public Policy Occasional Papers Series*, No. 33. University of Maine, The Canadian-American Center, 1998 (with T. McDorman and R. McKelvey).

The Economics of Fisheries Management: A Survey. In Tom Tietenberg and Henk Folmer (eds) *The International Yearbook of Environmental and Resource Economics 1998/1999*. Edward Elgar, Cheltenham, 1998 (with T. Bjørndal).

The Economics of Overcapitalisation and Fishery Resource Management: A Review. In Aaron Hatcher and Kate Robinson (eds) *Overcapacity, Overcapitalisation and Subsidies in European Fisheries*. CEMARE, University of Portsmouth, Portsmouth, 1999.

Current International Work on Subsidies in Fisheries. In Aaron Hatcher and Kate Robinson (eds) *Overcapacity, Overcapitalization and Subsidies in European Fisheries.* CEMARE, University of Portsmouth, Portsmouth, 1999 (with R. Steenblik).

Overcapitalisation and Excess Capacity in World Fisheries: Underlying Economics and Methods of Control. In *Managing Fishing Capacity: Selected Papers on Underlying Concepts and Issues*. FAO Fisheries Technical Paper No. 386, Food and Agriculture Organization of the United Nations, Rome, 1999 (with D. Gréboval).

An Economic Review of the UN Agreement of 1995. In T. Bjørndal, G. Munro and R. Arnason (eds) *Proceedings from the Conference on the Management of Straddling Fish Stocks and Highly Migratory Fish Stocks, and the UN Agreement*. Centre for Fisheries Economics, Bergen, Paper on Fisheries Economics, 38, 2000.

The Management of High Seas Fisheries. *Annals of Operations Research* 94: 183-96, 2000 (with T. Bjørndal, V. Kaitala and M. Lindroos).

On the Economic Management of 'Shared Fishery Resources'. In A. Hatcher and D. Tingley (eds) *International Relations and the Common Fisheries Policy*. CEMARE, University of Portsmouth, Portsmouth, 2000.

The United Nations Fish Stocks Agreement of 1995: History and Problems of Implementation. *Marine Resource Economics* 15: 265-80, 2000.

Climate Uncertainty and the Pacific Salmon Treaty: Insights on the Harvest Management Game. In R. Johnston and A. Shriver (eds) *Proceedings of the Tenth Biennial Conference of the International Institute of Fisheries Economics and Trade*. Corvallis, International Institute of Fisheries Economics and Trade, 2001. (with K. Miller, R. McKelvey and P. Tyedmers).

The Management of High Seas Fisheries and the Implementation of the UN Agreement of 1995: Problems and Prospects. In R. Johnston and A. Shriver (eds) *Proceedings of the Tenth Biennial Conference of the International Institute of Fisheries Economics and Trade.* Corvallis, International Institute of Fisheries Economics and Trade, 2001 (with T. Bjørndal).

The 1999 Pacific Salmon Agreement: A Sustainable Solution? *Canadian–American Public Policy Occasional Paper Series*, No. 47. The Canadian-American Center, University of Maine, Orono, 2001 (with K. Miller, T. McDorman, R. McKelvey and P. Tydemers).

The Effect of Introducing Individual Harvest Quotas Upon Fleet Capacity in the Marine Fisheries of British Columbia. In Ross Shotten (ed.) *Case Studies on the*

Effect of Transferable Fishing Rights on Fleet Capacity and Concentration of Ownership, FAO Fisheries Technical Paper 412. Food and Agriculture Organization of the United Nations, Rome, 2001.

The Problem of Overcapacity. *Bulletin of Marine Science* 70: 473-84, 2002 (with C.W. Clark).

The Impact of Subsidies Upon Fisheries Management and Sustainability: The Case of the North Atlantic. *Fish and Fisheries* 3: 233-90, 2002 (with U.R. Sumaila).

The Management of High Seas Fishery Resources and the Implementation of the UN Fish Stocks Agreement of 1995. In H. Folmer and T. Tietenberg (eds) *The International Yearbook of Environmental and Resource Economics 2003/2004.* Edward Elgar, Cheltenham, 2003 (with T. Bjørndal).

On the Management of Shared Fish Stocks. In Food and Agriculture Organization of the UN. *In Papers Presented at the Norway-FAO Expert Consultation on the Management of Shared Fish Stocks, Bergen, Norway, 7-10 October 2002.* Food and Agriculture Organisation Fisheries Report No. 695, Supplement. FAO, Rome, 2003.

The Management of Shared Fish Stocks and High Seas Fisheries: The Case of the Mediterranean. In *Conference on Fishery Management and Multi-Level Decisional Systems: The Mediterranean Case.* Istituto Richerche Economiche Per La Pesca e L'Acquacoltura, Salerno, 2003.

Fishing Capacity and Resource Management Objectives. In S. Pascoe and D. Gréboval (eds) *Measuring Capacity in Fisheries*, Food and Agriculture Organization, Fisheries Technical Paper 445. FAO, Rome 2003 (with C.W. Clark).

On the Management of Shared Fish Stocks: Critical Issues and International Initiatives to Address Them. In A.I.L. Payne, C.M. O'Brien and S.I. Rogers (eds) *Management of Shared Fish Stocks*. Blackwell, Oxford, 2004 (with R. Willmann and K. Cochrane).

The Whole Could be Greater Than the Sum of the Parts: The Potential Benefits of Cooperative Management of the Caribbean Spiny Lobster. In A.I.L. Payne, C.M. O'Brien and S.I. Rogers (eds) *Management of Shared Fish Stocks*. Blackwell, Oxford, 2004 (with K. Cochrane and B. Chakalall).

Climate and Cooperation: A New Perspective on the Management of Shared Fish Stocks. *Marine Resource Economics* 19: 367-93, 2004 (with K. Miller).

The Conservation and Management of Shared Fish Stocks: Legal and Economic Aspects, Food and Agriculture Organization, Fisheries Technical Paper 465. FAO, Rome 2004 (with A. van Houtte and R. Willmann).

Climate, Competition and the Management of Shared Fish Stocks. In *Proceedings of The Twelfth Biennial Conference of the Institute of Fisheries Economics and Trade: What are Responsible Fisheries?* International Institute of Fisheries Economics and Trade, Corvallis, 2004 (with K. Miller and T. Bjørndal).

Subsidies, Buybacks and Sustainable Fisheries. *Journal of Environmental Economics and Management* 50: 47-58, 2005 (with C.W. Clark and U.R. Sumaila).

The Incentive-Based Approach to Sustainable Fisheries. *Canadian Journal of Fisheries and Aquatic Sciences.* Forthcoming (with R.Q. Grafton, R. Arnason, T. Bjørndal, D. Campbell, H.F. Campbell, C.W. Clark, R. Connor, D.P. Dupont, R. Hannesson, R. Hilborn, J.E. Kirkley, T. Kompas, D.E. Lane, S. Pascoe, D. Squires, S.I. Steinshamn, B.R. Turris and Q. Weninger).

Decommissioning Schemes and Their Implications for Effective Fisheries Management. In J. Nielson (ed.) *Reconciling Fisheries with Conservation: Proceedings of the 4th World Fisheries Conference.* American Fisheries Society, Bethesda. Forthcoming (with C.W. Clark and U.R. Sumaila).

Section 1

Property rights and fisheries management

2

Phases in the Evolution of Property in Sea Fisheries

Anthony Scott

It takes a great deal of history to produce a little literature
— Henry James

2.1 INTRODUCTION

Four hundred years ago the great lawyer Grotius declared that because offshore fisheries were inexhaustible there should be international freedom of access.[1] Apparently there was little disagreement. Roughly speaking, for the 400 years after the English Civil War, those who thought at all about the public right of fishing or the freedom of the seas felt secure in taking no action to conserve the sea fisheries. In the 1880s that confidence was re-stated by the great scientist Huxley.[2] But by that time some fisheries were already being regulated, sometimes for biological-management reasons.[3] Two generations later, Huxley's optimism was forgotten. Conservationist regulation was widespread.

---

[1] This was Grotius (1916) thesis *Mare Liberum*. He argued that ownership of the seas was impossible *inter alia* because of the fact that resources in the oceans were inexhaustible. But why, it is asked, does the secondary law of nations which brings about this separation when we consider lands and rivers cease to operate in the same way when we consider the sea? I reply, because in the former case it was expedient and necessary. For every one admits that if a great many persons hunt on the land or fish in a river, the forest is easily exhausted of wild animals and the river of fish, but such a contingency is impossible in the case of the sea.

[2] T.H. Huxley (1882) Inaugural Address at the Fisheries Exhibition 'I believe that it may be affirmed with confidence that, in relation to our present modes of fishing, a number of the most important sea fisheries, such as the cod fishery, the herring fishery, and the mackerel fishery, are inexhaustible.'

[3] In Britain, for example, see the *Sea Fisheries Act, 1868.* c. 45 31 & 32 Vict.; at the same time, the state of Connecticut passed an Act allowing private citizens to bring *qui tam* actions against any persons, partnerships or corporations that created pollution deemed to be deleterious to clams, oysters, eels or fish. This Connecticut statute of 1868 'for encouraging and regulating fisheries' is cited in *Blyndenburgh v. Miles*, 39 Conn. 484 (1872) at para. 1. The Act, reads, 'Every person, partnership or corporation that shall permit or allow any coal tar, or refuse from the manufacture of gas, or refuse from establishments operated to extract oil from white-fish, or other deleterious substances to clams, oysters, eels and fish, to run, flow, drain or be placed in any of the harbors, rivers, creeks, arms of the sea, or waters adjacent to this state, shall forfeit

During these 400 years fish markets and fish prices had not been forgotten. From early fishmonger holdings of property rights to later local shellfish monopolies, the price and quality of market deliveries played a role in influencing what fish might be landed and when they might be delivered. It is true that in our century, there emerged conservation regulation but many of its rules were gradually modified by price considerations. And, as fishermen acquired the present variety of quasi-exclusive rights over catching and over the stock, they were given the ability *individually* to take account of price effects of their operations.

In what follows I trace the cycling of institutions through five phases. Property had been abolished, and then it was regained.[4] The state of the fish stock had been disregarded, and then it was brought under management. Individualism had been everything, then it was suppressed, then it recovered and now it may become an element in joint action. At every turn the market for fish had been an important influence.

2.2 PHASE ONE: VERY EARLY DAYS — FISHERY PROPERTY, MONOPOLIES AND SOVEREIGNTY

I will start with medieval regimes emerging in the English common law. England had been subject to Roman property concepts, applying to lands, rivers and fisheries.[5] This was modified in a regime of Saxon concepts that flourished on the continent, the result of a quite different history of sea fishing rights. For example, for centuries in France, Holland, Spain, Scandinavia, and the Baltic, coastal landowners could claim to have ownership and control of fishing just offshore. This rule survived, in a few countries, until today.

To understand developments in England, we must first understand the fishing technology. Line and net fisheries were unimportant offshore, where few fishermen ventured offshore unless they were hurrying across to other coasts. Most harvesting was in tidal rivers, estuaries and inlets. Much of it was done with fixed gear: weirs, traps and barriers ('kiddles') in streams and around the beaches. And property rights were well adapted to this fixed-gear technology. The abstract principles of titles to fishery ownership were the same for tidal waters and for fresh waters. Derived from their holders' titles to adjoining land, they governed their encroachment disputes. (Customary local law was then used to settle disputes among owners and feudal tenants.) Who might fish, and when, were questions settled by property law, certainly not by reference to feared over-fishing of the stock.

---

the sum of one hundred dollars, one half to him who shall prosecute to effect, and one half to the treasury of the town within which the offense is committed.'

[4] This refers to the so-called 'Grotius doctrine.' Grotius argued that the seas may not be subject to appropriation by any nation because it is, by nature, inalienable, and because the abundance of fish resources deprives any individual of exclusive access rights. Grotius's doctrine was a response to the Dutch East Indies Company's effort to monopolise access to the oceans; it resolved the vacuum left by the failure of Spain and Portugal to agree on a division of the maritime trade routes.

[5] See, for example, Plunkett (1956: 294-300) and Hunter (1897). At 310 Hunter states, 'The right of fishing belongs to all men.'

2.2.1 The public right of fishing in tidal waters

For almost 200 years after the Norman Conquest the holder of land next to both freshwater and salt-water location fisheries was entitled to the fishing and to the power to grant the holder's rights to others. The holder was already bound by customary fishing obligations to feudal tenants. We may say that this entitlement to fisheries in tidal waters was land-based,[6] as it was in rivers and lakes.

That these land-based titles over salt-water (tidal) fisheries disappeared is, surprisingly, due to the concessions made to the barons by King John.[7] They had complained of John's propensity to make money by granting rights to place fixed gear in tidal rivers, rights that over-rode the local baron's land-based rights. Probably John had been granting these new fixed-gear rights to those who supplied town-based fishmongers and to other large-scale buyers. The barons forced the king's agreement, (*Magna Carta,* 1215) not to grant such fixed fisheries in named tidal waters. (Furthermore, the barons complained, John's grantees' fixed gear was blocking river navigation.)

The surprise lies in this personal undertaking by King John becoming transformed into general law, and subsequently, a basic social principle. Soon, in the courts' interpretation of the concessions in *Magna Carta,* neither the king nor *any other landholder* might thenceforth grant (exclusive) fishing rights in the named tidal rivers.[8] Within a century this view was interpreted as banning *anyone's* granting of (exclusive) fishing rights in any tidal waters. In a few decades it followed that no one, from the king down, might even *hold* title to exclusive fishing rights. This doctrine quickly led to that which prevails today: there is a general public right of fishing in tidal waters.

Did this new entitlement doctrine emerge from the new sea-fishing technology then appearing? Probably the causation was the other way around: the new public right of fishing permitted the new technology. The profitable way to exercise the old granted fishing rights had been to place fixed gear in tidal rivers. The competition allowed by a public right provided an incentive to abandon fixed gear and to adopt vessels and gear with more capacity.

---

[6] In 1995 in our 'Evolution of Water Rights' Georgina Coustalin and I developed the idea of alternating periods of 'land-based' and 'use-based' rights. Referring to non-tidal waters, Lord Hale explained in *De Jure Maris,* that lands covered by water were the same as lands not so covered and therefore susceptible to private ownership upon a grant from the Crown. Until such grant, they remained in the royal demesne. See Hale (1787) c. 1 & 3.

[7] *Magna Carta,* s. 33

[8] A granted and privately held fishery was and is, confusingly, called a free fishery, just as title to a private land holding may be referred to as 'freehold'. By the mid-nineteenth century the House of Lords held that *Magna Carta* did indeed prevent Crown grants of several fisheries in tidal waters, and that to establish a lawful exclusive fishery in such waters one had to produce either a Crown grant 'not later than the reign of Henry II' or evidence of 'long enjoyment' of the fishery from which it might be inferred that such a grant had been made. See *Malcomson v. O'Dea* (1863), 10 H.L. Cas. 593 at 618. The common law came to conclude that after the reign of Henry II, the public right of fishing in tidal waters could not be limited or abrogated except by Act of Parliament. See, *Neill v. Duke of Devonshire* (1882), 8 App. Cas. 135 at 177-83.

2.2.2 Capture

At the same time, the common-law doctrine known as the 'law of capture' was gradually applied to ocean fisheries.[9] No one could own a fish still swimming at large, it must be captured first. This law of capture, taken with the right from *Magna Carta*, reinforced the public nature of the right of fishing in tidal waters.

2.2.3 Europe

Coastal Europe clung to the doctrine that tidal fisheries could be private, until the seventeenth century. Then, in France, Colbert's attempts to create a fishing labour force (from which he could lure recruits into Louis XIV's navy) introduced open fishing for all.[10]

Elsewhere in Europe various kinds of land-based monopoly and territorial fishing rights survived longer. Legal writers are vague about the duration, extent, and the enforceability of this doctrine, under which great nobles continued to claim the fishing out to sea from their estates. Indeed, there are still in Nordic countries vestiges of private ownership, or jurisdiction, over adjoining sea fisheries.[11]

2.2.4 Testing the public right of fishing

Once established, the English public right of fishing was not often tested or modified in the courts. The few cases on record deal mainly with shoreline disputes in tidal rivers or over oyster beds. Without cases there could be no judge-made law refining fisheries' ownership, grants or bequests, and no nuisance law. So the freedom of fishermen in salt water was not changed, not even defined, until the nineteenth and twentieth centuries when it was constrained by regulatory laws.[12]

2.2.5 Mercantilism: monopolies and legislation

Under their charters some early local governments had been able to close fisheries, set gear and size limits, ration access to the best fishing places and make rules to regulate fish quality.[13] Their bylaws and regulations were not directed at protecting the stocks but at protecting local fishermen and merchants from unwanted competition. In the

---

[9] Under the doctrine of the law of capture, landowners do not own migratory substances underlying their land, but have exclusive rights to drill for, produce or otherwise gain possession of such substances, subject only to restrictions and regulations pursuant to police powers. See *Black Law Dictionary* (St. Paul: West Publishing, 1998) p. 613.

[10] This may well have anticipated the modern Warming/Gordon economic theory of fishing by predicting that open access would lead to an expansion of fishermen employed in fishing. (For clarity, when referring to coastal Europe I am referring to those continental nations bordering along coastal waters).

[11] For early examples of this see Moore and Moore (1903: 26). For a current example, see Lappalainen *et al.* (2002).

[12] I should mention inshore oyster beds and other sedentary fisheries, established perhaps by Royal Charter, and placed under private or guild control, or under that of local government. Their arrangements were tested in the courts.

[13] See, for example, *Placita Coronae for Cumberland,* 6 & 7 Edward 1 as cited in Patterson (1868) pp. 35-6. Here Patterson outlines a statute 'concerning fishery, which is a local Act for the county of Cumberland'. Included in the statute was authority to restrict gear limit sizes as well as prohibit fishing altogether.

same mercantilist spirit, Queen Elizabeth gave the Newfoundland fishery to Sir Humphrey Gilbert (a monopoly she could not have given in England.)

These are both exceptions to the idea of a universal right of fishing. In spite of them I would emphasize that in general no arm of government exercised any kind of jurisdiction to make fishery policy or to change the public or government's fishing rights. Modern theorists who see the history of property titles as a continuing battle between private ownership and government cannot apply their approach to the centuries-long governmental neglect of the public right of fishing.

2.2.6 The international freedom of the high seas

A few words about inter-nation rights. While for centuries there were almost no changes in public rights of individual access to ocean waters, there was from the seventeenth century an evolution of national powers over the high seas. Everywhere, one gathers, pressure groups lobbied their governments to exclude 'foreigners' from 'their' fisheries. For example, the English herring fishermen encouraged war and diplomacy to protect their market niche.

The motive everywhere was to dominate a market by monopolising a coastal fishery. This motive led the English and the Danes actually to exclude the Irish and the Icelanders from fishing off the coasts of Ireland and Iceland. Fishing exclusion was probably less important than fish-trade protection. But when the debate arose about whether Spain or Britain or Holland could claim sovereignty over the high seas, the protection of fisheries did get mentioned as one important argument for national control. The weakness of that argument was that, in Elizabeth's time, Britain and Holland seem to have agreed on the inexhaustibility of marine resources. Apart from the vulnerability of such valuable species as sturgeon, whales, salmon and dolphins, the fish stocks of the ocean were regarded complacently.[14]

2.3 PHASE TWO: THE COMING OF BIOLOGICAL REGULATION DRAWS GOVERNMENT BACK INTO FISHERY POLICY

In the early nineteenth century the general complacency about the condition of salt-water fisheries was continuing because there appeared little reason to intervene. True, each nation did worry about the invasions of their inshore fisheries by foreign vessels, however the general belief in the north-east and north-west Atlantic was that the offshore fish stocks were resilient and their annual yields unaffected by the fishing then going on.

It was in the 1880s and 1890s that the political world began to hear the fishing industry's grumbles about mature fish being hard to find, grumbles that coincided with new technology: the steam-powered Danish or 'otter' trawler. Realisation of the steam trawler's ability to scrape along at great depths soon converted some in Britain to the view that in the north-east Atlantic, at any rate, over-fishing was a possibility. That

---

[14] As noted earlier, in 1609 Hugo Grotius, the great formulator of the law of nations had used the 'fact' of inexhaustibility to strengthen his view that there was no need for *exclusive* fishing rights. The doctrine of the seas, the public right of fishing, and the law of capture complemented each other.

view was given official recognition by the Huxley commission of 1883. But what were the policy ramifications of this?[15]

The authorities had little regulatory experience to fall back on and most fishermen could still get a full net if they tried long enough. So it was not until 1920 that fishery interests were confronted with very suggestive evidence. A recovery of north-east Atlantic fish stocks during the war was due to the naval blockades preventing fishing. Therefore reductions in the stock of fish might be due to increases in fishing. The newly-coined word 'over-fishing' was heard. New European scientific evidence was confirmed by new overseas studies, especially of the Pacific halibut. With such glimpses of measurable effects of fishing, fisheries science came of age. With it came some quantitative understanding of a biological rationale for government controls to reduce fishing pressure.

The evidence led to support for not one but two dominating types of control: gear-regulation (including mesh-size regulation) and seasonal closures. Administratively, there were already local versions of both of these: protection of migrating salmon especially in Scotland (and later in eastern Canada) and riparian owners' regulation of river sport angling (the latter known to MPs who fished on their holidays.) Furthermore a few coastal shellfish and crustacean fisheries were still under special local regulations. There was also support from parts of the fishing industry, weary of fishing-ground competition.

But with all these supporters, not much regulation emerged. Gear regulation and seasonal closures were hard to implement and enforce, even where the stock areas did not sprawl into the high sea. True, the amazingly successful Bering Sea *Fur Seal Convention* 1911 and the later Pacific Halibut Treaty 1923 had shown that international management was possible. But few European or North American governments wanted to go further[16] until after World War II. Then, to make regulations in their home waters more effective, they did agree that national fishing limits should be pushed out to 12, 50 and eventually 200 miles.

2.3.1 *Property rights at the end of the first two phases*

These new regulatory regimes had two unwelcome effects. First, closed seasons and gear regulation did not halt the growth of fleets. Indeed, the case can be made that it often encouraged fleet expansion. More and more vessels raced each other to find land and deliver fish to port. This raised costs, stimulated the use of more destructive technologies and drew fishing communities closer to the fishing grounds. Administrators began to shorten openings and to forbid the more effective technologies. Second, the closures had been directing everyone's attention miles away from demand and price. With the new technologies and vessels, fish were landed and dumped on the market as long as there was an opening.

Both effects touched off new ideas. Some governments realised that proliferation of vessels and gear (effort) had to be stopped, and some of these began to adjust the conditions on vessel fishing licences to control technology and enforce closures. In beginning to consider changes in the number and characteristics of licences, they were trimming the public right of fishing. From being a purely administrative (and revenue-

---

[15] Thomas Huxley was actually appointed to three commissions. Mark Kurlansky argues that these commissions established a tradition of governments ignoring the observations of fishers. See Kurlansky (1997).
[16] See Christy and Scott (1965) for the timing of various international fishery treaties.

earning) instrument, almost invisible to some experts and 'incomplete' and 'attenuated' to others, the humble vessel licence was taking on a few of the characteristics of private property.[17]

2.4 PHASE THREE: THE LIMITATION OF LICENSING AUGMENTS THE 'CHARACTERISTICS' OF A PROPERTY RIGHT

In Canada, limiting the number of licences was phased in during the 1960s. This limitation was soon widespread.[18] To reduce the effort pressing on a fishstock there was to be control of the *number* of licensed vessels. This gave the licence some importance among the instruments of fishery management.

Limitation caused the licence to acquire property-right type characteristics. All real interests in land and resources are somewhat attenuated with none of them completely equipped with the most wanted amount of each of six characteristics. It may have limited duration and yet be nonrenewable. It may be transferable and yet indivisible. Here is a list of the six characteristics.

The first is *duration*. The more enduring the holder's right, the greater the opportunities to increase the productivity and value of land or resource. Easy *renewability* may be a substitute for duration.

The second is *quality (or security)* of title. The more a right has of this characteristic, the smaller the likelihood that its holder will lose possession of the resource and his improvements to it.

The third is *exclusivity*. The more a right has of this characteristic, the more the holder can call on the resources of the legal system to free him from spillovers and damage arising extrinsically.

The fourth is *transferability*. The more a right has of this characteristic, the greater its value derived from its value to potential users; heirs, buyers and renters.

The fifth is *divisibility*. With this characteristic, parts of the land or resource may be acquired by and placed under a different holder, in search of a higher value.

The sixth is *flexibility*. This characteristic is most useful in contracts and leases. It provides the parties with powers, in the future, to agree on different terms and indeed characteristics of their bargain.

2.4.1 *The property characteristics of government's simple fishing license*

With this list of characteristics, we can now scrutinise the primitive vessel licence. Because of the public right of fishing, the issue of licences was unlimited. Did the governments' third-phase fishing regulations affect the licences' characteristics?

Having a public right of fishing had meant the licences completely lacked *exclusivity*. Their holders may have been given secure titles to go fishing, perhaps even for a long *duration*. Legalistically, the licenses may have had some

---

[17] See generally, Scott (1989: 11-38). In *Re: Bennett* (1988: 24) B.C.L.R. (2d) 346 (B.C.S.C.) Ryan J. declared that a commercial fishing licence was property in relation to the Bankruptcy Act.
[18] See Wilen (1989), *supra* note 17 p. 250.

transferability, and some *divisibility*. But the complete lack of *exclusivity* meant that only their *quality of title* and *duration* had any practical value to the holder.[19]

2.4.2 Some licence characteristics benefit the holder

Economic specialists remind us that, as an isolated policy measure, the licence limitation begun in the 1960s was a poor way to save the fishstock or to reduce the fishing costs. It could not remove the fisherman's old incentive to race, out-invest and out-fish his rivals. It could not remove the existing regime of concentrated landings during the brief opening, and of concentrated marketings as well, no matter how far the price fell. And it could not remove the incentive to outwit the regulators by concealing and evading, free riding on the conformity of others.[20]

Nevertheless, licence limitation must be applauded for a good effect (in nations where there were limitations) putting the fishery on the road toward the conversion of the licence into real property rights in the fishery. Having fewer vessels to interfere with each other had brought about a glimpse of *exclusivity*. True, with racing behaviour and over-investment, the vessels rarely enjoyed really exclusive fishing but the ideas of exclusivity and of fixed shares were implanted. The vessel owner, no longer faced with the likelihood of an indefinite fleet expansion, might even take an interest in the fashioning of regulations and in the size of the fishstock.

As well, the perceived idea of licence exclusivity cast light on other characteristics. The licence's *transferability* took on a little meaning, because if regulation did increase the value of the catch, the increase would be reflected in the value of each licence accruing to licence holders when they sold out to new entrants. And the licence's duration and *renewability* were also reflected in the licence value.

2.4.3 The security characteristic: alternating land-based and use-based title systems

I digress briefly to consolidate earlier remarks about introducing an individual title to use a common-property resource. Well-known questions about doing this are, from whom, or what, is the holder's title derived? There have been two answers. In the case of water in inland rivers and streams one answer was that the holder's title arose from extending into the river the boundaries of his lands. The holders' rights may be labelled 'land-based' (or, perhaps, 'territorial'). The alternative answer is to apply the 'first come - first served' rule. The users acquire an exclusive title to a flow of water because they are already taking it (for, say, their mills along the river.) Where all flows have been acquired in this way, the holders' water titles are 'use-based'.[21]

The two systems, land-based and use-based title, are found in inland fishing. Are they found in sea fisheries as well? As seen above, there were feudal land-based rights entitling the landowner to the fishing area offshore, rights that disappeared with the arrival of the public right of sea fishing. That had to await the arrival of licences and

---

[19] A designation should be made between 'limited licences' on the one hand, and 'free licences' on the other. With the former it may be argued that there accrues a limited amount of exclusivity, while in the latter there is virtually no exclusivity whatsoever. However, it is important to recognise that this exclusivity dealt only with the number of boats on the water and did not, in effect, provide any exclusivity with regard to the fish itself.

[20] This is the fisheries version of the now famous 'The Tragedy of the Commons' (Hardin, 1968).

[21] Scott and Coustalin (1995).

permits, hundreds of years later. Under these feeble instruments, vessels (making 'use' of the resource) acquired title to fishing or to land a catch, or even to the fish stocks.

If one system of fishing entitlements replaced another, is this replacement part of a regular reciprocation of fishing titles? Probably not, although the systems of rights to shellfisheries did wobble to and fro, from land-base to use-base and back again. The amplitude of such cycles would be mild compared to those of water titles. But the advents of sea fishing quotas (below) and of aquaculture rights suggest that innovation and cycling of title bases is not at an end.

2.4.4 A note on exclusivity of titles in international zones

Only in the *international* allocation of fishing rights have there been complete swings of right systems. Early medieval landowners' rights had been good against local and foreign fishers – they were land-based. In the following period, fishermen's rights arose from the freedom of the seas and the public rights of fishing. These were use-based. Then, new naval technology allowed three-mile fishing limits to be enforced. Fishery rights were therefore localised and land-based. Today in European and other treaty waters quota-type rights indicate another swing toward use-based titles.

2.5 PHASE FOUR: THE QUOTA: A USE-BASED PROPERTY RIGHT, DESCENDED FROM REGULATORY LICENSES

It is now well known how, with industry consent, governments in the 1970s and 1980s introduced numerical quota-licences to particular fisheries. Iceland and New Zealand were the leaders, with Canada, Australia and eventually several others following into the 1990s.[22] The quota's limit on a vessel's landings was first seen as a strengthening of the biologists' instruments of management.[23] This policy innovation meant that the regulatory instruments (above) would hardly be needed, so that enforcement costs could well be reduced. For vessel owners, quotas meant less desperate racing and so less incentive to install every novelty in power or fishing gear.

This was not the first time property rights had been invented unwittingly. In the common law, those landowning litigants who had taken their land-based fishery disputes to court had not realised that their suits were creating precedents that would serve as bases for clearer fishing powers, equipped with stronger property characteristics. So it was with quota-licences in the sea fishery. Cabinet members and fishery managers who seized on quotas as this year's gimmick to plug leaks in last year's regulatory system did not fully realise that they were on the verge of introducing a new and powerful managerial technique. Even less did they grasp that they were introducing quota *rights,* with some amount of each of the five characteristics of property rights.

---

[22] For early views on quotas see Christy (1973), Maloney and Pearse (1979) and Neher *et al.* (1989).
[23] James Acheson notes that quota limits on the lobster fishery clearly improved management of the stock (Acheson, 1987). At 52 Acheson states, 'We obtained unmistakable evidence that reduced fishing effort in the perimeter defended areas had both biological and economic effects.'

2.5.1 Transition to exclusivity

The biggest transitional difficulty was surmounting struggles among active licence-holding vessel owners, fishing companies and the wider public over the distribution of the new quota rights. Some have favoured auctions. Many have favoured 'grandfathering' the amount of a holder's previous years' landings into the amount of quota. Some favoured giving weight to individual native and aboriginals' claims. Differences about these methods have been a major aspect of each country's history of quotas. The fishermen who survived the distribution found that the rights they were holding had features strikingly like those of quantitative irrigation water rights, quantitative oil-field rights and quantitative grazing rights.[24] Legal experts learned that title to the fishing quota, like those of existing resource rights, was not land-based but use-based.

Typically quota rights had long *duration*, simple *transferability* and *divisibility, accepted security of title,* and, above all, an *exclusivity* that was far greater than that of any ocean fishing right-holder had been since AD 1300 and that enabled the holder to recapture the vessel's pricing-setting powers. Medieval fishmonger guilds had been aware that price depended on amounts supplied and they endeavoured to hold down these amounts when price was low. Many years later fishing regulation changed this. It became the aim of all connected with marketing to get the fish landed while the fishery was open, and to sell as much of this as they could, regardless of price. Now, individual transferable quota (ITQ) holders with their exclusive right to a certain landed amount could behave rather like those earlier guildsmen: reduce catches when prices are low and save the rest of the quota till price rises again. These were the marketing aspects of harvesting that had been emphasised by Crutchfield/Zellner and again by Crutchfield, but were ignored by most managers and other experts until quotas made price-oriented landings a reality.[25]

2.6 PHASE FIVE: PROPERTY RIGHTS EXCLUSIVITY—FROM OWNERSHIP TO 'MEMBERSHIP'

In 2005 we have entered the fifth phase of fishery property-rights development. Quota-licensing is already found world-wide. The owners hold quotas that do have some of each of the five main characteristics of property rights, although they are deficient in the security (quality of title) characteristic and, especially, in the exclusivity characteristic, as will be discussed in the next two sections. This discussion

---

[24] See Stollery (1988), Scott (1999) and Devlin and Grafton (1996).
[25] See Crutchfield and Zellner (1963) and Crutchfield (1981). In Arnason (1996) there is a useful survey of the ITQ systems introduced by then to various extents in five nations: Iceland, New Zealand, Greenland, Holland, and Australia. All five reported a decline in fishing costs and a related increase in productivity and profitability, but did not mention improvements in quality or price. This is surprising for Holland, where the flatfish industry, somewhat similar to the Pacific halibut industry, was one of fisheries improved by the introduction of ITQs. Many of the other fisheries with new ITQs had typically supplied undiscriminating bulk markets; perhaps we will learn soon that their catches are now finding their way to quality-conscious markets. For a discussion of market control as illegal 'collusion' see Adler (2004).

will lead into a re-defining of quota licences as 'membership cards' for collective fishery ownership.

2.6.1 Weakness in the security (or quality of title) characteristic

Turn first to security. The vessel-owner is exposed to two risks: that the quota gives its holder a secure claim to the catch, and that the quota regime will not give the holder a secure title or claim to the quota. We may defer the first risk till later. The likelihood of the second risk depends on the firmness with which the government upholds the basis on which quota-licences were originally handed out. Little prevents the government, under political pressure, from legislating to re-distribute them on a different basis. The likelihood of the first eventuality also depends on whether the government can be relied on to maintain its police-type protection of the owners' quota holdings against trespasses and invasions by outsiders.

These risks suggest that in spite of similarities, the fishing quota-holder's title has less 'quality' and is less secure than farmers' and miners' land titles. The quota-holder relies on steadiness in politics, goodwill and custom. Parliament may cave in: witness Iceland where the well-established quota system that has been attacked and there is political pressure on the government to take back the 'undeserved enrichment' of those to whom quotas were given. The quota-holders' security against trespass by outsider vessels and fleets also depends on politics, goodwill and custom. In short, title and security are weaker than that of property in land because they are not part of the traditional property system, standing outside government. Therefore, it seems probable, one of the main reasons for some countries' failure to switch over to the quota-licence system is that their vessel owners and fishers, having long experience with the sudden changes of mind of government regulators, have insufficient faith in governments' reliability as guardians of their titles and quotas and will strenuously oppose the whole system. They need more than a licence holding to be won over to Phase-Four ITQs let alone to Phase-Five changes.

We will see below that converting their ownership to 'membership' may be the Phase-Five answer to weak quota titles. Meanwhile, we consider the parallel weakness in the exclusivity characteristics.

2.6.2 Weakness in the exclusivity characteristic: co-operative membership

Consider the effect of adopting quota licences. At the time of their inauguration most quota systems do not do much for the holders' security and title (as I have suggested) but they do a lot for their exclusivity. Wherever they can count on their licence holding being secure, they also can confidently rely on the *exclusivity* of their quota rights to a guaranteed catch at a constant cost, a vast improvement over earlier phases.[26]

But the exclusivity characteristic is only relative to a set of specific invasive threats, not to all possible invasions. For example, an exclusive fishing quota for Coho salmon would be good only against other vessels seeking the same variety of salmon. It would not be good as against vessels fishing for other species. And if this were remedied, it would be good only against those authorised to seek and land *fish*, not against those

---

[26] This is a brief way of putting what they may expect. Of course a quota is the maximum allowed catch, not the guaranteed minimum. And their costs can be *predicted* to remain constant, not guaranteed.

vessels using the same fishing grounds for transportation, recreation or waste disposal. The vessels' fishing ground is legally 'common property' having open access to all types of vessel and user, and becoming more and more congested. Even if *every* type of user had a 'fully exclusive' right, users of that type are almost certain to suffer many kinds of spillover from the activities of users of other types. In isolation, there will be nothing much a fishing owner can do to exclude these outside-source interferences, apart perhaps from appealing to government for relief.

Knowing this, owners can be predicted to act collectively, forming what I will call a co-operative. Note that while many modern co-operatives involve the pooling of work, as in a collective farm or kibbutz, a vessel co-operative would not have the aim of pooling members' fishing activities. It would be more like a buying co-operative, in that its vessel-owner membership would merge their persons and knowledge in making representations to government, in debating controls with scientific fishery managers, contracting with those representing fishers of other species; and contracting also with municipal and commercial enterprises conducting activities and using equipment in the fishing area.[27]

There are already a few such co-operatives in operation. Except perhaps for independent minded groups like the lobster catchers in Maine, they nearly all are made up of holders Phase 4-dated quotas. Most co-operatives start off by participating in the day-to-day and year-to-year management of 'their' fish stock. Since the quotas for a particular fish stock or species add up to the total allowable catch (TAC) of that species, it follows that by just organising that species' quota holders, the co-operatives unite 100 percent of the interested vessel owners. The co-operative may debate freely, among its membership and with government, the management of the landings or fish stock. There is no remaining outside vessels, either to free ride or to be harmed by whatever the co-operative decides.

Note that the co-operative has the potential to overcome some of the weaknesses of title and of the exclusivity titles in individual quota, mentioned above. The potential weakness of an owner's quota title is made up for by its recognition by all members with the same title and by their combined efforts to make their titles something more than a government fishery manager's administrative convenience. The weakness of the owner's quota's exclusivity characteristic is made up for by the capacity of the cooperative to enter into battle, or to make agreements that strengthen the recognition of its sole proprietorship over the TAC. It should be noticed that the members' individual quota-licences retain their other characteristics: transferability, divisibility and duration.[28]

The co-operative's capacity to negotiate and, especially, to contract, makes one hopeful that the conflict and congestion coming on all multi-use fishing grounds can be swept away. Japanese experience has shown that when village fishermen are organised as a co-operative, they can successfully and victoriously negotiate over airport, seaport, and industrial fishing ground developments over land and water

---

[27] There would be opportunity for discussion of the decision-making rules in a co-operative. (Existing farm co-operatives often weight votes according to member's acreages, purchases, or production). In this case, the weights could be the size of the quotas. Note however, that because many fishers are linked to adjoining communities, the members might select a one-man-one-vote instead of a one-vessel-one-vote rule. For an example on how such a system works within the farming industry see Ellickson (1993).

[28] In these respects a cooperative is like a club, of the kind analysed by J.M. Buchanan (1965). See also Stollery (1988) and Staatz (1989).

users.[29] One strategy is to resist the outside developers. Another, perhaps equally valuable, is to go into joint planning and to *contract* the sharing of the available water space.

2.7 CONCLUSION: LAND-BASED RIGHTS AND THEIR CHARACTERISTICS

For 200 years the Norman and Plantagenet kings of England, with their barons, held land-based rights to tidal fisheries. Then for 700 years, in tidal waters, individuals had free access, a use-based public right of fishing. Then for 50 years, in the name of official fish-stock management, individual access rights were gradually reduced and government regulators took increasing control.

Viewing this long history we can see that although property in the fishery started with high exclusivity and good title, it fell to open access, without either of those characteristics present. Then, through the efforts of government, there was at first a very gradual recovery of exclusivity, though not of title. As concerns about fishstock survival deepened, the licence was given increasing amounts of both characteristics, until, with the ITQ, it had features similar to common-law property right. But that was as far as the evolution of the individual right could go. To give the holder more exclusivity and better title, the individual right is being transformed into membership of a co-operative.

We can see that the individual sea-fishery right has wandered between the extremes of being completely land-based, as in 1066 and again today. In contrast, with the beginning of open entry and a public right of fishing, the right could be called use-based. Whoever fished had a right to do so. It was to remain use-based until government, to reduce the pressure of effort on the stock, began to issue licences, and to link these to particular places. The right was becoming location-based or land-based. There was to be one more alteration. The typical membership *title* will stem from ownership of a preceding quota licence just as that probably stemmed from a preceding simple fishing licence. From that point of view, the membership will be use-based; originally issued to those who were already fishing, not to those who had rights to adjoining land.

Acknowledgements

It was a privilege to join those honouring Gordon Munro. Sometimes alone, sometimes with co-authors, he has been influential among fishery economists. His work has ranged from practical policy options to abstract theory. Throughout the range he has been a real leader.

---

[29] Area, location and territory are important characteristics of the collective rights of Japanese fishing communities. On this see Yamamoto (1995).

References

Acheson, J. 1987. Lobster fiefs revisited. In B. MacKay and J. Acheson (eds) *The Question of the Commons: The Culture and Ecology of Communal Resources.* Tucson: University of Arizona Press.

Adler, J.H. 2004. Conservation through collusion: Antitrust as an obstacle to marine resource conservation. *Washington and Lee Law Review* 3-78.

Arnason, R. 1966. Property rights as an organizational framework in fisheries: The cases of six fishing nations. In B.L. Crowley (ed.) *Taking Ownership.* Halifax: Atlantic Institute for Market Studies.

Buchanan, J. 1965. An economic theory of clubs. *Economica* 32: 1-14.

Christy, F.T. and A. Scott. 1965. *The Common Wealth in Ocean Fisheries: Some Growth and Economic Problems.* Baltimore: Johns Hopkins Press.

Christy, F.T. 1973. *Fisherman Quotas: A Tentative Suggestion for Domestic Management.* Occasional paper # 19. Rhode Island: Law of the Sea Institute.

Crutchfield, J. 1981. *The Pacific Halibut Industry.* Technical report # 17. Economic Council of Canada.

Crutchfield, J. and A. Zellner. 1963. Economic aspects of the Pacific halibut fishery. *Fishery Industrial Research* Vol. 1 April 1962. Bureau of Commercial Fisheries, Fish and Wildlife Service, US Department of the Interior, Washington: GPO.

Devlin, R.A. and R.Q. Grafton. 1996. Marketable emission permits: Efficiency, profitability and substitutability. *Canadian Journal of Economics* 29: s260-64.

Ellickson, R. 1993. Property in land. *Yale Law Journal* 102: 1315.

Grotius, H. *Mare Liberum.* trans. R. Van Deman Magoffin (ed.) 1916. with an introductory note by J. B. Scott. New York: Oxford University Press.

Hale, L. J. 1787. De Jure Maris et Brachiorum Ejusdem. In F. Hargrave (ed.) *A Collection of Tracts Relative to the Law of England*, Vol. 1. London: T. Wright.

Hardin, G. 1968. The tragedy of the commons. *Science* 162: 1243-48.

Hunter, W.A. 1897. *A Systematic and Historical Exposition of Roman Law in the Order of a Code* (3rd ed.) London: Sweet and Maxwell.

Huxley, T.H. 1883. Inaugural address: The Fisheries Exhibition literature. *International Fisheries Exhibition* London 4: 1-22.

Kurlansky, M. 1997. *Cod: A Biography of a Fish That Changed the World.* Toronto: A.A. Knopf Canada.

Lappalainen, A., S. Pekka and R. Varjopuro. 2002. Management of Baltic coastal fisheries, a background report: 10-11. Finnish Game and Fisheries Research Institute. Accessible online http://www.rktl.fi/www/uploads/pdf/raportti241.pdf

Moloney, D.G. and P.H. Pearse. 1979. Quantitative rights as an instrument for regulating commercial fisheries. *Journal of Fisheries Research Board of Canada* 36: 859-66.

Moore, S.A. and H.S. Moore. 1903. *The History and Law of Fisheries.* London: Stevens and Haynes.

Neher, P., R. Arnason and N. Mollett. 1989. *Rights Based Fishing.* Dordrecht: Kluwer.

Patterson, J. (Chair) 1868. *Special Commission For English Fisheries.* London: Her Majesty's Stationary Office.

Plunkett, T. 1956. *A Concise History of the Common Law.* Boston: Little, Brown and Co.

Scott, A. 1989. Conceptual origins of rights based fishing. In P. Neher, R. Arnason and N. Mollet (eds) *Rights Based Fishing*. Dordrecht: Kluwer: 5-38.

Scott, A. and G. Coustalin. 1995. The evolution of water rights. *Natural Resources Journal* 35: 821-979.

Staatz, J. 1989. *Farmer Co-operative Theory: Recent Developments*. USDA Agricultural Co-operative Services, Research report No. 84 June.

Stollery, K. 1988. Co-operatives as alternatives to regulation in commercial fisheries. *Marine Resource Economics* 4: 289-305.

Wilen, J. 1989. Rent generation in limited entry fisheries. In P. Neher, R. Arnason and N. Mollet (eds) *Rights Based Fishing*. Dordrecht: Kluwer: 249-62.

Yamamoto, T. 1995. Development of community-based fishery management system in Japan. *Marine Resource Economics* 10: 21-34.

3

Property Rights Quality and Economic Efficiency of Fisheries Management Regimes: Some Basic Results

Ragnar Arnason

3.1 INTRODUCTION

Property rights are a social institution that specifies the rights of social agents to objects. The object may be physical or non-physical. Clearly, the rights in question can be more or less extensive. If they are sufficiently extensive, the object is customarily referred to as 'property' and the holder of the rights as 'owner'. By defining the relationship between objects and agents, property rights also determine a part, often a critical part, of the social relationship between social agents in matters having to do with the object.

In this chapter we investigate the relationship between property rights and economic efficiency. The first and fairly obvious observation is that without property rights there will be (i) virtually no accumulation of capital and (ii) relatively little production.

Considering point (i) first, it is obvious that no one is going to save valuables in the form of physical capital, natural resources or even human capital unless he enjoys adequate property rights over his accumulation. Accumulation of capital necessarily means sacrifice of current consumption. As a result accumulation must be sufficiently rewarded. Without property rights, this is not possible.[1] Moreover, even if some people decided to accumulate nevertheless – possibly for altruistic reasons, this accumulation would be seized by other, less altruistic persons, and, in order to avoid a similar fate, quickly consumed. So, we may safely conclude that without property rights there will be (a) virtually no accumulation and (b) what capital there might exist will be quickly seized and squandered.

Regarding point (ii), relatively little production, Smith (1776) explained how division of labour, i.e. specialisation in production, is a prerequisite for a high level of production. Specialisation obviously requires trade. Trade, however, is exchange of property rights. Thus, without property rights there can be no trade. So without property rights, there will be no specialisation in production and, consequently, poor use of productive resources and relatively little production.

---

[1] This assumes a sufficient number of less than perfectly altruistic individuals.

It is interesting to note in this context that property rights are really more fundamental to economic progress than markets. Assuming only that people look after their interests, if property rights exist markets for them will automatically arise. The reason is that trading – made possible by property rights – offers opportunities for mutual gain. On the other hand, without property rights, markets cannot exist. So, quite clearly, property rights are both necessary and sufficient for the existence and operation of markets. In this sense property rights are more fundamental than markets.

Having argued that no property rights, inevitably lead to a low level of production and little economic efficiency, what about full and complete property rights? The basic observation is that full and complete property rights allow the operation of a perfect market system which, under certain conditions, is known to be economically efficient (Debreu, 1959; Arrow and Hahn, 1971). Note, moreover, that with perfect property rights there can be no externalities, so that particular source of inefficiency will not occur.

In reality property rights are seldom perfect. Even the most solid property rights such as the ones in private land, houses, cars and even personal items are circumscribed in various ways. For instance, typically, non-owners can use the land of others for travel, rest and often several other purposes. Usually there are also restrictions on how one can modify one's house. In most cultures, even a person's most private property may be used by others, if the latter's need is deemed to be great enough.

So the question arises; what is the efficiency of less than perfect property rights? What happens to efficiency along the interval between perfect property rights and no property rights? Scott (1989, 1996), recognising that the rights in question can be more or less continuously reduced or increased likes to talk about the quality of property rights. In those terms; what is the relationship between the quality of property rights and the associated efficiency?

This chapter explores this issue. Its main hypothesis is that the relationship between the quality of property rights and economic efficiency is monotonically increasing. More precisely, the higher the quality of a property right, the more efficient is the associated economic activity, where the phrase 'associated economic activity' refers to economic activity wholly or partially based on this property right.

The chapter is organised as follows: In the next section, the basic theory of property rights and their main constituent characteristics will be reviewed. The following section attempts to delineate the relationship between the main dimensions of property rights and economic efficiency. The final section summarises the results of the paper and presents an example of the application of the theory.

3.2   THE BASIC THEORY OF PROPERTY RIGHTS

A property right is not a single variable. As pointed out by Alchian (1965), Demsetz (1967) and Scott (1989, 1996), any property right consists of a collection of different attributes or characteristics. The number of distinguishable characteristics that make up property rights is very high. However, according to Scott (1996, 2000) the most crucial property rights characteristics are:

- Security, or quality of title
- Exclusivity
- Permanence
- Transferability.

More specifically, the content of these characteristics are as follows.

3.2.1 Security, or quality of title

A property right may be challenged by other individuals, institutes or the government. Security here refers to the ability of the owner to withstand these challenges and maintain his property right. It is perhaps best thought of as the probability that the owner will be able to hold on to his property right. Probabilities range from zero to one. A security measure of one means that the owner will hold his property with complete certainty. A security measure of zero means that the owner will certainly lose his property.

3.2.2 Exclusivity

This characteristic refers to the ability of the property rights holder to utilise and manage the resource in question (his property) without outside interference and to exclude others from doing the same. An individual's personal belongings, such as his clothes, generally have a very high degree of exclusivity. A right to the enjoyment of a public park has almost zero exclusivity. The right of a fisherman to go out fishing has exclusivity reciprocal to the number of other fishermen with the same right. A holder of an individual transferable quota (ITQ) has a right to a specified volume of harvest from a given stock of fish over a certain time period. However, when it comes to the actual harvesting, the question of exclusivity refers to his ability take this harvest in the way he prefers and to prevent others from interfering with this ability. Any government fishing regulations clearly subtract from this ability. The same applies to the actions of other fishermen that may interfere with his ability to harvest his quota in various ways. Thus, an ITQ right generally provides substantially less than 100% exclusivity to the relevant asset, i.e. the fish stock and its marine environment. It should be noted that *enforceability*, i.e. the ability to enforce the exclusive right, is an important aspect of exclusivity.

3.2.3 Permanence

Permanence refers to the time span of the property right. This can range from zero, in which case the property right is worth nothing, to infinite duration. Leases are examples of property rights of a finite duration. By verbal convention, the term 'ownership' usually represents a property right in perpetuity or for as long as the owner wants. Note that there is an important difference between an indefinite duration, which does not stipulate the duration of the property right, and property right in perpetuity, which explicitly stipulates that the property right lasts forever. The duration of a property right may seem related to security; if a property right is lost then, in a sense, it has been terminated. Conceptually, however, the two characteristics are quite distinct. Thus, for instance, a rental agreement may provide a perfectly secure property right for a limited duration.

3.2.4 Transferability

This refers to the ability to transfer the property right to someone else. For any scarce (valuable) resource, this characteristic is economically important because it facilitates the optimal allocation of the resource to competing users as well as uses. An important feature of transferability is *divisibility*, i.e. the ability to subdivide the property right into smaller parts for the purpose of transfer. Perfect transferability implies both no restrictions on transfers and perfect divisibility.

3.2.5   Quality of property rights: the Q-measure

As suggested by Scott (1989), it is helpful to visualise these characteristics of property rights as measured along the axes in four-dimensional space. This is illustrated in Figure 3.1. Obviously, if more than four characteristics are needed to describe a property right, the number of axes in the diagram would simply be increased correspondingly as in Scott (1989).

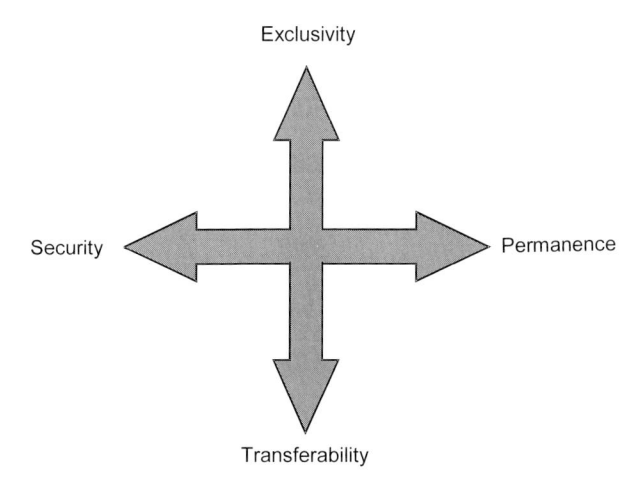

Figure 3.1   Characteristics of property rights.

A given property right may exhibit the different property rights characteristics to a greater or lesser extent. It is convenient to measure this extent on a scale of 0 to 1. A measure of zero means that the property right holds none of the characteristic. A measure of unity means that the property right holds the characteristic completely. Given this we can draw a picture of a perfect property right in the space of the four property rights characteristics as the perfect diamond illustrated in Figure 3.2.

We refer to the map of the property rights characteristics, of which Figure 3.2 is an example, as the *quality map* or *characteristic footprint* of a property right. Obviously, the characteristic footprint of a perfect property right represents the outer bound for the quality map of all property rights. It follows that the characteristic footprint of any actual property right must be completely contained within the characteristic footprint of a perfect property right. This is illustrated in Figure 3.3.

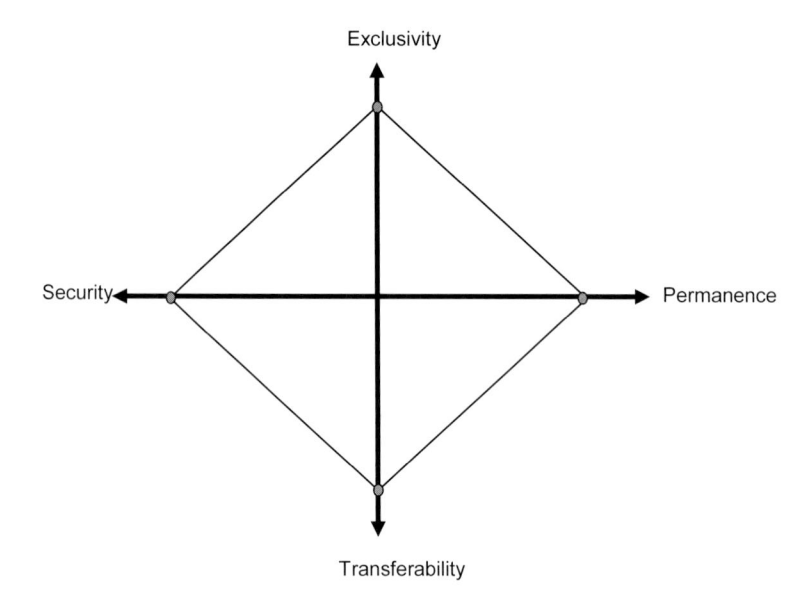

Figure 3.2 A perfect property right.

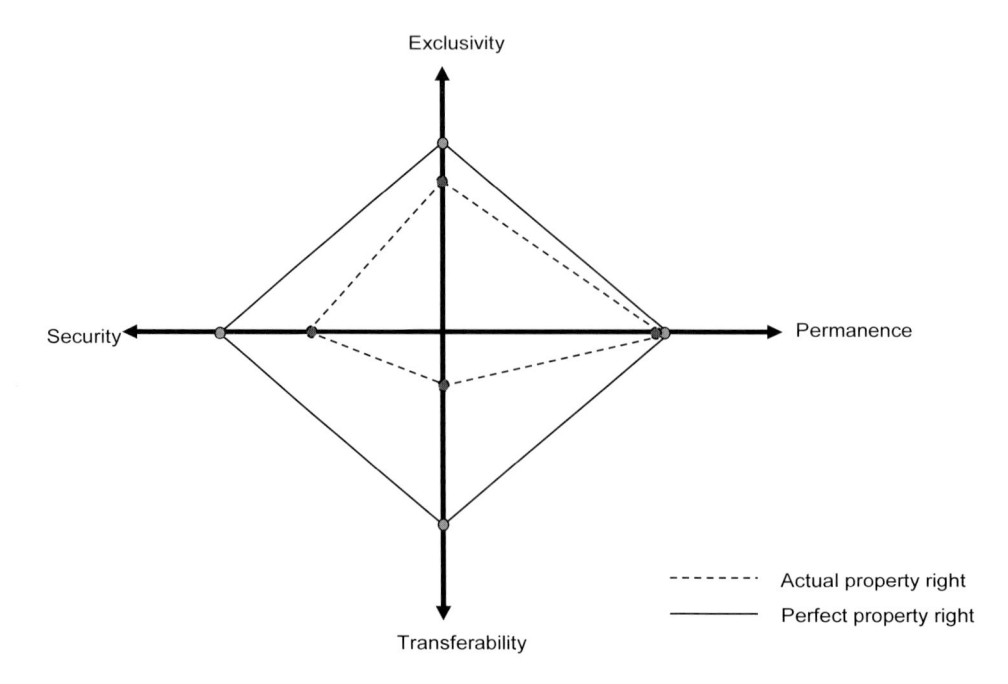

Figure 3.3 The quality map of a property right.

Figure 3.3 illustrates the characteristic footprint of some actual property right within the characteristic footprint of a perfect property right. The ratio between the two areas enclosed by the two quality maps provides an idea of the relative quality of the actual property right. Obviously the closer the characteristic footprint of a property right is to that of a perfect property right, the higher is its quality.

Given the multi-dimensional nature of property rights, it is obviously useful to have a uni-dimensional numerical measure of the quality of a property right. One such measure is the so-called Q-measure of property rights quality (Arnason, 2000). In the case of the above four property rights characteristics, the Q-measure may be defined by the expression

$$Q \equiv S^\alpha \cdot E^\beta \cdot P^\gamma \cdot (w_1 + w_2 \cdot T^\delta), \quad \alpha, \beta, \gamma, \delta, w_1, w_2 > 0 \text{ and } w_1 + w_2 = 1 \qquad (3.1)$$

where S denotes security, E exclusivity, P permanence and T transferability; α, β, γ and δ are parameters and w_1 and w_2 are weights. The Q-measure takes values in the interval [0, 1]. A value of zero means that the property right has no quality; it is worthless. A value of unity means that the property right is perfect. Note that in the formula in (3.1), the first three property rights characteristics are considered essential. If any one of them is zero, the overall property right quality is also zero. The fourth characteristic, transferability, by contrast, is not essential. Even when there is no transferability, the quality of the property right may still be positive.

3.3 PROPERTY RIGHTS AND EFFICIENCY: BASIC RESULTS

We examine the relationship between property rights and economic efficiency in fisheries with the help of the simplest possible fisheries model. In spite of its simplicity, this model includes non-renewable resource utilisation and production independent of natural resources as special cases.

Consider production opportunities (or technology) that can be summarised by the instantaneous profit function $\Pi(q, x)$, where q represents the level of production and x the existing stock of a natural resource, and both variables, as well as the function itself, may vary over time. Although not explicitly stated, this profit function generally also depends on a number of parameters such as prices. We assume this function to be non-decreasing in x – the resource is not detrimental to profits – and concave in both variables. To be economically interesting we also assume that the profit function has a maximum at some positive level of production.

It should be noted that this specification of the profit function accommodates both conventional production where profits are independent of natural resources, i.e. $\Pi(q, x) \equiv \Pi(q)$, and any natural resource dependent industry.

In what follows we will sometimes find it useful to assume that there exists a collection of production technologies that can be described by the same functional relationship but with quantitatively different parameters. Importantly, these profit functions may vary over time. When necessary we will refer to different functions by an index $i = 1, 2, \ldots I$.

The natural resource evolves according to the differential equation:

$$\dot{x} = G(x) - q, \qquad (3.2)$$

where the concave function $G(x)$ describes the natural growth or renewal function of the resource. This formulation is also quite general. In the case of a nonrenewable resource, the natural growth function is simply identically zero, $G(x) \equiv 0$. In the case of a renewable resource, there exists a resource level $\bar{x} > 0$ and a (non-empty) range of resource levels defined by $\bar{x} > x \geq 0$ such that $G(x) > 0$.

Both functions, $\Pi(q,x)$ and $G(x)$, are only defined for non-negative levels of the variables. For mathematical convenience we assume that both functions are continuously differentiable to the extent required in the analysis.

3.3.1 Socially efficient production

One of the fundamental results of economic theory is that if all prices are true, i.e. reflect marginal social benefits (this excludes externalities, monopolistic pricing and lack of information) and if we ignore complications having to do with uncertainty, then profit maximisation is necessary for Pareto efficiency.[2] In what follows we will assume that all market prices are true in this sense.

Now let $\Pi(q,x)$ refer to the most productive technology (highest profit function) at each time. In that case, socially efficient utilisation of the resource requires solving the following problem:

$$\underset{\{q\}}{Max}\ V = \int_0^\infty \Pi(q,x) \cdot e^{-r \cdot t} dt , \qquad (3.3)$$

$$\text{subject to: } \dot{x} = G(x) - q ,$$
$$q, x \geq 0 ,$$
$$x(0) = x_0 .$$

The functional V in equation (3.3) measures the present value of profits from the production path $\{q\}$ with r representing the opportunity cost of capital per unit time, i.e. the rate of discount.

For subsequent reference it is useful to list some of the necessary (in this case also sufficient) conditions for solving this problem (Seierstad and Sydsaeter, 1987; Léonard and Long, 1993):

$$\Pi_q = \lambda,\ \forall t \text{ provided } q > 0 , \qquad (3.4)$$

$$\dot{\lambda} - r \cdot \lambda = -\Pi_x - \lambda \cdot G_x,\ \forall t , \qquad (3.5)$$

$$\dot{x} = G(x) - q,\ \forall t \qquad (3.6)$$

$$\lim_{t \to \infty} e^{-r \cdot t} \cdot \lambda \cdot (x^\circ - x^*) \geq 0 , \qquad (3.7)$$

where in the last (transversality) condition x^* represents the resource volume along the optimal path and x° any other feasible path.

---

[2] This is basically just an implication of the First Welfare Theorem (see e.g. Debreu, 1959; Arrow and Hahn, 1971).

In the above conditions, the new variable λ represents the shadow value of the resource at each point of time and equation (3.5) describes its evolution over time. If the profit function is independent of the resource, λ will obviously be identically zero at all times and the above conditions collapse to the familiar one, $\Pi_q = 0$, $\forall t$. If the profit function depends on a non-renewable resource, the dynamics of the shadow price of the resource are simplified to $\dot{\lambda} - r \cdot \lambda = -\Pi_x$, $\forall t$.

In the case of a renewable resource a profit maximising equilibrium may exist[3] defined by the equations:

$$G_x + \frac{\Pi_x}{\Pi_q} = r, \qquad (3.8)$$

$$G(x) = q. \qquad (3.9)$$

In what follows we will generally assume that a profit maximising equilibrium exists and is attainable in a profit maximising way.

The nature of the solution to problem (3.3) may be illustrated by a phase diagram as in Figure 3.4.[4] The dark heavy arrowed curve indicates the optimal approach path to equilibrium. The lighter curves (with arrows at the end) indicate other possible profit maximising paths. Note that paths ending with zero harvest (bending downward) can never satisfy the appropriate transversality conditions and therefore cannot maximise profits.

The profit maximising path obviously depends on the initial level of the resource, the parameters of the problem and the form of the profit and biomass growth functions. Let us write these solutions in a general form as:

$$x^*(t) = X(x(0); r, p; t), \qquad (3.10)$$

$$q^*(t) = Q(x(0); r, p; t), \qquad (3.11)$$

where p is there to remind us of other parameters such as prices.

Corresponding to these paths, there is a maximum present value of the program which we write as:

$$V(x(0); r, p). \qquad (3.12)$$

To calculate $V(x(0); r, p)$ the profit maximising paths in (3.10) and (3.11) must be worked out. That, however, is generally quite difficult.

---

[3] Given our specifications, an equilibrium always exists, but production in equilibrium may not maximise profits.

[4] This phase diagram is worked out on the basis of a very simple example of the functions, namely $G(x) = a \cdot x - b \cdot x^2$ and $\Pi(q, x) = q - c \cdot q^2/x$.

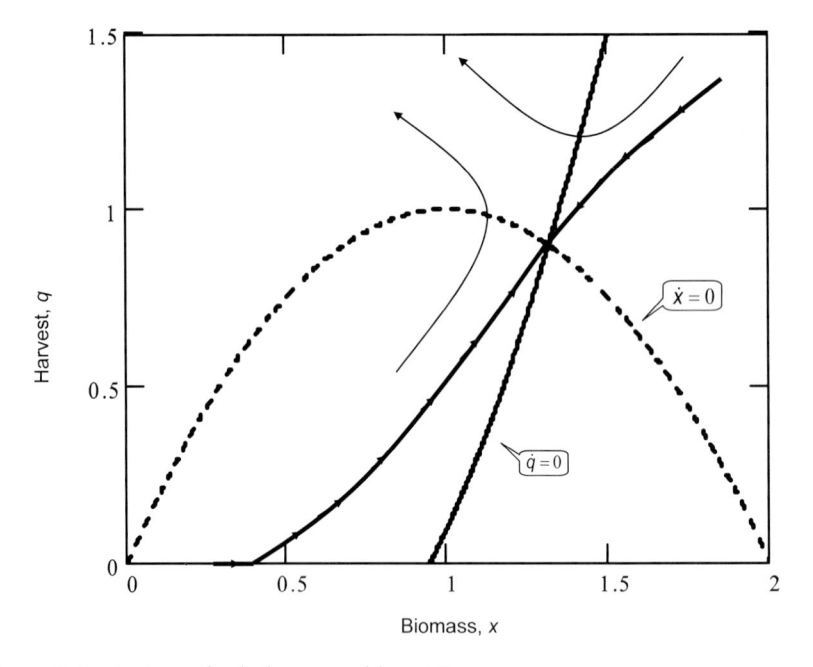

Figure 3.4 Nature of solution to problem 3.3.

It is sometimes convenient to examine the impact of exogenous changes, such as those in the quality of property rights, on efficiency by looking at the shadow value of the resource λ, along the solution to problem (3.3). The main reason is that λ is a measure of average resource rents (resource rents per unit of production) at each point of time. To see this, note that along the solution to (3.3) $\lambda = \Pi_q$ at all points of time. Resource rents, however, are defined as $R = \Pi_q \cdot q$ (Arnason, 2002). Second, note that resource rents are pure profits and thus constitute the lower limit of the contribution of the production activity to gross domestic product (GDP). In fact, if all firms are producing at the maximum of their average profit curves (a feature often associated with perfect competition), there will be little or no intramarginal rents and resource rents will be identical to the contribution to GDP. Third, note that λ may be regarded as the equilibrium price between the supply and demand of harvest. An independent owner of the resource would allow q to be taken from the resource at a price λ and an independent user would be willing to pay λ. This is particularly clear in equilibrium where the supply curve is $\lambda^s = \Pi_x/(r - G_X)$ according to (3.5), an upward sloping function of q and the demand is $\lambda^d = \Pi_q$, a downward sloping function of q.

3.3.2 Decentralised production

We now turn our attention to how decentralised, profit maximising firms would deal with the production problem. To do this we first have to specify the nature of their property rights.

The basic property right of relevance in this context is the firm's ability to produce/operate. As in most cases, this is a bundle of rights. Looking at the specification of the model, it appears that it contains four main subjects for property rights: (i) the quantity of production q, (ii) the volume of the resource, x, (iii) the resource renewal process, $G(x)$, and (iv) the production technology, $\Pi(q,x)$. Each of these is subject to the various dimensions of property rights discussed in section 3.2 including *exclusivity, durability, transferability* and *security*. Exclusivity, in this context, measures to what extent the firm can select production levels, keep control of the output and keep others from extracting from the natural resource or interfere with its renewal process. Security refers to the certainty to which the above rights will be held. Durability means that the firm can keep these rights and, therefore, carry on its operations as long as it wants. Finally, transferability means that the firm can sell off the selected operation if it so chooses.

3.3.3   Perfect property rights

We are now in a position to consider the role of property rights in the efficiency of our firm. We first consider the case of perfect property rights.

Under perfect property rights, the firm has control over all the variables in the profit maximisation problem subject only to limitations of technology, the laws of nature and the property rights of others as reflected in market prices. In terms of the various dimensions of property rights discussed in section 3.2 and above, the property rights in question are perfectly exclusive; no outside interference is possible without the owner's permission. They are also fully secure; they are held with full certainty. They are permanent; they last until the owner transfers them to someone else. Finally, they are fully and completely transferable.

Under these conditions, it should be clear that the problem facing a profit maximising firm holding these property rights is identical to problem (3.3). Hence, it will do its best to solve the social problem as expressed by (3.3). Moreover, note that to the extent that other firms can solve this maximisation problem better, a profit maximisation firm will seek to sell the operation to these more efficient firms.

3.3.4   The role of security

Security, as already stated, refers to the certainty to which property rights are held. For simplicity we will here regard only two alternatives; (I) that the firm retains its property rights intact and (II) that that the firm completely loses its property rights. The level of security is the probability that alternative (I) applies. The level of insecurity or risk is the probability that alternative (II) occurs.

Clearly property rights, here primarily the ability to utilise the resource, can be removed in many different ways. For instance, the government may simply abolish the firm's right to operate. The government may also pass a law making it impossible for the firm to operate profitably. A poor legal environment may generate a risk that other private agents usurp the firm's ability to use the resource and so on. The important point is that less than full security of property rights is not a feature of the world. It is a man-made arrangement.

Let the symbol ρ represent the probability that the firm will lose its property right during a period of unit length. It follows that $(1-\rho)$ is the probability it will retain the property during the same period. We refer to $(1-\rho)$ as the security parameter. Moving to continuous time, the probability that the firm will retain its property over a period of length t is $e^{-\rho \cdot t}$. Thus, for this simple case, reduced security, i.e. $\rho > 0$, works as an increased rate of discount.

Now, writing $i = r + \rho$, the firm's profit maximisation becomes

$$\underset{\{q\}}{Max}\ V = \int_0^\infty \Pi(q,x) \cdot e^{-i \cdot t} dt\ , \qquad (3.13)$$

$$\text{subject to:}\quad \dot{x} = G(x) - q\ ,$$
$$q, x \geq 0\ ,$$
$$x(0) = x_0$$

As we have already seen (equation (3.12)), profit maximising production depends on the rate of discount, in this case $i = r + \rho$. It immediately follows that less than full security, $\rho > 0$, will affect the production decisions of the firm compared to what would apply under full security. Since, as discussed above, perfect property rights lead to full efficiency, this establishes that, compared to perfect property rights, less than fully secure property rights are economically damaging. We formulate this as Result 1.

Result 1

Less than fully secure property rights are economically damaging.

The complete relationship between efficiency, in terms of the present value of profits, and the insecurity parameter, i.e. ρ, is more difficult to work out. However, a simple argument shows it must be monotonically declining, i.e. the greater the insecurity parameter, ρ, the less the economic inefficiency. We have already established in Result 1 that compared to $\rho = 0$, $\rho > 0$ is economically damaging. Clearly, for any level of ρ and r, there exists an equivalent level of rate of discount; $r°$, such that $r° = r + \rho$. Now, obviously $r°$ could have been the true rate of discount, in which case the production path selected by the private operator would have been profit maximising. It then follows from Result 1 that any increase in the uncertainty parameter would reduce profits. This establishes monotonicity.

Result 2

Economic efficiency declines monotonically with the level of uncertainty.

This monotonic relationship is illustrated in Figure 3.5 based on numerical calculations (see Appendix 3.1 for details) of profit maximising paths under different levels of security.

While the negative impact of less secure property rights on economic efficiency is monotonically increasing, the impact on the quantity of production and the resource level at each point of time is more complicated. In the case of renewable resources, insecure property rights generally lead to initially increased production rates and reduced biomass at each point of time compared to what would otherwise be the case. This is illustrated in the phase diagram in Figure 3.6, which illustrates the shift in profit maximising equilibrium curves when security is reduced. Obviously the optimal adjustment paths will be similarly shifted.

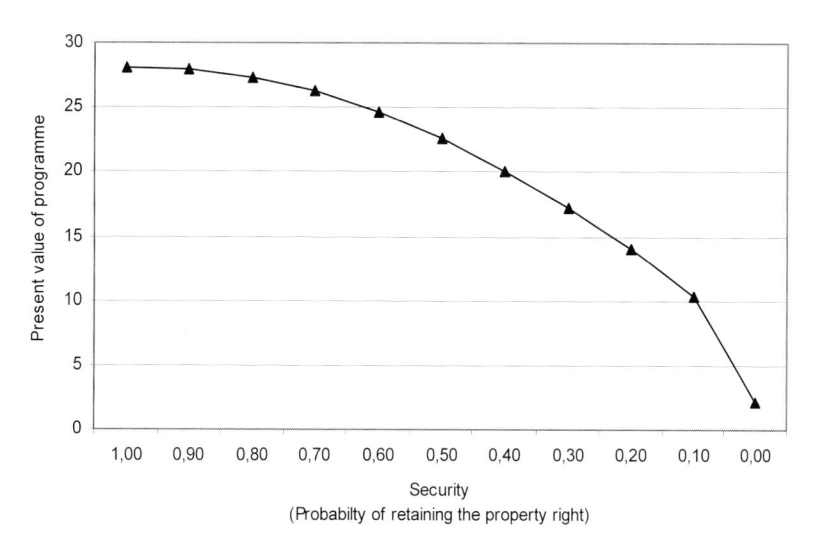

Figure 3.5 Relationship between security and efficiency.

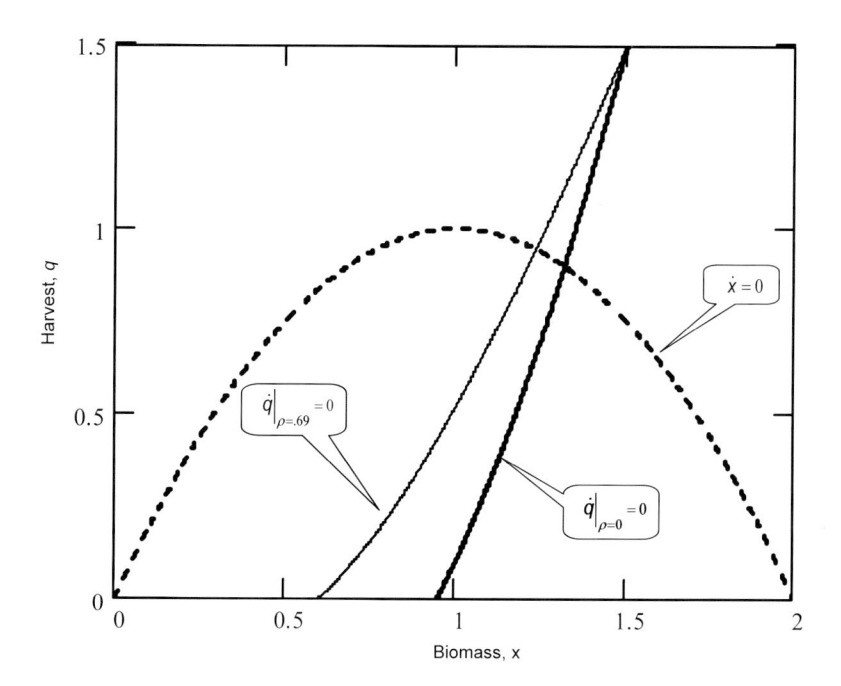

Figure 3.6 Impact of security on equilibrium curves:
($\rho = 0$ full security, $\rho = 0.69$ risk of losing property is 0.5 per period).

Finally, let us observe that reduced security also reduces the shadow value of biomass, λ, both in equilibrium and along adjustment paths. Indeed as security goes to zero ($\rho \to \infty$), $\lambda \to 0$. To see this note that according to (3.5) λ in equilibrium is:

$$\lambda = \frac{\Pi_x}{(i-G_x)} = \frac{\Pi_x}{(r+\rho-G_x)}.$$

Since both Π_x and G_x are bounded it must be the case that

$$\lim_{\rho \to \infty} \lambda = 0.$$

Similarly, as stated in (3.5) along the optimal path:

$$\dot{\lambda} = (r+\rho-G_x) \cdot \lambda - \Pi_x. \tag{3.5'}$$

But according to (3.4) $\lambda = \Pi_q \ \forall t$, which is obviously bounded above. So, clearly for (3.5') to hold, λ=0 along the optimal path.

So, if there is no security, λ=0 at all times. This means that the fishery behaves like a competitive fishery and economic rents ($R = \Pi_q \cdot q = \lambda \cdot q$) go to zero. This is verified by the calculations illustrated in Figure 3.5.

3.3.5 The role of exclusivity

Exclusivity refers to the ability of the property holder to use his property without interference by others. Within the framework of our basic fisheries model, (3.3), lack of exclusivity may take several forms including:

- Seizure of output
- Taxation
- Non-exclusive access to the natural resource
- Restrictions on activities.

The first two basically remove output from the property rights holder. The third reduces the availability of resource inputs. The final one puts limitations on how the property rights holder can use his property. All, however, alter the firm's opportunity set and therefore, in general, modify its behaviour.

Seizure of output, taxation

Output seizure and taxation are essentially the same operation. Both represent a subtraction from the instantaneous profit function of the firm. Let us write the modified profit function as $\Pi(q,x) - \tau \cdot q$, where $\tau \in [0,1]$ denotes the rate of taxation or confiscation. Obviously, a positive τ is going to affect the firm's behaviour compared to what would be socially optimal. Thus, for instance, under this regime the first equilibrium condition, equation (3.8), is altered to

$$G_x + \frac{\Pi_x}{\Pi_q - \tau} = r. \tag{3.8'}$$

So, the higher the fraction of output that is expropriated, the higher the marginal stock effect (Clark and Munro, 1975) and, consequently, the equilibrium biomass level.

A corresponding effect may be seen outside equilibrium. The following figure, Figure 3.7, draws a phase diagram for this case. As shown in the diagram, the impact of taxation is to turn the harvest equilibrium curve clockwise to the right. As a result the equilibrium is shifted toward a higher biomass level as stated in equation (3.8'). Moreover, most likely the optimal approach paths will involve lower extraction at all levels of biomass.

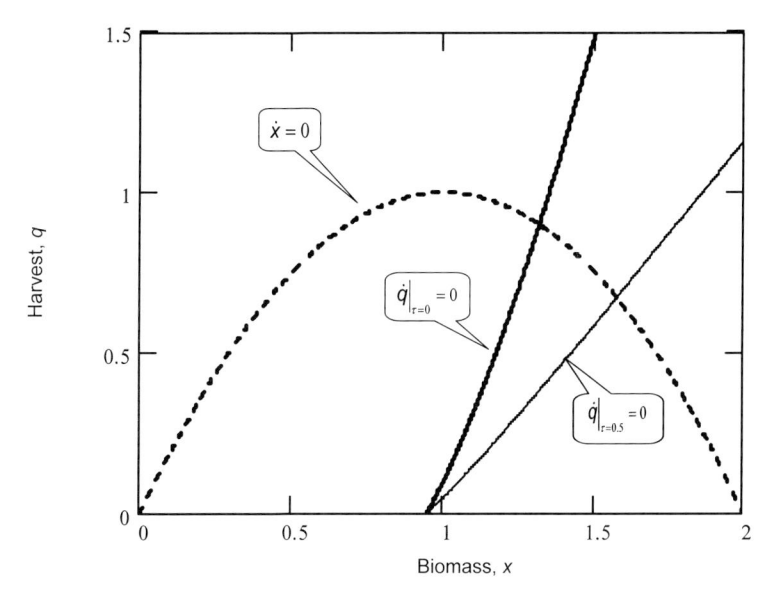

Figure 3.7 Effects of taxation on equilibrium curves.

By its impact on the economic return to the property rights holder, output taxation or expropriation will divert production from the socially optimal level. Consequently it is socially costly. How costly will depend on the situation and the level of taxation. Generally, however, taxation beyond a certain level, i.e. exclusivity below a certain minimum, will generally lead to termination of production and, therefore, no social benefits whatsoever.

An example of the relationship between the value (including the expropriated value) of the production path and the level of taxation is given in Figure 3.8. The relationship (based on the linear model defined in Appendix 3.1) illustrates the key results discussed above. The present value of the programme is monotonically declining in taxation and it converges to zero if the rate of taxation is high enough. Note that production ceases before 100% taxation because a certain minimum share of output is needed to cover the cost of production.

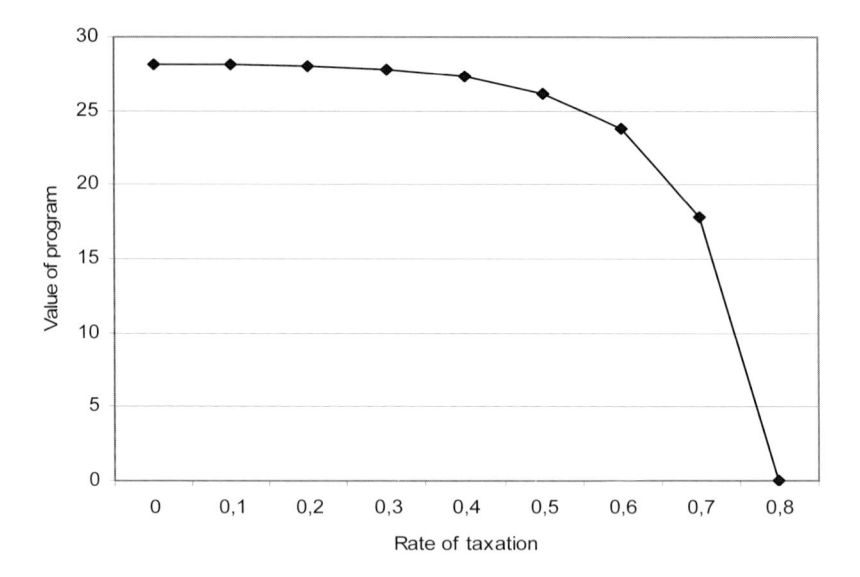

Figure 3.8 Present value of production (incl. taxation income) for different levels of taxation, τ.

Non-exclusive access to the resource: the common property problem

Exclusivity may also be reduced if other firms have access to the same resource. This is usually referred to as an external effect. One firm's production will reduce the availability of the resource to all other firms. This, as is well known, leads to a reduction in overall efficiency unless the rights holders can somehow co-ordinate their activities (Gordon, 1954; Hardin, 1968).

One simple way to see this is to note that the shadow value of the resource as seen by the firms is monotonically increasing with its share in total the resource use. More formally, it can be shown (Arnason, 1990) that in equilibrium:

$$\lambda(i) = \lambda^* \cdot \alpha(i), \qquad (3.14)$$

where $\lambda(i)$ represents firm i's shadow value of biomass, λ^* is the socially optimal shadow value, i.e., the one that a single owner would use, and $\alpha(i)$ is the share of company i in the resource use. Obviously, the lower a firm's share in the resource (low exclusivity) the lower will the firm's assessment of the shadow value of biomass be. In the limit, when the number of firms approaches infinity $\alpha(i) \to 0$ and, therefore, $\lambda(i) \to 0$. This means that the quality of the property right has been totally eroded and the fishery behaves like an open access, common property fishery. Since, as we have seen, total resource rents always equal $\lambda \cdot q$, where q is the production level, this situation is one where there are no resource rents.

The above argument establishes that the present value of production is (*ceteris paribus*[5]) monotonically declining in the number of companies exploiting the resource. It does not tell us what happens along the dynamic approach paths to this equilibrium. It is possible, however, to obtain some perspective of what is going on in disequilibrium, by looking at the phase diagrams in production-biomass space for different number of firms. This type of diagram is shown in Figure 3.9. For simplicity it is assumed that all the firms are identical. Thus, an increased number of firms automatically implies reduced production share for all companies.

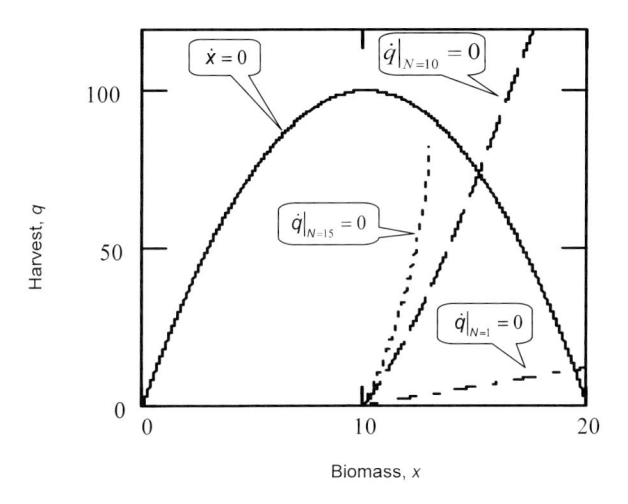

Figure 3.9 Effects of the number of firms (N): phase diagramme.

As shown in the diagram, the production equilibrium curve ($\dot{q} = 0$) is shifted to the left as the number of rights holders, denoted by N, increases. As a result, the equilibrium production is increased and biomass reduced compared to what would be optimal. Similarly, along the adjustment paths to equilibrium, for every level of biomass, production would exceed the socially optimal level. Moreover, it would appear from the diagram that in the limit, when the number of firms approaches infinity, the fishery converges to the common property fishery with zero economic rents in equilibrium. As indicated by equation (3.14) above, as well as Figure 3.10, the fishery would find itself in the neighbourhood of this situation for any number of firms exceeding 20 or so.

An example of the numerical relationship between the aggregate present value of production and the number of firms is illustrated in Figure 3.10. As in the corresponding Figures 3.5 and 3.8, this present value was calculated on the basis of the simple linear fisheries model in the Appendix 3.1 for a certain initial biomass level. Figure 3.10 is drawn on the basis of identical firms. Since the firms are identical the number of firms may be calculated as the reciprocal of the share.

---

[5] Since an increased number of companies does not necessarily imply reduced shares for all firms, the *ceteris paribus* simplification is invoked.

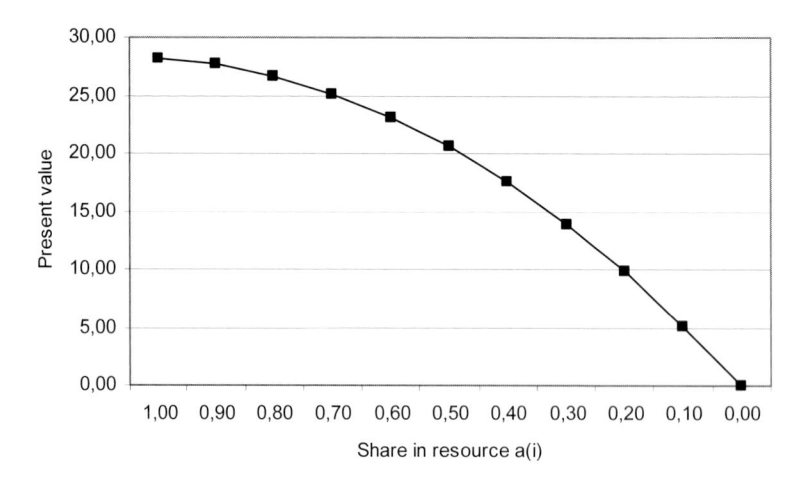

Figure 3.10 Present value of production for different share in the resource.

The above analysis has established the following result:

Result 3

Less than full exclusivity is economically damaging.

Although not formally proven, the above results also provide strong arguments for the following monotonicity result.

Result 4

Economic efficiency is monotonically increasing in exclusivity.

3.3.6 The role of permanence

Let us now consider the impact of a finite duration of the property right. Let this duration be T. So, in this case the firm knows with certainty that its property right is going to elapse at time T. Consequently, it attempts to solve the limited time problem:

$$\underset{\{q\}}{Max}\ V = \int_0^T \Pi(q,x) \cdot e^{-r \cdot t} dt, \qquad (3.15)$$

subject to the same constraints as before.

The necessary (and sufficient) conditions for solving this problem will be the same as for problem (3.3), except the transversality condition (3.7) is replaced by

$$\lambda(T) \geq 0,\ x(T) \geq 0,\ x(T) \cdot \lambda(T) = 0. \qquad (3.16)$$

So, if the resource is not completely exhausted, the profit maximisation period will end with the shadow value of the resource being equal to zero. This, of course, is as expected. Since the firm will lose its property right at *T*, from its perspective the resource will have no value from that time on. Note that since along the optimal path $\Pi_q = \lambda$, at the terminal time $\Pi_q = \lambda(T) = 0$. Therefore, at the terminal time, resource rents are zero just as in the common property case.

So, basically, a limited duration property right will induce the firm to move off the socially optimal production path in order to maximize its present value of profits over the duration of its property right. The impact of limited duration may be illustrated with the help of the phase diagram in Figure 3.11.

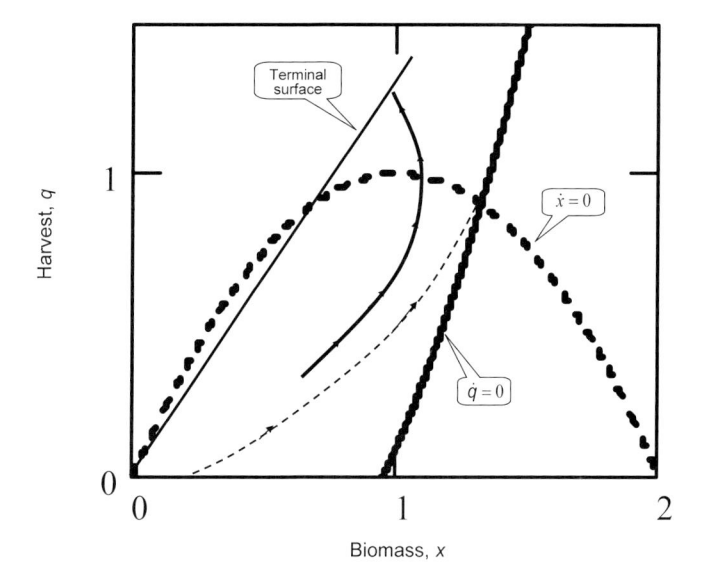

Figure 3.11 Limited Time: example of a profit maximising path.

As shown in Figure 3.11, the limited duration of the property right moves the optimal production-biomass path off the one to the long term equilibrium to a new path that terminates at a point satisfying the transversality conditions in (3.16). Note that the longer the duration of the property right, i.e., the higher *T*, the more similar to the socially optimal path will the production path be and vice versa. When $T \to 0$, the production will take place right at the terminal surface where $\Pi_q = \lambda = 0$. But $\lambda = 0$ is the defining characteristic of common property behaviour. It immediately follows that a sequence of very short-term property rights will operate very much like a common property regime.

Now to make further progress, let *V*(x(0), T)* denote the true maximised value function starting with biomass *x(0)* and ending at time *T*. This function is defined by:

$$V*(x(0),T) = \underset{\{q\}}{Max} \int_0^T \Pi(q,x) \cdot e^{-r \cdot t} dt + V*(x(T),\infty) \cdot e^{-r \cdot T}, \text{ s.t. } \dot{x} = G(x) - q \quad (3.17)$$

where $V*(x(T),\infty) \cdot e^{-r \cdot T}$, which corresponds to *V*(x(0), T)*, is the maximum value function starting with biomass *x(T)* and having an infinite time horizon. This function is the appropriate terminal (or scrap) function at time T for the true maximisation problem. Similarly, let $V(x(0), T)$ denote the maximum value for the firm faced with a limited duration property right of T. More precisely:

$$V(x(0),T) = \underset{\{q\}}{Max} \int_0^T \Pi(q,x) \cdot e^{-r \cdot t} dt; \text{ s.t. } \dot{x} = G(x) - q. \qquad (3.18)$$

Note that the difference between the maximisation in (3.17) and (3.18) is that the firm in the latter case, having a limited horizon, ignores the terminal value of the programme, $V^*(x(T), \infty) \cdot e^{-r \cdot T}$. It is precisely because of this that the firm will act suboptimally from a social perspective.

With these definitions, the following observations are now straightforward:

- $V(x(0), T) \leq V^*(x(0), T)$.

  To see this, it is sufficient to note that V^* is the maximum value function.

- $\lim_{T \to \infty} V(x(0),T) \to V^*(x(0),T)$.

  When $T \to \infty$ the two maximisation problems become identical.

- $\dfrac{dV(x(0),T)}{dT} > 0$

To see this, note that V cannot decrease when the restriction on the time horizon is reduced. Moreover if it is not optimal to end the programme before T is reached (in which case the optimal programme will be ended too and the two value functions will coincide), the value function must be increasing.

This shows that efficiency is monotonically increasing in the duration of the property right (or alternatively monotonically falling with reduced duration), asymptotically reaching the maximum attainable as duration or property rights permanence approaches infinity.

The above results are illustrated in Figure 3.12 which draws the present value of profits obtained from infinitely repeated limited duration periods of equal length as a function of the length of duration. The underlying model as before is the linear one specified in Appendix 3.1. The initial biomass level in all cases is one that can generate no profits (i.e. is on the intersection of the terminal surface and the horizontal axis, see Figure 3.11).

As predicted by the theory, the value of the programme converges to zero as duration of the property right approaches zero. Alternatively, if duration approaches infinity, the value of the programme approaches the maximum upper bound. Note that in this example over 99% of the maximum value of the programme is attained if duration of the property right is 30 periods (years) and about 85% if the duration is 10 periods. For shorter duration, the value falls very fast being only about 12% when the duration is one period.

The key point being illustrated in Figure 3.12 is the monotonicity of the relationship between duration and value of production. Different conditions such as initial biomass, biomass growth functions, maximum harvest levels and the rate of discount will affect the curvature of the relationship but not the basic monotonicity property.

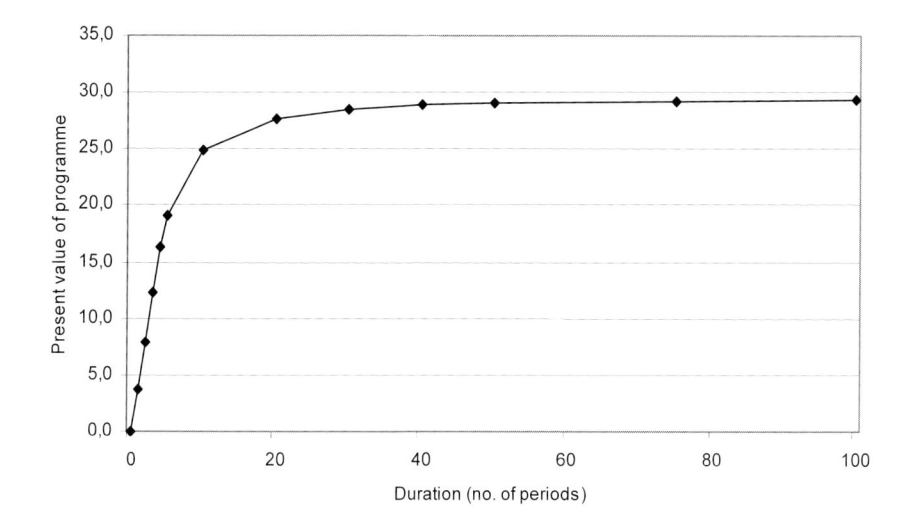

Figure 3.12   Value of production under a repeated limited duration property rights.

To summarise, we have established the following two results:

Result 5

Limited duration property rights are economically damaging.

Result 6

Economic efficiency is monotonically increasing in duration.

3.3.7   The role of tradability

To attain economic efficiency, only the firms having the best technology (highest profit function) should carry out the production at each point of time. If the property rights, on which firms base their production decisions, are transferable, private profit maximisation will tend to ensure that this will be the case.

Consider the firm's bundle of rights as one consolidated property right. If markets are effective, there will at each point of time be a market price for this property right. This price will, among other things, depend on the quantity of the resource. It will also evolve over time with general and firm specific technical progress and other developments. Seen from the present, let us write this price as:

$$F(x(t), t) = P(x(t), t) \cdot e^{-r \cdot t} \tag{3.19}$$

Under these conditions the property rights holder will at each point wonder whether or not he should sell his property. This problem can be formally written as follows

$$\underset{\{q\},T,x(T)}{Max} \quad V = \int_0^T \Pi(q,x) \cdot e^{-r \cdot t} dt + P(x(T),T) \cdot e^{-r \cdot T}, \qquad (3.20)$$

$$\text{subject to:} \quad \dot{x} = G(x) - q,$$
$$q, x \geq 0,$$
$$x(0) = x_0.$$

The solution to this problem is the same as in problem (3.3) except the transversality condition, (3.7) is now replaced by the two transversality conditions

$$\lambda(T) = P_x(x(T),T) \cdot e^{-rT}, \qquad (3.21)$$

$$H(q(T),x(T)) = r \cdot P(x(T),T)) - \dot{P}, \qquad (3.22)$$

where $H(q(T),x(T))$ represents the Hamiltonian function at time T and $\dot{P}$ denotes the firm's expected rate of price growth. So $\dot{P}$ is basically the firm's speculative component. For simplicity of exposition we ignore this variable (assume it equals zero) in what follows.

Conditions (3.21) and (3.22) basically stipulate when to sell the property right and at what level of biomass, $x(T)$. Under perfect market conditions, when trades take place, the price, $P(x(T),T))$, equals the maximum value function of the other firm, i.e. $P(x(T),T)) = V_2(x(T),T)$. But, as is well known, the maximum value of the (present value) Hamiltonian function is just the interest on the maximum value function.[6]

$$H^*(x(T),T) = r \cdot V(x(T),T), \qquad (3.23)$$

where the symbol * indicates the maximised present value Hamiltonian. It immediately follows that the trading condition, (3.22), is the simple rule that the property right should be transferred to the firm with the higher value function. More formally, we may restate (3.22) as:

$$\text{Sell if } V^1(x(T),T) \leq V^2(X(T),T), \qquad (3.22')$$

where $V^1(x(T),T)$ denotes the value function of the first firm, i.e., the one holding the property right, and $V^2(X(T),T)$ the value function of the second firm, i.e. the one willing to buy the property right.

On our previous assumptions, basically all prices being true, the rules expressed in (3.21) and (3.22) are also socially optimal. It immediately follows that any limitations on tradability can only reduce the social benefits derived from the resource use.

---

[6] This is easily seen from basic equation of dynamic programming, namely:
$r \cdot V(x(T),T) = \underset{q}{Max}[H(q,x)]$. See e.g. Kamien and Schwartz, 1981.

A priori, however, there is not much more of general validity that can be said of the social costs of limitations on trade. If the current property rights holder is the most efficient firm from now to eternity, restrictions on his ability to trade will not lead to loss of social benefits. Indeed, the limitations on trade will turn out to be non-binding. If, on the other hand, there are or will be more efficient firms, there will be social costs of trading limitations. These costs will obviously be monotonically increasing in the efficiency differential and the sooner in time it emerges.

Economic experience shows that there is ample reason to expect existing firms to lose their advantage over newer firms over time and gradually fall behind in efficiency. The reasons for this are many and it is beyond the scope of this paper to discuss them. Let it suffice to say that this is the main reason why most firms do not last for a long time and none indefinitely.

To give some idea of the efficiency impact of this, let us assume a constant productivity growth differential between existing property rights holders and other firms not holding the property right. Assume the productivity growth is neutral and denote the productivity growth differential by the symbol ε. Thus the profit function of the most efficient firm evolves according to the expression $\Pi \cdot e^{\varepsilon t}$, where Π is the profit function. Now assume that at some point the maximised Hamiltonian function of the property rights holder is equal to the maximised Hamiltonian function of the most efficient firm. Then, according to (3.23)

$$H_1^*(x(T),T) = r \cdot V^1(x(T),T) = (r-\varepsilon) \cdot V^2(x(T),T) = H_2^*(x(T),T),$$

where the index 1 refers to the property rights holder and 2 to the other most efficient firm. It immediately follows that the relative efficiency is

$$\frac{V^1}{V^2} = \frac{r-\varepsilon}{r}.$$

Note that this efficiency ratio is time independent. It compares the value of the total programme if the property right is kept by a single company with the value of the programme if the most efficient company always has the property right.

On this basis we can illustrate the loss in efficiency due to non-tradability as in Figure 3.13. Obviously, the loss in relative efficiency is monotonically increasing in the productivity growth differential. Limited tradability would reduce this loss. Full tradability would eliminate it. Note also that if the rate of productivity growth exceeds the rate of discount, the present value integral does not even converge and the relative efficiency is not defined.

From this we can surmise that efficiency is most likely monotonically increasing and certainly non-decreasing in tradability. However, even with no tradability, there would normally be substantial economic rents. This is different from the other dimensions of property rights as we have seen.

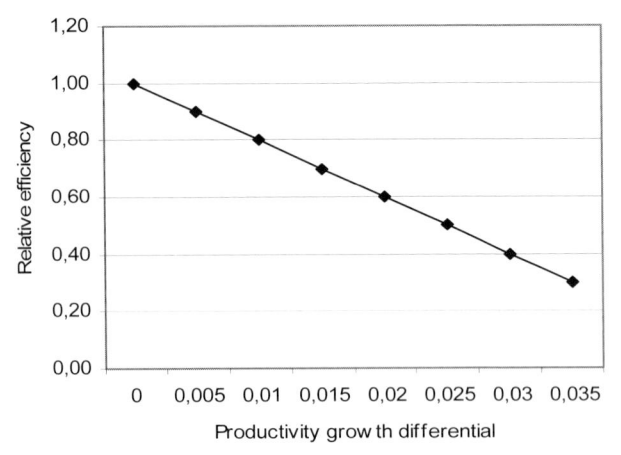

Figure 3.13 Non-tradability: efficiency loss.

3.4 CONCLUDING REMARKS

Above, we have established that economic efficiency is monotonically increasing in three of the four main dimensions of property rights, i.e. security, exclusivity and duration. We have also shown that efficiency is non-decreasing in the fourth dimension, tradability. This means that if security, exclusivity or duration are reduced, even minutely, there will be a reduction in the efficiency of the associated economic activity. Moreover, if any of these variables are reduced to zero, the activity will become wholly inefficient in the sense that it will not produce any economic rents. The outcome may actually be even worse. It is entirely possible, even likely, that the activity in question will simply cease. Certainly, investment in physical and human capital, barring subsidies and other public interventions, will be greatly reduced as well as distorted. In the case of natural resources, the resource may even be exhausted beyond its ability to regenerate itself.

The case of reduced tradability is much less dramatic. Even with no transferability, there is every reason to believe that economic rents will continue to be generated. Moreover, if the agent holding the property right is reasonably efficient, the cost of non-tradability will be comparatively small. However, over time the relative efficiency of any firm or agent tends to decline. Therefore, at least in the long run, the cost of limited tradability can be high.

All this we have been able to deduce by purely analytical means. What the analysis has been unable to tell us, though, is the quantitative relationship between property rights and economic efficiency between the two extreme point of no and complete property rights. All we know about this relationship is that it is monotonically increasing in the first three dimensions of property rights and non-decreasing in the third. To go much further, not to mention determine the exact quantitative relationship, as we did in the examples provided in section 3.3, we must know the profit function in question, the natural renewal function, the parameters of the situation and the initial state of the resource. This problem, not surprisingly is fundamentally empirical.

This, of course, implies that it is not possible to construct a simple measure of property rights quality, such as the Q-measure in equation (3.1) above, that can also serve as a measure of the associated economic efficiency. In other words, a cardinal measure of this kind does not seem to be obtainable. However, to the extent that there is a monotonic relationship between property rights quality and economic efficiency, it should be possible to construct an index of property rights quality such that the efficiency of the associated economic activity is monotonically increasing in the index. The Q-measure, perhaps, comes close to being such an index. However, since, as we have seen, efficiency is not necessarily increasing in tradability, this basic monotonic property does not necessarily hold for tradability. It holds for the other three main dimensions of property rights, however.

The property rights theory that has been developed has certain clear applications to the theory of fisheries management.

- Fisheries management methods that are not based on property rights are unlikely to work except in cases where there is basically no room for individual firms to manoeuvre (like 100% taxation, complete control of the fishery and so on).
- The better the property rights, certainly along the first three dimensions, the greater the efficiency. Thus, all deviations from perfect property rights will be economically costly. The only question is how costly.
- Property rights quality indices such as the Q-measure can provide a useful indication of the efficiency of the fisheries management system in question.
- The invention of better property rights in fisheries and improved protection of existing property rights can be extremely valuable. Economically speaking there is a reason to encourage activity and enterprise in these areas.

Let us conclude by illustrating the use of this theory by looking at particular cases of property rights in fisheries.

Iceland, New Zealand and Norway all base their fisheries management on individual quota property rights. In Iceland and New Zealand the regime is a fairly complete ITQ system. Norway, by contrast, operates an IQ system, i.e. an individual quota system with very limited transferability of the quotas. In all three countries, the security of the property right is fairly high. However, in Norway, in the most important fisheries, new vessels may be allocated quotas, thus subtracting from the quota shares of the other fishing vessels. Clearly this reduces the security of the Norwegian property right. In all three countries the exclusivity of the harvesting right is pretty high, really limited only by government fisheries regulations which in the case of Iceland and in particular Norway are more extensive than those in New Zealand. Permanence of the property right differs greatly between the countries. In New Zealand the quota rights are explicitly perpetual. In Iceland they are of indefinite duration but there are non-trivial socio-political threats to the continuation of the system. In Norway individual quota rights are explicitly non-permanent, allocated only for a year at a time. However, since quotas are customarily allocated to the previous recipients in more or less the same proportions, it may be claimed that the associated property right has gained a degree of permanence. Finally, transferability in New Zealand is close to perfect (only foreigners are excluded). In Iceland, transferability is only slightly more restricted. In Norway, as mentioned above, there is virtually no transferability of the quotas.[7]

A rough numerical estimate of the values of the property rights characteristics for these three countries based on the above description is provided in Table 3.1.

Table 3.1   Estimated quality of quota property rights: Iceland, New Zealand and Norway: Q- values.

Characteristics	Iceland	New Zealand	Norway
Security	1.00	1.00	0.90
Exclusivity	0.90	0.95	0.70
Permanence	0.80	1.00	0.50
Transferability	0.90	0.95	0.10
Q	0.86	0.96	0.44

Q-values calculated as: $Q \equiv S^{\alpha} \cdot E^{\beta} \cdot P^{\gamma} \cdot (w_1 + w_2 \cdot T^{\delta})$, where $\alpha=\beta=\gamma=\delta=1$; $w_1=0.6$, $w_2=0.4$

According to the Q values reported in Table 3.1, the quality of the New Zealand quota property right, $Q=0.96$, is near perfect. The property rights quality of Iceland's quota rights, $Q=0.86$, is considerably lower but still quite high. The property rights quality of Norway's fishing rights, $Q=0.44$, is much lower than that of both New Zealand and Iceland. Thus, although of substantially higher quality than common pool property rights, for which Q-values are typically in the range $Q=[0.05-0.2]$, Norway's IQs must be regarded as comparatively weak property rights.

How do these Q-measures correspond to the efficiency of the respective fisheries? The available evidence suggests that the Icelandic and New Zealand fisheries, which score much more highly on the Q-measure than Norway, also have much more efficient fisheries. Moreover, the indications are that New Zealand, which scores higher than Iceland on the Q-measure, also has a slightly more efficient fishing industry.

While calculations of this kind are fairly rough, they nevertheless suggest that overall measures of property rights quality such as the Q-measure, can serve as a short-hand assessment of the economic efficiency of the fisheries management systems in question. Needless to say, this theory generalises to other areas of resource use including the environment.

---

[7] Fishing vessels can be sold with their quotas attached and two vessels can combine their quotas provided one of them leaves the fishery. In the latter, however, the combined quota is reduced slightly and can only be kept for 13 years. This rule is now in the process of being changed so that 100% of the combined quota is retained and it can be kept indefinitely.

APPENDIX 3.1
Fisheries models used in calculations

A3.1.1 *For generation of phase diagrams*

The following basic fisheries model was used to generate the phase diagrams in the text:

$$\Pi(q,x) = p \cdot q - c \cdot \frac{q^2}{x}, \tag{A.1}$$

$$\dot{x} = G(x) - q = \alpha \cdot x - \beta \cdot x^2 - q, \tag{A.2}$$

where q represents production (harvest) and x biomass.

A3.1.2 *For calculating the value of profit maximising paths*

For the calculation of the value of profit maximising paths the following linear version of the basic model was used:

$$\Pi(q,x) = p \cdot q - c \cdot \frac{q}{x} \tag{A.3}$$

$$\dot{x} = G(x) - q = \alpha \cdot x - \beta \cdot x^2 - q \tag{A.4}$$

$$q \in [0, e_{max}] \tag{A.5}$$

This kind of a model implies a bang-bang profit maximising behaviour with the control variable, q, jumping between its upper and lower bound and its singular value, if it exists. The singular value for q (and x) is found by solving the two equations for q:

$$\alpha - 2 \cdot \beta \cdot x + \frac{c \cdot q}{(p \cdot x - c) \cdot x} = r, \tag{A.6}$$

$$\alpha \cdot x - \beta \cdot x^2 - q = 0, \tag{A.7}$$

where r is the rate of time discount.

The value of the programme (the present value of optimal paths) depends on the initial biomass level and any constraints (such as limited time and so on) as well as the parameters of the problem. The calculation of the value of the programme consists of determining the optimal paths under the given conditions and calculating the corresponding present values.

References

Alchien, A. 1965. Some economics of property rights. *Il Politico* 30: 816-29.

Arnason, R. 1990. Minimum information management in fisheries. *Canadian Journal of Economics* 23: 630-53.

Arnason, R. 2000. Property rights as a means of economic organization. In R. Shotton (ed.) Use of Property Rights in Fisheries Management. *FAO Fisheries Technical Paper* 401/1. Food and Agriculture Organization of the United Nations, Rome.

Arnason, R. 2002. Resource rent taxation: Is it really nondistortive? A paper presented at the 2nd World Congress of Environmental and Resource Economists, Monterey.

Arrow, K.J. and F.H. Hahn. 1971. *General Competitive Analysis*. Holden-Day. San Franscisco.

Clark, C.W. and G.R. Munro. 1975. The economics of fishing and modern capital theory: A simplified approach. *Journal of Environmental Economics and Management* 2: 92-106.

Debreu, G. 1959. *Theory of Value*. Yale University Press, New Haven.

Demsetz, H. 1964. The exchange and enforcement of property rights. *Journal of Law and Economics* 3: 1-44.

Demsetz, H. 1967. Toward a theory of property rights. *American Economic Review* 57: 347-59.

Gordon, H.S. 1954. Economic theory of a common property resource: The fishery. *Journal of Political Economy* 62: 124-42.

Hardin, G. 1968. The tragedy of the commons. *Science* 162: 1243-47.

Kamien, M.I. and N.L. Schwartz. 1981. *Dynamic Optimization: The Calculus of Variations and Optimal Control in Economics and Management*. North Holland, Amsterdam.

Léonard, D. and N. van Long. 1992. *Optimal Control Theory and Static Optimization in Economics*. Cambridge University Press.

Scott, A.D. 1955. The fishery: The objectives of sole ownership. *Journal of Political Economy* 63: 116-24.

Scott, A.D. 1983. Property rights and property wrongs. *Canadian Journal of Economics* 16: 555-73.

Scott, A.D. 1989. Conceptual origins of rights based fishing. In P. Neher, R. Arnason and N. Mollett (eds) *Rights Based Fishing*. Kluwer Academic Publishers, Dordrecht.

Scott, A.D. 1996. The ITQ as a property right: Where it came from, how it works and where it is going. In B.L. Crowley (ed.) *Taking Ownership: Property Rights and Fisheries Management on the Atlantic Coast*. Atlantic Institute for Market Studies, Halifax.

Scott, A.D. 2000. Introducing property in fisheries management. In R. Shotton (ed.) Use of Property Rights in Fisheries Management. *FAO Fisheries Technical Paper* 404/2. Food and Agriculture Organization of the United Nations, Rome.

Seierstad, A. and K. Sydsaeter. 1987. *Optimal Control Theory with Economic Applications*. North Holland, Amsterdam.

Smith, A. 1776. *An Inquiry into the Nature and Causes of the Wealth of Nations*. Edition by E. Cannan 1977. University of Chicago Press, Chicago.

4

How to Resolve the Class II Common Property Problem? The Case of British Columbia's Multi-Species Groundfish Trawl Fishery

R. Quentin Grafton
Harry W. Nelson
Bruce Turris

The market sends out incorrect signals to the participants in the fishery. Input controls constitute an attempt to address the problem by making it difficult for participants to respond to incorrect market signals. Output controls, on the other hand, change the market signals themselves.

— G.R. Munro and A.D. Scott (1985: 661)

4.1 INTRODUCTION

In what is now a classic in the fisheries economics literature, Munro and Scott (1985) divided the economic problems of fisheries into class I problems, where the absence of regulation results in dissipation of rent, and class II problems where authorities set a total allowable catch (TAC), but fail to prevent crowding and the race to fish that result in rent dissipation. After reviewing the outcomes of fisheries regulations, especially in Canada, their insight was to suggest the use of individual transferable quotas (ITQs) that they foresaw would both change the incentives of fishers and help prevent rent dissipation.

In the 20 years since their work, ITQs have been applied in a large number of fisheries in several different countries, including Canada (Kaufmann et al., 1999), New Zealand and Iceland (Hannesson, 2004). Despite the apparent success of ITQs in delivering substantial economic benefits (Fox *et al.*, 2003; Grafton, 1996; Grafton *et al.*, 2000; Shotton, 2001; Weninger, 1998) these advantages have largely been identified in so-called 'single-species' fisheries, where fishers are able to target particularly species of fish. What is less clear is whether the difficulties and complexities of managing multi-species fisheries (Holland and Maguire, 2003) mean that ITQs can not be adopted for such fisheries (Boyce, 1996) with similar success.

One of the perceived problems associated with multispecies fisheries includes discarding at sea (Anderson, 1994; Arnason, 1994; Boyce, 1996). Although this problem also exists even in the absence of ITQs, quota systems can also induce discarding at sea (Copes, 1986; Turner, 1997). Indeed, even in so-called single species fisheries this problem exists, although the regulatory system simply ignores the issue by excluding bycatch and discards from the management regime. Perhaps more problematic is that *if* fishers have a jointness-in-inputs technology (Kohli, 1983), such that they have no ability to separately target species, their catch mix may not match their quota allocation. A limited ability to target individual species could also contribute to undesirable quota underages and overages, and may severely constrain economic returns if total harvests are limited to protect vulnerable species, or lead to unsustainable harvesting of some stocks if managers ignore the bycatch or discarding issues (Squires *et al.*, 1998). Although some types of technology or gear, such as longlines, provide greater flexibility in catching individual species than other types of gear, such as trawls (Squires and Kirkley, 1991), all fishers can to some extent adjust their fishing behaviour to increase their ability to target fish.

Using insights from the British Columbia (BC) groundfish trawl fishery, managed by ITQs since 1997, we show that with the appropriate incentives fisher behaviour can be modified to meet the twin goals of sustainability and economic efficiency. We also argue that an ITQ system can be effectively implemented in a multi-species fishery that includes over 55 distinct quotas while also addressing the discarding issue, equity issues and improving economic returns and sustainability. The experience of the BC groundfish trawl fishery demonstrates that the common property problem can be overcome if the incentives can be changed. The approach adopted in this fishery has been to change the incentives that, in turn, modify fisher behaviour. The end result has been a transformation from a 'race to the fish' mentality towards one oriented towards reducing costs, maximising value, and improving management.

The implications for multi-species fisheries management are significant. If the basic management approaches used in the BC groundfish trawl industry can be effectively applied elsewhere they have the potential to both mitigate sustainability challenges and raise resource rent. More generally, we argue that the lessons learned apply to single-species fisheries as well, and that the issue of discards and bycatches should be incorporated into the management regime to improve both the sustainability as well as the economic performance of the fishery.

The chapter is divided as follows. First, a description of the BC groundfish trawl fishery and the management regime in place prior to the introduction of the ITQs in 1997 is provided. This is followed with a detailed discussion of the current ITQ system and how the management programme has helped to develop the 'right' incentives for promoting sustainability and enhanced economic benefits from fishing. The experiences of this fishery show that ITQs have contributed to conservation objectives, led to substantially improved profitability, and improved the collection of scientific information to help regulate the fishery. Several unique features of the ITQ system designed to address concerns over concentration of quota ownership and the associated distributional effects on processors, coastal communities, and crew members are also reviewed.

4.2 THE BRITISH COLUMBIA GROUNDFISH TRAWL FISHERY

To understand what the incentive approach to fisheries management has achieved in the Groundfish Trawl Fishery (GTF) we first review the major features of the fishery and its history.

4.2.1 Background

The GTF is the largest wild fishery in terms of both the value and total catch in British Columbia (BC) (Gislason et al., 2004) that lands about 100,000 tonnes (t) of fish per year worth some CDN$ 60-70 million (see Table 4.1). The harvest consists of dozens of different species caught exclusively off the BC coast. Important species in terms of catch are rockfish, hake, Pacific cod, thornyheads, sole and lingcod (halibut are excluded as they fall under a separate licensed fishery). In terms of volume, the single largest species caught is hake that accounts for about twice as much in terms of landed weight than the catch of all other species combined. However, its price of 7 to 10 cents per pound is much less than the average 50-60 cents per pound received for other species. Other fish harvested by the trawl fleet also include sablefish, dogfish, turbot, skate, flounder and other groundfish. Management is also complicated by the fact that many species caught in the fishery have location–specific populations although some, such as sablefish and hake, can be harvested over the entire BC coast.[1]

Table 4.1   Quantity (thousands of tonnes) and value of landings (nominal values in Canadian dollars) in the British Columbia trawl fishery 1993-2002.

	1993	1994	1995	1996	1997	1998	1999	2000	2001	2002
Landings (000 t)	122.6	164.6	121.3	138.0	129.3	129.5	124.0	58.3	92.9	98.1
Landed value (CDN$ millions)	42	50	45	42	48	57	62	58	57	66

Source: G. S. Gislason & Associates

Fish are caught in trawl nets that may extend as much as 1500 metres behind the vessel. Depending upon the species, fishers may harvest along the ocean floor for many rockfish and other groundfish, while hake, pollock, and some rockfish species (i.e. greenies and brownies) are targeted with mid-water trawls. All species can be harvested year round, although there can be seasonal variations as well as annual fluctuations in relative abundance.

---

[1] For example, there are five distinct stocks of Pacific Ocean perch fished; one on the lower west coast of Vancouver Island and another on the upper west coast, Hecate Strait, the west coast of Queen Charlotte's and Queen Charlotte Sound.

The large number of species that can be caught poses various difficulties for the industry. First, there is the complexity inherent in determining stock sizes and abundance for dozens of species. Second, the challenge exists to reconcile, within an ITQ system, the actual catches to initial fisher allocations plus net trades. Third, the problem of ensuring stocks are protected from unreported discarding, while at the same time allowing fishers to harvest up to the sustainable catch levels.

4.2.2 Historical development of the fishery

A commercial groundfish trawl fishery has existed in BC for over 60 years. In the early 1960s the fishery consisted of 80 trawlers, of which about half were operated on a full-time basis. In the mid-1960s foreign fleets arrived in BC waters, initially targeting Pacific Ocean perch, and then hake. By the early 1970s, the catch of foreign fleets was significantly higher than the Canadian harvest for a number of key species. In 1977 Canada extended its fisheries jurisdiction to 200 miles, and since then foreign fishing has been phased out.

In 1960 a key report was released on the state of BC fisheries that recommended the use of vessel licensing to help prevent the further expansion of fishing effort (Sinclair 1960). In 1969 limited-entry licensing was implemented in the commercial salmon fishery. In 1976 limited licensing was introduced into the GTF in the form of groundfish trawl *T* licences. *T* licences were allocated to 142 vessels that were allowed harvest multiple species, and permitted fishers to catch groundfish anywhere along the Canadian west coast. The licences also created categories of prohibited species (halibut, salmon and herring) that fishers were not allowed to keep, as well as permitted species. A limited entry *L* license (by hook & line gear) for halibut was implemented in 1979 followed by limited entry *K* licences (by hook & line and trap gear) in 1981 for sablefish. The groundfish trawl fleet was allocated 8.75% of the annual sablefish TAC. During the late 1970s the fishing fleet experienced a significant expansion due to favourable tax treatment and price subsidies designed to replace the effort formerly undertaken by foreign vessels in Canada's exclusive economic zone.

Despite the significant increase in landings in the 1970s, low prices, persistent unprofitability, and a significant amount of idle capacity characterised the GTF. The proposed remedy was given in a 1982 Royal Commission Report that recommended that 10-year quotas be established for groundfish species, individually where they were targeted separately, and issued by zone, with temporary permits issued on a year-to-year basis to handle fluctuations in stock abundance (Pearse, 1982, 130-132). Unfortunately, none of the recommendations were implemented, and the groundfish trawl fleet continued to expand its harvesting capacity and overall harvest levels as predicted by Munro and Scott (1985) in their description of the class II common property problem.

Prior to the establishment of ITQs in the fishery in 1997, the TACs for individual species in the fishery were specified for the whole of BC by adding individual sub-stock TACs together. The annual species TACs were also divided into four quarters, corresponding to a 12-month fishing season. Within the overall season were monthly fishing periods, and vessels could choose different fishing options (2, 4 or 15 trips per month) with vessel trip limits for each species. The trip limits were calculated by estimating fishing effort for the quarter and were reduced accordingly as the total landed catch approached the quarterly TAC. The trip limit was also related to the fishing option chosen by fishers such that more trips per month option had smaller trip limits. These individual trip allocations were non-transferable. Indeed, the incentives for misreporting and discarding were magnified by the fact there were individual species limits and, thus, fishers were obliged to curtail their trip if they exceeded just one of their individual limits out of the many species harvested on a trip. As trip limits were reduced, the problems of misreporting and discarding worsened.

In addition to trip limits and TACs on individual species, each fisher also faced various vessel and gear restrictions that included regulations on vessel size, gear and the ability to combine multiple *T* licences on one vessel. Fishers were also required to maintain daily fishing logs recording their catch, and the regulator introduced 100% monitoring of all landed catch by independent dockside monitors.

Despite these controls harvests consistently exceeded coast-wide TACs for a number of species in the 1990s, and fisheries managers increasingly had to specify shorter trip limits in order to control effort. As fishing trip duration declined, the ability of fishers to modify their fishing operations to avoid reaching individual species limits diminished, magnifying the incentive to discard overages and bycatch. In addition to the impact on the sustainability of some stocks, the costs of fishing were also rising and increased the 'race to fish'. Moreover, the landed value of fish declined because of reduced quality and because an increased proportion of fish were sold as a frozen product due to the ever shortening fishing season.

The key difficulty faced by fishery managers was how to manage the stocks in the absence of reliable information on stock specific harvesting at sea and the level of discarding (Walters and Bonfil, 1999). One of the principal problems identified at the time was that stocks could not be managed on a stock specific basis as reporting by fishers had little or no credibility. Moreover, there was no ability to assess whether or not harvests were being taken out of a specific area, potentially endangering vulnerable sub-stocks. Given the lack of information, there was also much uncertainty over stock assessments, and in response to this uncertainty, fishery managers increasingly took a precautionary approach to management by setting lower TACs in the belief that the current TACs were likely to be exceeded. By 1995, the official catch exceeded the coastwide TACs for several species. As a result, fishery managers took the unprecedented action of closing the entire GTF in September 1995.[2]

The groundfish trawl fishery reopened in February of 1996 with 100% at-sea observer coverage.[3] This addressed the concerns of stock specific management and discarding as fisheries managers now had, for the first time, a reliable means of quantifying discards and identifying where the fish were coming from, reducing the main sources of uncertainty in managing within TACs. The fishery could now be managed on a stock-specific basis rather than on a coast-wide basis.

---

[2] It should be noted that other high profile fisheries in Canada, most noticeably the east coast cod fishery and the west coast salmon fishery, had recently run into highly publicised problems with low returns and in fact, the Fraser River sockeye salmon run (the most important in the province) had been shut down in August, only a month earlier.
[3] Some fisheries have only partial observer at sea coverage, such as halibut, sablefish, schedule II fisheries and a localised in-shore fishery within the trawl fishery (Option B fishery).

4.3 ESTABLISHMENT OF ITQs IN THE BRITISH COLUMBIA GROUNDFISH TRAWL FISHERY

To address the economic problems of the fishery, fishers agreed to implement what was initially a trial of ITQs in April 1997. As part of the negotiations over allocation of quota that included a range of interests (crew, shoreworkers, processors, fishing communities and licence holders) only 80% of the species TACs were allocated to *T* licence holders. The remaining 20% was placed under the purview of a newly created non-profit society called the Groundfish Development Authority (GDA) charged with promoting regional development, market and employment objectives, sustainable fishing practices and fair and safe treatment of fishing crews.

4.3.1 Allocation and transferability

The TACs were first divided between trawl and hook and line gear for rockfish, lingcod, and dogfish. The allocation of ITQ within the TACs for trawlers was then made on the basis of catch history and vessel length. All groundfish were converted into groundfish equivalents (GFEs) in order to make different species comparable.[4] Trawlers then received proportionate shares across all species and stock combinations.[5] This resulted in 55 different ITQ allocations all expressed as a percentage of the respective stock TAC (DFO, 2004).

Under the ITQ rules, vessels are permitted to fish up to their allocation (which can also include carryover from the previous year) of a species within an area for stock specific allocations. If they exceed their allocation for the area they will not be permitted to continue fishing unless additional ITQs for that species are transferred onto the vessel to cover the overage. If the species in question is delineated as a coast-wide ITQ then the vessel is not permitted to bottom trawl anywhere and for any species until enough quota for the species in question is transferred to cover the overage.

Individual vessels are also permitted to retain fish caught in excess of their allocation and apply it to their next year's ITQ, although there are annual overage limits. They may also carry over underages into the next year. However, this underage expires at the end of the next year, and thus cannot be accumulated. The maximum underage and overage is 37.5% for all groundfish species other than hake and halibut for which the limit is 15%. There are also overall individual species cap set (between 4% and 10%) and a total holdings cap set limiting the amount of quota (based on groundfish equivalents) any vessel owner can accumulate in a particular species or in general on the vessel.[6]

Fishers also have individual halibut bycatch quotas even though they are not permitted to retain halibut. There is an overall mortality bycatch cap assigned to the fishery of one million pounds for which each individual vessel owner receives a proportionate share. This is freely transferable although no licence holder can

---

[4] Pacific Ocean perch (POP) was used to establish the baseline (so the price of POP is set to equal 1.0).

[5] Hake were excluded, but followed the same principles although the weighting – 70% history for catch and 30% for vessel length was the same (although it was for hake vessels only).

[6] The total holding cap is determined using groundfish equivalents to compare different stock–species allocations.

accumulate more than 4% of the total and, as is the case for individual species allocations, fishers cannot continue fishing bottom trawl when they have exceeded their cap until they acquire more. In addition to the halibut bycatch, fish discarded at sea that are considered marketable are deducted from each vessel allocation, while non-marketable fish are recorded, but not counted against the ITQ. Mortality rates for fish caught and discarded are calculated based on the species and other factors, such as towing time.

There is an active market in transferring quota with over 2700 trades made among 70 boats annually. The transferability rules are determined on an ongoing basis and are reviewed every three years. The main concern addressed in designing the transferability rules has been the prevention of concentration of quota and an expressed desire to see quota stay on 'active' vessels. Quota can only be transferred between licensed trawl vessels. While quota is freely transferable, there are restrictions on how much can be transferred between vessels because of limits on how much can be accumulated on any one boat due to the individual species caps and total holdings limit.

The current rules require that 25% of the ITQ be 'locked' on to the vessel for three years (2004-2006) unless permanently transferred off, while 75% of the ITQ is freely transferable on a temporary basis within the year (this quota returns to the vessel owner at the start on the next fishing season). Each vessel licence is allowed 2 one-way permanent transfers of locked-on ITQ during a 3-year period. As is the case for total holdings cap, this is measured using groundfish equivalents to make different species quotas comparable.

4.3.2   Addressing distributional concerns

In addition to quota concentration, there were also concerns expressed about the impact of quota trading on both crew members and those communities that were either home ports to fishers or had processing facilities. These concerns were addressed through the development of the Code of Conduct Quota (CCQ) and the Groundfish Development Quota (GDQ).

The CCQ consists of the 10% of each TAC that is automatically allocated to individual trawl-licensed vessels based on the ITQ allocation formula, provided the Groundfish Development Authority (GDA) has not been advised that a specific vessel receive less than their full CCQ allocation. The CCQ is designed to shelter crewmembers from unfairly absorbing any of the costs associated with the introduction of the ITQ system and to limit, to some extent, the ability of vessel owners to reduce crew size. In addition, licence owners of vessels are required to satisfy 'safe vessel' criteria to prevent changes in crew size or maintenance due to ITQs that might compromise the safety of the vessel. Any crew member who feels that their rights have been violated can complain to the GDA, but to date there have been no complaints filed. The lack of complaints, however, may be partially explained by the fact that crew members who report a vessel owner will also suffer financially as the total allocation for the vessel will be reduced by 10%, thereby reducing their crew share.

The GDQ has also 10% of each groundfish trawl TAC allocated by the Groundfish Development Authority (GDA) based on proposals jointly prepared by processors and trawl licensed vessel owners. The GDA administers both the CCQ and GDQ. The GDA was the compromise negotiated in response to requests by processors, unions and coastal communities for direct ITQ allocations. Its Board of Directors is drawn from those stakeholder groups, but vessel owners active in the fishery are excluded. There is also a steering group that provides information and expertise to the board made up of processors, vessel owners, a government representative, and a First Nations'[7] representative.

The GDQ is allocated based on the amount of ITQ fish in the proposal, processor production history, and the rating of the proposal. Criteria used for evaluation include market stabilisation, maintenance of existing processing capacity, employment stabilisation, benefits to local communities, increasing the value of groundfish production, job-training and sustainable fishing practice.

4.3.3 Financial responsibilities and rent capture

The shift towards the ITQ system has also seen a change in the roles and obligations of industry and government. Overall direct costs to industry of managing the fishery are approximately CDN$4 million annually (this excludes government expenditures on management, enforcement and science), or about 6% of the total landed value of fish harvested in the GTF. The government currently picks up one-third of the cost of data management with the industry paying two-thirds of the cost. The expense of maintaining on-board observers at sea, approximately CDN$300/day, is paid for by vessel owners. The annual cost of at-sea and dockside monitoring is approximately CDN$3 million while annual industry-funded science activities cost CDN$800,000, and GDA expenses are CDN$80,000 per year. The federal government also collects licence fees worth, in total, some CDN$800,000 annually while prior to ITQs the government annually collected CDN$1,420 in licence fees (DFO, 2004).

4.4 ECONOMIC AND SUSTAINABILITY EFFECTS OF ITQs

The potential economic benefits of ITQs include fresh fish year round, increased values, less loss of gear, lower quota overage and bycatch discards, and increased safety at sea (Dupont and Grafton, 2001). Offsetting these benefits are increased management and enforcement costs and distributional concerns around the reduction in crew sizes. In their study of the Alaska sablefish fishery in which ITQs were introduced, Sigler and Lunson (2001) found that the fishing season lengthened and that catching efficiency improved. As well, selective fishing techniques were introduced, reducing bottom damage, and there was a reduction in catching of smaller immature fish and related discards.

As yet, there has not been a detailed economic assessment of the impact of ITQs on the BC groundfish trawl fishery. The available evidence, however, points to both improved profitability and changed fishing patterns and effort. Prior to the introduction of ITQs there were 142 limited entry licences of which 115 to 135 boats would be active, depending upon the year. After the introduction of the ITQs, there still remain 142 *T* license holders, but there are now only about 60-70 boats operating. Both the very small boats (under 50 feet) and larger boats have exited the fishery. The smaller boats exited because of the cost of an observer at sea meant smaller-scale fishing was no longer profitable, while the larger boats exited because they had been designed for larger volumes and longer trips, but were not as profitable operating with smaller volumes taken over a long period of time.

---

[7] Native Canadian Indian.

The most significant change with the advent of ITQs has been in terms of fisher behaviour. This has manifested itself in several different ways. First, under the earlier regime, in order to maximise their catch fishers were forced to fish in all areas. This occurred because there was no incentive to individually withhold effort from an area given the possibility of the area TAC being reached and then closed to fishing. Thus harvesters would try to fish every area to ensure that they gained some catch before the limit was reached. Under the ITQ system, however, fishers have chosen to specialise both in regions and species. Figure 4.1 illustrates the significant reduction in nominal effort (total hours fished at sea) that has taken place since the introduction of the quota. There has also been the development of customised 'shopping lists', based on market demand, and fishers have been making shorter trips to improve the quality of the fish.

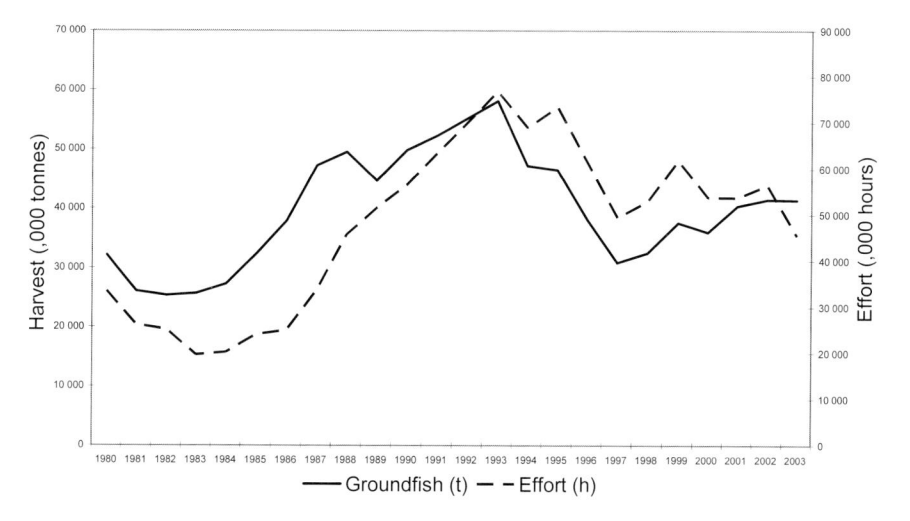

Figure 4.1   Groundfish harvest (000 tonnes) and effort (thousand hours) in the BC Groundfish Trawl Fishery 1980-2002.

4.4.1   *Sustainability*

One of the most important changes in terms of sustainability has been the change in fishing practices. Prior to the ITQ system, reliable data on discards was not available. Under the new management system the presence of on-board monitors, introduced independently of ITQs, ensures that discards are reliably estimated. As a result, stock assessments and the setting of TACs can be made with much more accurate data.

In addition to improved information, mortality rates on quota species are also assessed on fish released at sea that are considered marketable, and these count against the quota owned or leased by fishers. As a result fishers now face an economic disincentive to discard catch, and economic incentives to minimise bycatch and avoid ITQ overages. Fishers have several alternatives when they do exceed their limits — they can purchase quota, borrow from their quota next year, or shut down, but all of these cost money that provide an economic signal to avoid catching unwanted species.

This policy of counting discards against quota allocations has given fishers the impetus to be much more selective in their fishing practices. For example, many fishers no longer target Pacific Ocean perch, sablefish, silvergray, and canary rockfish as such species are caught incidentally when targeting other species. Consequently, fishers reserve their fishing effort and quota for preferred species that can be targeted effectively.

Table 4.2 shows discard ratios for selected species that are the at-sea releases (both marketable and unmarketable) divided by the landed weight. For all five species shown, this ratio drops as fishers have learned over time to fish more selectively. In some cases the drop in at-sea releases to retained catch is dramatic such as for spiny dogfish where discards as a proportion of the retained catch were in 2003/04 about 5% of what they were in 1997/98. This change is *not* because of the observer monitoring at sea as full monitoring began in 1996, but is a direct result of adjustments by fishers as to when, where, and for how long they trawl so as to ensure that bycatches of non-targeted species do not prevent them from fishing. However, the at-sea monitoring is critical to ensure that the proper economic incentives are in place.

Table 4.2 Proportion of at-sea releases to retained catch for TAC-managed species in the BC groundfish trawl fishery, selected species for years 1997-2004

Species	**97-98**	**98-99**	**99-00**	**00-01**	**01-02**	**02-03**	**03-04**
Pacific Ocean perch	0.028	0.024	0.016	0.010	0.008	0.007	0.012
Yellowmouth	0.008	0.008	0.003	0.003	0.003	0.002	0.002
Redstripe Rockfish	0.316	0.342	0.206	0.122	0.113	0.097	0.132
Shortspine Thornyheads	0.065	0.062	0.049	0.049	0.043	0.042	0.030
Spiny Dogfish	0.46	0.123	0.30	0.09	0.044	0.034	0.025

The 2003/2004 data only includes reports up to February 3, 2004

In addition to helping the sustainability of the groundfish species that fall within the GTF management system, changed fishing practices in response to economic incentives, have also reduced the annual bycatch mortality for halibut to about 15% of it's previous level, dropping from around 900 t to a little over 100 t since the introduction of ITQs.[8]

---

[8] We note here, however, that despite the presence of another market for halibut quota, that this halibut is discarded, forgoing potential revenues, suggesting that there is still room for improvement in terms of integrating the two quota systems.

Fishers have been able to more selectively harvest in several different ways. First, there is greater communication between vessels advising each other not to fish in certain areas where there is a high incidence of unwanted species.[9] Second, fishers have changed their behaviour, using shorter tows and more frequent checking of the net, as well as test tows to see what they encounter before actually fishing. Third, fishers have also invested in technology (electronic equipment that allows them to vary the net opening while trawling) as well as net mensuration gear (remote sensors that transmit what is being caught and how much is in the net). Finally, harvesters are experimenting with their gear (nets, bridles, footropes, headropes, lengthening pieces, doors, and codends) to improve selectivity.

The better information generated by the industry in terms of discard mortality also provides for improved stock assessments that, in turn, feed backs into better management. For example, the BC longspine thornyhead quota was initially established at an unsustainable level because of insufficient information regarding the status of the fish stocks. Since the introduction of ITQs, however, managers have been able to work closely with industry to collect improved information that has allowed managers to redo their stock assessments and reduce the TAC to sustainable levels. Indeed, the ability to use better information to improve the management is one of the most important outcomes of the ITQ system, and is recognised as such by fishers (Haigh and Shute, 2003).

4.4.2 Economic outcomes

The introduction of ITQs into the fishery has allowed more profitable fishers to purchase quota from less profitable operators. This has led to a consolidation in quota holdings and increased catches per vessel as shown in Table 4.3 along with about a 50% reduction in 'active' fishing vessels.

Table 4.3 Distribution of landings by size of harvest, selected years 1994-2000.

Landings per vessel (lb)	1994	1998	2000
> 1 million	53%	68%	82%
0.5-1.0 million	36%	21%	13%
< 0.5 million	11%	11%	5%

In addition to a change in the overall number of vessels, the composition of the fleet has also changed with the smallest and largest vessels exiting and greater specialisation for remaining vessels in terms of their use of mid or bottom trawls, harvesting in deep or shallow water, landing fresh versus frozen fish, and choice of fishing locales.

Greater specialisation and quota consolidation has led to improved economic outcomes for vessels within the fleet. These benefits are manifested in a number of ways, such as increased output prices, because fish is now landed over a much longer

---

[9] This information sharing also appears in other fisheries (Platteau and Seki, 2002). It is interesting to speculate to what extent such co-operation emerges in response to changes in incentives from moving to different management regimes.

period of time and in better quality or form (fresh rather than frozen). Figure 4.2 shows the significant difference in trends between landed values and volumes after the introduction of quota, with overall fleet revenues increasing despite the reduced harvest. The market has recognised these increased returns with a doubling in the average lease price for quota and in quota values over the period 1997/98 to 2003/2004, as shown in Table 4.4 These increases have come despite a fully funded at-sea and dockside monitoring programme and increased licence fees. In addition, the 10% set-aside of the TACs each for Code of Conduct Quota and Groundfish Development Quota has allowed some of the benefits of ITQs to accrue to crew, some of whom own quota, as well as processors.

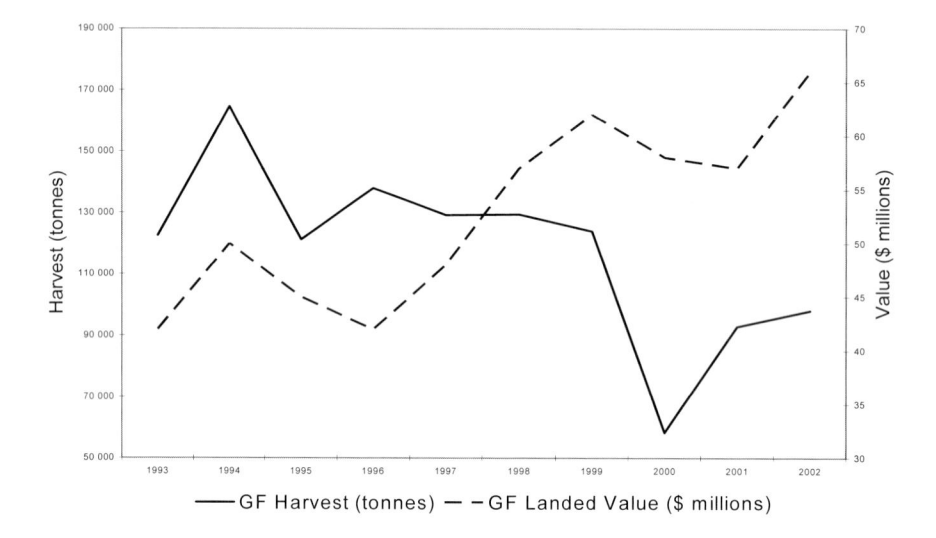

Figure 4.2 Groundfish harvest (tonnes) by the trawl fishery and landed value (nominal CDN$ millions), 1993-2002.

Table 4.4 Lease and quota prices in British Columbia trawl fishery.

	Lease price (CDN$ per lb)	Quota price (CDN$ per lb)
1997 - 1998	$0.10	$1.50
1998 - 1999	$0.12	$1.50
1999 - 2000	$0.14	$1.75
2000 - 2001	$0.20	$2.00
2001 - 2002	$0.18-0.20	$2.00-$2.50
2002 - 2003	$.20	n.a.
2003 - 2004	$.20	$3.00

n.a. = not available

4.5 CONCLUDING REMARKS

It is some twenty years since Munro and Scott (1985) identified the key factors in successful fisheries management as the market signals and incentives faced by fishers. Their insight has been adopted and applied in a wide range of fisheries with the use of individual output controls.

Using the experiences of the British Columbia groundfish trawl fishery — a multispecies fishery with 55 separate quotas — it is clear that the 'incentive principle' of Munro and Scott can help address some of the most vexatious problems in fisheries management: discarding at sea, overages of bycatch species, and how to obtain reliable catch information for stock assessment purposes. By creating a quota for bycatch, even though the species have no or little market value, fisheries managers did give them an economic value. Fishers have then responded to the economic incentives that were created. Under an effective monitoring system, they have adjusted their behaviour as to when, where and how they use their fishing gear to mitigate the discard problem. As a result, the fishery is much better managed than prior to the introduction of individual harvesting rights. Equally as important, transferability of individual quota has allowed more profitable operators to increase their share of the total catch. It has also allowed for greater specialisation by fishers that has contributed to a doubling in quota values since the introduction of incentive-based management in 1997.

The key lesson from the experiences of the British Columbia groundfish trawl fishery is to adaptively manage fisheries and set incentives such that fisher behaviour matches the goals required for a profitable and sustainable industry. The experiences of the British Columbia groundfish trawl fishery show that the insights of Munro and Scott regarding incentive-based management, implemented with adequate monitoring and operational tools, provides a powerful combination to help resolve the class II common property problem and the challenges of multi-species fisheries management.

Acknowledgements

The authors are grateful for helpful comments and suggestions provided by two anonymous referees.

References

Anderson, L. 1994. Highgrading in ITQ fisheries. *Marine Resource Economics* 9: 209-26.
Arnason, R. 1994. On catch discarding in fisheries. *Marine Resource Economics* 9: 189-208.
Boyce, J.R. 1996. An economic analysis of the fisheries bycatch problem. *Journal of Environmental Economics and Management* 31: 314-36.
Copes, P. 1986. A critical review of the individual quota as a device in fisheries management. *Land Economics* 62: 278-91.
DFO. 2004. Pacific region integrated fisheries management plan-groundfish trawl (April 1, 2004 to March 31, 2005). Department of Fisheries and Oceans. Accessible online http://www.dfo-mpo.gc.ca/

Dupont, D.P. and R.Q. Grafton. 2001. Multi-species individual transferable quotas: The Scotia-Fundy mobile gear groundfishery. *Marine Resource Economics* 15: 205-20.

Fox, K.J., R.Q. Grafton, J. Kirkley and D. Squires. 2003. Property rights in a fishery: Regulatory change and firm performance. *Journal of Environmental Economics and Management* 46: 156-77.

Gislason, G.S. and Associates. 2004. *The BC wild fisheries*. Mimeograph, 2004.

Grafton, R.Q. 1995. Rent capture in a rights-based fishery. *Journal of Environmental Economics and Management* 25: 48-67.

Grafton, R.Q. 1996. Individual transferable quotas: Theory and practice. *Reviews in Fish Biology and Fisheries* 6: 5-20.

Grafton, R.Q., D. Squires and K.J. Fox. 2000. Private property and economic efficiency: A study of a common-pool resource. *The Journal of Law and Economics* 43(2): 679-712.

Groundfish Special Industry Committee. 2002. Review of the Groundfish Trawl Management Program: Summary Report, March. Fisheries and Oceans Canada, Pacific Region.

Groundfish Special Industry Committee. 2003. Review of the Groundfish Development Authority: An Element of the IVQ / GDA Program in the BC Groundfish Trawl Fishery Final Report and Recommendations. Fisheries and Oceans Canada, Pacific Region.

Haigh, R. and J.T. Shute. 2003. The longspine thornyhead fishery along the west coast of Vancouver Island, British Columbia, Canada: Portrait of a developing fishery. *North American Journal of Fisheries Management* 23: 120-40.

Hannesson, R. 2004. *The Privatization of the Oceans.* The MIT Press, Cambridge, MA.

Holland, D. and J. Maguire. 2003. Optimal effort controls for the multispecies groundfish complex in New England: What might have been. *Canadian Journal of Fisheries and Aquatic Sciences* 60(2): 159-70.

Kaufmann, B., G. Geen and S. Sen. 1999. *Fish Futures: Individual Transferable Quotas in Fisheries*. Fisheries Economics, Research and Management Ltd, Kiama, Australia.

Kohli, U. 1983. Non-joint technologies. *Review of Economic Studies* 50: 209-19.

Munro, G.R. and A.D. Scott. 1985. The economics of fisheries management. In A.V. Kneese and J.L. Sweeney (eds) *Handbook of Natural Resource and Energy Economics* Volume II. North Holland, Amsterdam.

Pearse, P.H. 1982. Turning the Tide: a New Policy for Canada's Pacific Fisheries. The Commission on Pacific Fisheries Policy Final Report, Catalogue No. Fs23-18/1982E. Supply and Services Canada, Ottawa.

Plateau, J. and E. Seki. 2002. Community arrangements to overcome market failure: Pooling groups in Japanese fisheries. In M. Aoki, and Y, Hayami (eds) *Communities and Markets in Economic Development.* Oxford University Press.

Shotton, R. 2001. *Case studies on the Effects of Transferable Fishing Rights on Fleet Capacity and Concentration of Quota Ownership*. FAO Technical Paper 412. Food and Agricultural Organization, Rome.

Sigler, M. and C. Lunsford. 2001. Effects of individual quotas on catching efficiency and spawning potential in the Alaska sablefish fishery. *Canadian Journal of Fisheries and Aquatic Sciences* 58(7): 1300-12.

Sinclair, D. 1960. *Licence Limitation — British Columbia*. Report to the Department of Fisheries and Oceans, Pacific Region.

Squires, D., H. Campbell, S. Cunningham, C. Dewees, R.Q. Grafton, S.F. Herrick Jr, J. Kirkley, S. Pascoe, K. Salvanes, B. Shallard, B. Turris and N. Vestergaard. 1998. Individual transferable quotas in multispecies fisheries. *Marine Policy* 22(2): 135-59.

Squires, D. and J.E. Kirkley. 1991. The potential effects of individual transferable quotas in a pacific fishery. *Journal of Environmental Economics and Management* 21: 109-26.

Turner, M.A. 1997. Quota-induced discarding in heterogeneous fisheries. *Journal of Environmental Economics and Management* 33: 186-95.

Walters, C.J. and R. Bonfil. 1999. Multispecies spatial assessment models for the British Columbia groundfish trawl fishery. *Canadian Journal of Fisheries and Aquatic Sciences* 56: 601-28.

Weninger, Q. 1998. Assessing efficiency gains from individual transferable quotas: An application to the mid-Atlantic surf clam and ocean quahog fishery. *American Journal of Agricultural Economics* 80: 750-64.

5

Auctions of IFQs as a Means to Share the Rent

Daniel D. Huppert

5.1 INTRODUCTION

This chapter explores options for sharing rents from individual fishing quota (IFQ) systems by auctioning IFQs. It reviews the experience of past and existing fishery quota auctions and draws on examples from other natural resource auction systems. The intent is to capture for public use some portion of the returns over and above opportunity costs accruing to the fishing industry. In the past, a systematic review of options for capturing fishery rent by Grafton (1995) analysed the following: (a) annual IFQ rental charges, (b) annual charges on fishing profits, (c) *ad valorem* royalties on landings, or (d) annual lump sum fees. The relative merit of the four options is reviewed below. Any of these four payment options could conceivably be implemented by either administrative fee setting or quota auctions. This discussion focuses on the relative merits of using auctions to capture rents from the industry.

Rent capture differs from cost recovery. Many nations have levied fees intended to recover annual management costs (Hannesson, 2004: 58; National Research Council, 1999). Because management costs are relatively predictable and independent of variations in fishery costs and returns, cost recovery is often pursued via administrative fees on fish harvested – either annual rental charges or *ad valorem* landing fees. In Iceland, the permanent, divisible and transferable IFQs are subject to 'a small annual charge to cover enforcement costs' (Runolfsson and Arnason, 2001). In the USA, the Secretary of Commerce is required to 'collect a fee to recover the actual costs directly related to the management and enforcement of any ... individual fishing quota programme' (Magnuson-Stevens Act, Section 304(d)(2)(A)). This fee is to be no more than 3% of the landed value of the harvest. For example, a landing fee of 2% of ex-vessel value was established for halibut/sablefish individual fishing quota (IFQ) holders in Alaska to cover estimated costs of $3,430,357 in 2001 (Alaska Region Office, National Marine Fisheries Service, 2002).

Capturing a chosen share of resource rent is more difficult than cost recovery, because annual rent is variable, uncertain, and hard to measure. Administrative fees to capture rent must vary with fishery earnings, and hence, must be re-calculated frequently (probably annually). This will require accurate information on rents being earned. Annual auctions of IFQ are a good alternative mechanism for rent capture, because they automatically produce information about rents and adjust payment accordingly. So the two dimensions in designing a rent capture system are choosing between administrative fee-setting and auctions, and choosing from among the four payment mechanisms mentioned above. Auctions for quotas should be considered in light of the advantages and disadvantages compared to the administrative fee-setting option. These are summarised in the chapter conclusions.

Why consider rent capture at all? Anderson (1989), Johnson (1995), and Lindner *et al.* (1992) warn that taxes, royalties, or quota auctions to extract rent from the fishery could reduce entrepreneurial vigor and technical innovation in the fishery, encourage non-optimal fish stock conservation, and unfairly divert short-run rents from harvesters during the transition from open access to an economically efficient fishing industry. A similar finding pertains to fees in offshore oil and gas sales in the USA (Mead *et al.*,1985), where royalty payments are typically equal to 1/6 of the well-head value. The fee or tax has the same effect on optimal firm decisions as an additional real cost. In the case of oil extraction, the royalty tends to cause premature abandonment of oil fields (production is stopped when real costs of extraction are still below the well-head price). In the fishery, a harvest-related fee could encourage firms to shift species targets or to retard investment in fishing technology or lower firm's contribution to fish-stock assessment and conservation efforts. Yet many observers support the notion that the fishery rent should support agency budgets or contribute to general government funds.

Two recent commissions on ocean policy in the USA (Pew Oceans Commission, 2004; and the US Commission on Ocean Policy, 2003) considered mechanisms to collect funds from IFQ holders to compensate for public costs and to capture some of the rent. The Pew Commission (p. 38) report suggested that 'those who wished to profit from fishing had to offer bids for the opportunity to fish — the same method used to apportion public timber and fossil fuels.' The US Ocean Policy Commission recommends that private sector use of public resources be contingent on providing a 'reasonable return' to taxpayers, (p. 408). While the Commissions supported sharing of fishery rent between IFQ holders and the broader public, they have not provided solid grounds for choosing a reasonable division of rents or a well-designed method for rent collection.

In the economics literature, Matthiasson (1992) examines the efficiency and equity of allocation rules for fishing rights, concluding that competitive auctions would assign IFQs to the most cost effective fishing firms and that the resulting income distribution would depend upon the form of auction bidding system used. These auctions could clearly capture much of the resource rent. Morgan (1995) evaluates quota allocation by (a) administrative decision, (b) lottery, and (c) auction; and he contrasts free allocation procedures in fisheries to auctions in the USA for rights to the electromagnetic spectrum for cellular phones, to airport landing slots, and to financial assets such as Treasury Bonds. Morgan concludes that an auction is more efficient than lotteries or administrative decisions in allocating rights to those who value them most. Stoneham *et al.* (2005) note that auctions effectively utilise information regarding costs and returns from fishing, information which may be difficult or costly for government agencies to obtain and interpret accurately. These assessments provide reasons to consider quota auctions along with the standard administrative fee-based systems of capturing rent.

This chapter does not argue for any particular division of rent between the industry and the government. Rather, it extends the discussion of the role that auctions could play in sharing resource rents. The first section below reviews experience with IFQ auctions. The following section focuses on the choices that must be made in designing administrative fees and IFQ auctions. The final section summarises the findings and adds some speculations concerning the usefulness of IFQ auctions.

5.2 EXPERIENCES WITH AUCTIONS FOR FISHERY AND OTHER RESOURCES

Auctions for rights to extract natural resources from the public domain are fairly common. In the United States, auctions are prevalent in public timber harvest allocation (Gorte, 1995) and in the assignment of oil and gas exploration and mining rights on the outer continental shelf (Mead *et al.,* 1985). Allocation by auction is rare in fisheries. However, recent examples include fish quota auctions in Estonia (Vetemaa *et al.*, 2002) and in the Russian far east (Anferova *et al.,* 2004). Both of these auctions have been halted for reasons explained below. Further, harvest quota auctions have been used to allocate IFQs in Chilean fisheries, and Washington State has been auctioning rights to harvest large clams, called geoducks, for three decades. Each of these is briefly described below.

Estonia adopted an individual fishing rights system in 2000, allocating 10% of the rights to the government and 90% to fishing firms based upon recent catch history (last three years). For each fishery right, 10% of the total allowable catch (TAC) was divided into units which are auctioned in open, public out-cry auctions with ascending prices. An auction was held in 2001-2002 for the open-sea fisheries (herring and sprat trawling) and Atlantic ocean fishing; while auctions were held in each of 15 counties for local, small-scale fishing. For example, in the herring fishery auction there were eight 10-tonne units, four 30-tonne units and nineteen 100-tonne units (Vetemaa *et al.,* 2002, p. 99). Every fisherman and fishing enterprise registered in Estonia was eligible to participate in the auction, after making an up-front payment equal to 50% of the starting price (set by the minister) of the object being auctioned. The auction brought a significant flow of financial resources to the government, but it also provoked some significant opposition from, not surprisingly, the fishing industry. The auction system was abandoned in Estonia after 2002.

In the Russian Far East, quotas were auctioned during 2001–03 for the purpose of 'improving the allocation of quotas to the industry..., [and] to provide a transparent access of enterprises to fishing, and to prevent corruption' (Anfernova *et al.,* 2004, p. 5). In 2001 53% of all the quota put up for auction was unsold due to lack of interest by both Russian and foreign firms (Anferova *et al.*, 2004, p. 7). Monitoring and enforcement of fishing rules was apparently very poorly executed as firms would purchase some quota to provide a legal base for entering a fishing zone, only to harvest species not covered under their quotas. Hundreds of firms reportedly exported fish directly and illegally. Compliance with TACs was not strongly controlled, and some estimates place the ratio of legal to illegal catches at an astounding 1:2 to 1:5. As a consequence, the annual catches exceeded the TACs, the fish stocks have been in decline since, and the quota-auction system is considered to be one contributing cause. Because payments for quota lots are an additional expense, firms tend to take even more illegal catch to avoid that expense. The Russian experience illustrates the importance of monitoring and enforcement to the success of IFQ programmes, with or without auctions.

In Chile, IFQs were adopted in accordance with the Fisheries Act of 1991. In March 1992 Chile successfully employed auctions in two rather small fisheries, a red shrimp fishery and the southern cod fishery. As explained by Peña-Torres (1997) the red shrimp TAC was auctioned in seventeen lots, each of which conferred a harvest right for 10 years. The harvest rights are reduced by 10% annually, and a new auction for the remaining 10% is carried out in December of each year. According to Gonzalez *et al.* (2001), auctions of IFQs have been introduced to more important Chilean fisheries, such as the offshore Patagonia toothfish fishery, since 1996. As with the red shrimp and southern cod fisheries, the successful bidders receive permits to harvest shares of the annual TACs. And each harvest right is reduced by some portion annually, releasing some of the TAC for regular annual auctions. This procedure forces all IFQ holders to participate in the auctions or to experience shrinking harvest rights.

Washington State Department of Natural Resources (DNR) has long employed auctions to allocate harvesting for geoducks on State 'aquatic lands'. Geoducks are large, long-lived clams found in calm, subtidal areas from California to south-east Alaska. Based upon surveys by the Washington Department of Fish and Wildlife that identify concentrations of geoducks and determine which tracts exhibit sufficient abundance for harvesting, commercial tracts are laid out between the 18 and 70 ft. isobaths in Puget Sound and connected waters. A TAC of about 1.3 million kg (about 2% of the estimated commercial abundance) is rotated among tracts for harvest by commercial divers. Over the period 1996-2005, the state received $60 million from winning bids in geoduck auctions. Half of the revenue is placed in the Aquatic Lands Enhancement Account, a fund created by the State legislature in 1984 to develop public access to aquatic lands and to restore native aquatic habitat (e.g. by purchase of shoreline property for public access, construction of docks). The other half goes into the Resource Management Cost Account, which funds sales activities, harvest management, and resource surveys.

Geoduck lease tracts generally range in size from 80 to 120 acres. The number of harvesting operations on a tract varies from four to six. The potential harvesters bid for the right to harvest in common on the tract. These tracts differ from IFQs as DNR does not promise a given weight of clams will be available for each harvester. Rather, it puts an upper limit on the amount that can be taken from each tract sale. But, based on the clam surveys and a one-day test fishery, the bidders have a fair idea of the quantity of clams that is likely to be taken. The tracts are generally harvested over a 3 to 6 month period, during which DNR personnel install buoys to demarcate the geoduck tracts and maintain continuous monitoring during harvest operations from an observer vessel. Harvesting is restricted to weekdays, 9am until 4pm, to ensure enforcement personnel the opportunity to weigh all clams harvested.

DNR runs a high-price, sealed-bid auction among 'responsible bidders' twice per year. Bidding firms are required to document their ability to complete the harvest and to honour terms of the contract, and they must possess state seafood-handling permits. Required disclosure of the bidders' financial records also provides DNR with information which is used to try to identify potential collusion among bidders. There are no special allowances for local or small firms and no restrictions, such as State citizenship requirements, which would limit the number of potential bidders. If the high bidder does not meet qualifications, the next highest bid is chosen.

In May 2004 DNR auctioned 127,000 kg in a tract off Freshwater Bay in the Strait of Juan de Fuca and 231,400 kg of harvest quota in Hood Canal portion of Puget Sound. In each tract the total harvest ceiling was divided into 12 segments (each called a 'quota'). Sealed bids for each 'quota' were opened sequentially, 20 minutes apart, starting at 10am. After each 20-minute period a new set of sealed bids is opened and the high bidder identified.

Each bidder offers a total cash price which is then divided into a bonus bid and a royalty rate. The state sets the royalty rate, based upon an estimate of 'fair market value', which is multiplied by the expected harvest from the tract to determine expected royalty revenue. That royalty revenue is subtracted from the cash bid, and the remainder becomes an up-front bonus payment. The winning bidder pays the cash bonus before the harvest agreement is final, and then he pays the state's fixed royalty rate as the clams are taken from the tract.

The geoduck auction system resembles forest timber auctions in the Pacific North-west region of the United States. Table 5.1 compares the key features of US Forest Service timber sales and the state geoduck sales. The current Forest Service sales system involves: (1) preparing timber sales by identifying the sale site, planning roads, appraising the timber to establish an advertised rate (value) per 1,000 board feet, preparing environmental reports, and advertising the sale; (2) awarding the contract to the qualified bidder who offers the highest bid, determined as the total value for the estimated timber volume at the bid rate for each tree species; and (3) administering the contract by checking that logging roads are built to standards, that other harvest standards are met, that all merchantable timber is removed, that only marked trees have been cut, and by spot-checking logs removed to assure that appropriate payments are being made (Gorte, 1995).

Revenue from forest sales is split up in various ways. Twenty-five percent of the gross value of timber sales is distributed to counties and states to compensate local governments for the tax base which is effectively removed by designating National Forests. Some of the revenue is placed in special trust funds, such as the Knutson-Vandenburg fund, to be used for reforestation, timber stand improvements, and salvage sales. Some of the sales revenue is credited to timber harvesters to compensate for the construction of required roads (Johnson, 1985). The remainder of the revenues (about 10% of gross sales revenue in recent years) goes to the Federal Treasury.

Recent timber sales in the Pacific North-west region have been conducted in two stages. In the first stage firms submit sealed bids at or above the advertised rate along with a deposit or bond. Bidding firms must declare whether they are a large or small company, must agree not to export the logs, must declare that they are not de-barred from Federal contracts, and must generally show that they are qualified bidders. This first stage is really a preliminary step in the bid process; very few bidders are disqualified based upon the sealed bid. The successful sealed-bidders then participate in an oral auction where the sale winner is determined. Each bidder lists prices offered for each tree species in the sale tract, and the Forest Service sums the total amounts based upon these offer prices and the estimated timber volume. The bidder offering the highest total value wins the auction. Contracts specify that timber must also be removed by a specified date, and they often list a number of additional restrictions on harvest methods depending upon the elevation and slope of the site.

The Washington State geoduck auction, the US Forest Service timber auctions, and the Chilean fishery auctions attempt to capture essentially the full rent from the resources, while the Estonian fish auctions sought to capture something closer to 10% of the rent. Auctions could be designed to capture any desired portion of the resource rent. The forest timber auctions deal with multi-species harvest, an element that can be important in many fisheries.

Table 5.1 Elements of Washington State geoduck auction and US Forest Service timber auctions

Elements	Washington State geoduck auction	US Forest Service timber sales
1. Bidding procedure	Sealed bid, high price auction among 'responsible bidders' for portions of permitted harvest in each tract.	Combination sealed bid, then oral bids for each sale tract. Bidder offers price/unit for each species. Total bid is sum over species of price times estimated volume.
2. Method of payment	Up-front bonus payment plus a royalty on harvests (recently $0.82/kg).	Lump-sum payment before cutting on each 'payment unit'.
3. Reserve price	Equals the 'fair market value' based on survey of processors and equals the royalty charged on harvests.	Reserve price based upon appraisal of past sales functions. Also, minimum bid established by law.
4. Packaging of units for sale	Single species, annual quotas divided into multiple tracts for sale	Each forest has 10-year harvest plan, divided into sale tracts of multiple species. Bidder must buy all species of timber on tract.
5. Bidder eligibility criteria or special subsidies	Bidders required to document capability and disclose financial records. No special allowances for small or local firms.	Firms with capacity to harvest timber are eligible. 'Set-aside' sales for small firms (<500 employees).
6. Miscellaneous elements	The auction revenue is divided between the Aquatic Lands Enhancement Fund and the Departments of Natural Resources and Fish and Wildlife.	Numerous forest practices dictated by USFS and other agencies; 25% of gross sales revenue is allocated to states and counties.

Source: Revised version of Table 3 in Huppert and Brubaker (1999).

5.3 CAPTURING RENTS WITH AUCTIONS AND FEE SYSTEMS

Given that some portion of the resource rent is to be captured by the government, two main options are: (1) to determine both payment option and size of fees in an administrative process; and (2) to combine an administrative selection of payment option with a market mechanism, such as auctions, to determine the payment size. The administrative fee and auction systems are not completely distinct. Administrators could use competitive market prices for IFQs in setting fees, for example. Under both systems, the agency or broader political process decides upon the overall rent capture objective. Further, in both systems the payment option is selected by an administrative process.

Aside from the similarities there are two important differences. First, the level of administrative fee depends upon information that the agency gathers and analyses. In the auction, the bidders are relied upon to process all relevant information in formulating bids, while the agency operates the mechanism by which bids are submitted and the winners are selected. The auction, like the secondary market in IFQs, marshals private knowledge. Consequently, the auction bidding system, as noted by Stoneham *et al.* (2005) and others, can be more effective than administrative procedure in allocating resources, and can reduce the costs of information collection and analysis by the agency in capturing a pre-determined share of the rent. This information cost savings may be small if the agency has ready access to current rental prices for IFQs, because they could set the fees based upon market price. Second, auctions involve the transfer of fishing rights or privileges, while administrative fees are levied on established IFQ holders. The following two sub-sections focus on (1) the extent to which the auction system could incorporate the four payment options discussed above and (2) the issues pertinent to the choice between auctions and administrative fees.

5.4 PAYMENT OPTIONS WITH AUCTIONS

As noted earlier, the list of payment options that can be used in either auctions or administrative procedures for capture of fishery rents includes: (1) a quota rental charge, which is levied annually and is proportional to the market value of the IFQs ; (2) a profit charge, which is simply a fraction of the profits (rents) earned in any time period; (3) the *ad valorem* royalty, which is a proportional share of the total revenue earned on IFQs, and (4) a lump-sum charge, which is an annual payment per firm (such as an annual licence fee) that is independent of actual rents earned or harvests taken in any year. Grafton (1995) shows that these four methods are not equally favourable in their economic effects on the fishery. Grafton's analysis envisions these fee systems as stemming from administrative decisions, but an IFQ auction system could use any of the four payments methods as well.

Suppose the government decides to collect a quota rental charge (fee proportional to market value of quota holdings) scaled to equal X% of the rent. Grafton (1995) suggests setting the quota rental charge per unit of quota equal to X% times the market price of quota times the competitive rate of interest. To calculate the appropriate fee level, the agency needs information on the market value of permanent quotas and the relationship between annual expected rents and the market price. The observed market price should reflect current private expectations of future rental values of IFQs –

including expected TAC levels, fish prices, and costs. Each quota holder would be willing to pay this amount for additional quota. Presumably, the government recalculates the appropriate rental charge each year based upon updated IFQ market prices and interest rates.

Alternatively, the government could allocate (1-X)% of the TAC as IFQs, holding the remaining X% for auction. This 'partial auction' could be designed as, for example, a uniform price sealed-bid auction as described by Milgrom (2004). The government divides the available quota into N units. A bid is an order to purchase q units at any price up to p. Each bidder can submit several such bids with various bid prices. The agency collects the bids and organises them into a demand curve. The demand curve shows how many units would be sold at each price level. The agency sets the price at the level which just sells the quantity available. If there are private markets for IFQ rentals, each firm can base its bids on those price signals, and it can adjust total holding of IFQs through the market after the auction. One could expect the auction prices and market prices for rentals to be roughly the same. Hence, an auction for X% of the TAC annually would bring in roughly X% of the annual short-term rental value generated by the fishery.

A major difference between the quota rental fee system and the auction is that the fee system burdens the government agency with collecting and interpreting market price information. This could be fairly easy and uncontroversial, or it could be difficult and divisive. One can imagine a vigorous debate about whether the private market rental price is reliable, or over what interest rate should be applied to calculate rental value from price of permanent IFQs as suggested by Grafton (1995). The competitive auction system automatically supplies the appropriate rental value for the quotas each year. On the other hand, the competitive auction system may burden the quota buyers with some additional risk as discussed in the next section.

If there is an active and competitive secondary market in both permanent IFQs and annual rentals, and if it is an open and well-documented market, then there is minimal or no advantage to using a partial auction rather than administrative calculation to establish the appropriate level of fees. The auction price for annual IFQ and the market rental would be equivalent measures of short-term rents. Both rent capture methods would yield about the same level of rent capture. Where market information is unreliable and costly to obtain, or where the results of the administrative calculation are subject to legal confrontations, or where the markets for IFQs are significantly constrained by IFQ programme elements, the auction system ought to capture the desired portion of the resource rent more accurately and with lower information costs than the administrative fee system.

Now consider the profit charge payment option. The administrative procedure calls for selection of a rate of payment applied to annual profits. With IFQ auctions the bid variable can be stipulated as a share of profits. In either case, the payments are made after the fish are caught and profit levels are established. Profit share bidding was once attempted in US Minerals Management Service in offshore oil and gas auctions (Meade *et al.,* 1985). Each bidder determines what fraction of their anticipated profits to offer, and the high bidders win the auctioned IFQs. Then the government agency collects the amount corresponding to the firm's profit for the year. This option is fraught with the accounting information difficulties inherent in monitoring and verifying firms' reported profits – especially when the fishing firms involved participate in other fisheries and/or are held by vertically integrated fishing/processing firms. There may be opportunities for shifting fixed costs or shared overhead costs among divisions within a firm to obscure the profits earned in a particular fishery. Hence, the profit share approach, whether the fee is determined by administrative calculation or auction bidding, has some disadvantage during the fee collection stage But, an auction of some fraction of the TAC, based upon profit share bidding, would be at least as effective at rent capture as an administratively determined profit share fee system.

The third payment option, an *ad valorem* royalty, sets the annual payment to equal some fraction of the annual revenue earned on fish sales under IFQs. Generally, the fee is collected after sales documents are available. In the Washington geoduck fishery a royalty payment (State determined reserve price) is charged on all clams harvested, but the overall bid in the auction covers both the expected royalties and an up-front bonus payment. A similar system is used in offshore oil and gas leases in the United States. Federal lease sales are generally awarded to the highest bonus bids, but the winners also must pay a 1/6 royalty rate on wellhead value of oil and gas extracted from Federal offshore tracts. In principle, the IFQ auction could adopt the royalty rate as a bid variable. That is, each bidder offers a percentage of sales revenue as a payment. If fishing costs and the market price for fish varies annually, rents as a fraction of sales value will vary widely over time. Assuming that firms can anticipate these variations, the IFQ auction system might generate essentially the same *ex ante* payments whether the bids are rentals (i.e. independent of earnings for the year) or royalties (proportional to sales revenue). If the royalties are collected after sales are complete, post-auction fluctuations in fish prices could generate some random movements in royalty payments. Also, with the *ad valorem* royalty payment the agency must accurately monitor the sales value of fish – whether the payment is based upon an administrative fee system or on a set of contracts settled in an auction.

Another possible difficulty with *ad valorem* royalty payments is some ambiguity about the relevant sales revenue. For example, in some fisheries there are a wide range of fishing and processing combinations. Some fish are landed as whole, fresh fish; some are headed-and-gutted and frozen at sea, while others are processed at-sea and on shore into products such as roe or surimi. The same fish species may be processed in two or three ways contemporaneously by different firms. Each product form entails unique costs, caters to a specific market, and receives a different price. So, there is no standard basis on which to calculate a 'sales value of fish'. The sales observed are not associated simply with harvesting fish, but also reflect investment in and operation of fish processing and packaging plants. An administrative rental rate could be based upon the price paid for the portion of fish that are sold fresh, for example, but this could create an incentive to shift the processing and fishery product mix. Hence, it may be difficult to design an *ad valorem* royalty rate that has desirable economic incentive effects.

The last payment option, a lump sum charge, could also be implemented through an auction. For example, where a permit must be held by anyone landing fish under IFQs, all IFQ holders could be required to engage in an annual permit auction. If there are N permits, these could be sold via the uniform price sealed-bid auction. Each fishing firm makes an equal lump sum payment. Grafton's (1995) analysis concludes that a lump sum charge distorts economic incentives by favouring firms with larger absolute profits (typically the larger firms) and penalises those with smaller profits. A lump sum charge per firm could induce an otherwise inefficient expansion in firm size, as average charge per unit of fish caught declines with volume caught. With IFQs firms could consolidate quota, reducing the fleet size to less than N and driving the auction price of permits to zero. Hence, auctioning of annual permits could be fruitless. So, even though the lump sum charge has the lowest information costs of the four options and is easiest to monitor, it is not considered further in this chapter as a means of capturing rent.

5.5 AUCTION OR ADMINISTRATIVE FEE

At least two of the payment options (quota rentals and *ad valorem* royalties) are worthy of consideration with either auctions or administrative fees. The choice between quota rental charge or an *ad valorem* royalty should be based upon the economic efficiency effects and administrative/monitoring costs as discussed, for example, by Grafton (1995). The choice of auctions or administrative action must focus on the relative efficiencies of the two approaches of establishing the actual rents captured under the given payment option. And if the auction attracts wide and competitive bidding, one expects that the auction would more accurately track the rents earned in the fishery than would an administrative procedure. That is, auctions can be a cost-effective means of revealing economic information about rents.

If auctions are efficient, why are auctions not always used instead of administrative fee-setting? For one thing, auctions add an element of risk and uncertainty to the fishery. Administratively set fees and royalties tend to persist for lengthy periods simply because the procedures for setting and changing them can be cumbersome. But auction prices can easily shift widely over short intervals. This could complicate firms' decisions on investment and hiring plans for upcoming fishing seasons. A higher auction price may dictate that some firms curtail harvests, while others (those offering higher bids) increase their harvests. This may be efficient in allocating IFQs, but it adds a noisy uncertainty that would not exist in a system of allocated IFQs with stable administrative fees. Also, the auction process itself introduces uncertainty for firms in making bids. Each bid is an attempt to forecast future profits, hedge against potential market shifts, and compete against similar bidders. The winning bidders may be saddled with economic losses due to the 'winner's curse' (Milgrom, 1989) – the high bidder pays too much due to mistakes in assumptions or calculations. In theory, the bidders may develop complex bidding tactics to avoid the curse. If bidders are uncertain of the outcomes from auctions, the government must also be uncertain of its revenue stream. If the rents are channelled into regular programme budgets, the government agencies also may be averse to the inherent risks. In these circumstances, the agency administrators and fishing firms may agree to a stable and predictable quota rental charge or royalty rate instead of entering the uncertain territory of auctions. On the other hand, if competitive secondary markets in IFQs (for both permanent sale and short term rental) exist, bidders could tie their auction bids to contemporaneous market prices, minimising the risks inherent in bidding. Further, use of a uniform price sealed-bid auction would minimise the risk of the winner's curse, as the highest bids (which may be mistakenly high) will exceed the price at which the IFQs are sold.

Another reason for avoiding auctions is that it shrinks the pool of IFQs for free allocation, and allocation of IFQs in established fisheries is often dominated by the tussle to obtain fishing rights for free. For auctions to be used as a substitute for annual fees, there must be periodic sales of publicly held IFQs, and this requires that some or all of the IFQs be retained by a government authority. Assuming that the basic resistance to this idea can be overcome, there are two broad possibilities here. First, the government could retain X% of the quotas (presumably allocating the remainder freely as permanent IFQs) and auction its share as annual (or longer-term) fishing rights. Second, the full TAC could be allocated freely as IFQs, but each harvest right would shrink by X% annually and the government would re-sell the expired IFQs as in the Chilean fisheries. Choosing the portion of IFQs to hold for periodic sale is similar to choosing the portion of the resource rent that the government will capture. One big difference between the partial auction and administrative fees, however, is that typical fee systems collect fees from all IFQ holders. The partial auction would collect only from those buying annual IFQs. Hence, instead of collecting a chosen percent of the rent from all IFQs, the partial auction would collect all or most of the rent from a chosen percent of the IFQs.

5.6 CONCLUSIONS

In considering a system to share resource rents between IFQ holders in the fishing industry and the government (or public) there are a series of decisions. The decision to capture some portion of the potential fishery rent comes first. In the process, we should consider the effects of reduced earnings on the economic incentives of fishing firms and subsequent effects on fishing communities. There have been objections by a number of economists to rent capture, basically focusing on the efficiency effects, which reflect the standard description of dead weight losses caused by sales and excise taxes. The additional payment per fish caught alters the marginal costs perceived by the fishing firm, causing inefficient adjustments of harvests and input decisions. On the other hand, if rents earned on the public fishery resources are to be shared with the public, then some sub-optimal fee or auction system will need to be implemented.

Second, a decision on payment option involves a selection from at least the four options discussed: (1) a quota rental charge, (2) a profit share, (3) an *ad valorem* royalty, and (4) a lump sum charge. These options have been evaluated in the literature (and reviewed herein) based upon their likely effects on economic performance of the fishery. The lump-sum fee option is rejected outright, because it distorts the incentives concerning firm size. The profit share could be a rather information-demanding option, and it raises difficulties for collection where fishing firms have complex operations that obscure the costs associated with the IFQ fishery. Also, as noted by Grafton (1995), the *ad valorem* royalty does not accurately track resource rents when prices are volatile. And, we have noted that the royalty on sales revenue requires a clear definition of fish sales revenue when multiple products and prices may be involved. Consequently, if the objective is to adopt a method of reliably collecting some agreed portion of the fishery resource rent, the quota rental charge – a simple charge per unit of IFQ – seems to have advantages.

Third, a decision must be made between capturing the rent via administrative fee or via auctions of some or all of the IFQs. Our review of experience of fishery auctions in Estonia, the Russian Far East, Chile, and Washington State shows that quota auctions are feasible but demanding. With the partial auction approach, a portion of the TAC is allocated to historical fishing participants, assuring that they can continue in the fishery. Auctions for the remaining IFQs assign them to high bidders, who are generally the most efficient harvesters. The fraction of IFQs allocated via partial auction determines the portion of the resource rents that are captured. The main difference between auctions and administrative fees concerns the efficient use of private economic information. When competitive, open markets for permanent transfers and annual rentals of IFQ are operating continuously, the essential information is contained in the market prices. The rent-capturing agency can set administrative fees as a fraction of the market price, or the individual bidders in an agency-sponsored partial auction of IFQs can use market price information selecting bid amounts. In either case, the amount collected should reflect the desired public share of the rent. The main advantage of the auction is that it will reflect resource rent from additional harvests, and it will adjust the payments to the public accordingly, regardless of whether there is a competitive and open IFQ market. And the auction obviates the need for administrators to determine whether the reported market prices accurately reflect resource rents. For these reasons, the partial auctions of IFQs deserve consideration as a means of distributing the fishery rent between the commercial fishing industry and the public at large.

References

Anderson, L.G. 1989. Conceptual constructs for practical ITQ management policies. In P.A. Neher, R. Arnason, and N. Mollett (eds) *Rights Based Fishing*. Kluwer Academic Publishing, Dordrecht.

Anferova, E., M. Vetemaa, and R. Hannesson. 2004 (in press). Fish quota auctions in the Russian Far East: a failed experiment. *Marine Policy*.

Gonzalez, E.P., R.C. Norambuena, and M.A. Garcia. 2001. Initial allocation of harvesting rights in the Chilean fishery for Patagonia toothfish. In R. Shotton (ed.) *Case Studies on the Allocation of Transferable Quota Rights in Fisheries.* Fisheries Technical Report 411. Food and Agricultural Organization, FAO Rome: 304-21.

Gorte, R.W. 1995. *Forest Service Timber Sale Practices and Procedures: Analysis of Alternative Systems.* Congressional Research Service Report 95-10. Washington D.C.

Grafton, R.Q. 1995. Rent capture in a rights-based fishery. *Journal of Environmental Economics and Management* 28: 48-67.

Hannesson, R. 2004. *The Privatization of the Oceans*. MIT Press: Cambridge, MA.

Huppert, D.D. and H.W. Brubaker. 1999. *Designing Individual Fishery Quota Auctions to meet Efficiency, Equity, and Political Feasibility Criteria*. School of Marine Affairs Working Paper 99-1. University of Washington, Seattle.

Johnson, R.N. 1985. US Forest Service policy and its budget. In R.T. Deacon and M.B. Johnson (eds) F*orestlands, Public and Private*. Ballinger Publishing Company, Cambridge, MA.

Johnson, R.N. 1995. Implications of taxing quota value in an individual transferable quota fishery. *Marine Resource Economics* 10(4): 327-40.

Lindner, R.K., H.F. Campbell, and G.F. Bevin. 1992. Rent generation during the transition to a managed fishery: the case of the New Zealand ITQ system. *Marine Resource Economics* 7(4): 229-48.

Marine Fish Conservation Network. 2004. *Individual Fishing Quotas: Environmental, Public Trust, and Socioeconomic Impacts*. Washington D.C. Accessible online http://www.conservefish.org

Matthiasson, T. 1992. Principles for distribution of rent from a 'Commons'. *Marine Policy* 16: 210-31.

Meade, W.J., A. Moseidjord, D.D. Muraoka and P.E. Sorensen. 1985. *Offshore Lands: Oil and Gas Leasing and Conservation on the Outer Continental Shelf*. Pacific Institute for Public Policy Research, San Francisco.

Milgrom, P. 1989. Auctions and bidding: a primer. *Journal of Economic Perspectives* 3(3): 3-22.

Milgrom, P. 2004. *Putting Auction Theory to Work*. Cambridge University Press.

Morgan, G.R. 1995. Optimal fisheries quota allocation under a transferable quota (TQ) management system. *Marine Policy* 19(5): 379-90.

National Research Council. 1999. *Sharing the Fish: Toward a National Policy on Individual Fishing Quotas*. National Academy Press, Washington D.C.

Peña-Torres, J. 1997. The political economy of fishing regulations: The case of Chile. *Marine Resource Economics* 12(4): 253-80.

Pew Oceans Commission. 2003. *Management of Marine Fisheries in the United States*. Proceedings of the Workshop on Marine Fishery Management, Arlington, VA.

Runolfsson, B. and R. Arnason. 2001. The effects of introducing transferable property rights on fleet capacity and ownership of harvesting rights in Iceland's fisheries. In R. Shotton (ed.) *Case Studies on the Allocation of Transferable Quota Rights in Fisheries*. Fisheries Technical Report 411. Food and Agricultural Organization, Rome: 66-102.

Stoneham, G., N. Lansdell, A. Cole, and L. Strappazzon. 2005. Reforming resource rent policy: An information economics perspective. *Marine Policy* 29: 331-338.

United States Commission on Ocean Policy. 2004. *An Ocean Blueprint for the 21st Century*. Final Report Pre-publication Copy. Washington D.C. Accessible online http://www.oceancommission.gov.

Vetemaa, M., M. Eero, and R. Hannesson. 2002. The Estonian fisheries: From the Soviet system to ITQs and quota auctions. *Marine Policy* 26: 95-102.

Wilen, J. and F. Homans. 1994. Unraveling rent losses in modern fisheries: Production, market, or regulatory inefficiencies? In E.K. Pikitch, D.D. Huppert and M.P. Sissenwine (eds) *Global Trends: Fisheries Management. American Fisheries Society Symposium 20*. Bethesda, MD: 256-63.

6

Shadow Prices for Fishing Quota: Fishing with Econometrics

Diane Dupont
Daniel V. Gordon

6.1 INTRODUCTION

The world's fisheries provide a fascinating area for economists to study how economic agents respond to changing regulations regarding their choices of inputs, production technologies and investment decisions. This has especially been the case since the widespread movement towards the adoption of individual transferable quotas (ITQs) during the last two decades (Grafton *et al.*, 1996). This new form of management was adopted in the face of failure by limited-entry programmes to prevent excess capacity and rent dissipation (Wilen, 1988; Squires, 1987; Dupont, 1996; Homans and Wilen, 1997). The ownership of quota provides the right to harvest a particular share of the total allowable catch (TAC). When this is combined with a well-functioning market, which allows the owner to sell part or all of his quota holdings and/or purchase additional units of quota, economic models predict a number of desirable outcomes. Most importantly, these include the orderly exit of marginal vessels and transfer of quota to more efficient vessels. In this way, a reduction in fleet size and excess harvesting capacity is encouraged through the efficient operations of the quota market.

Market-clearing prices for quota reflect the overall degree of scarcity within the market and this information shows changing trends in behaviour (Newell *et al.*, 2005). Perhaps more interesting, however, is how individual fishers value quota, as these shadow prices reveal the distribution of values across individual vessels. Knowledge of shadow prices allow predictions on the types of transactions that might take place in a quota market, thereby yielding information regarding the direction of fleet adjustment. This is important because there is a concern among regulators that large vessels will place a larger value on quota and come to dominate the fishery.

This chapter illustrates how to obtain shadow prices per unit of a species that are not only vessel-specific but also conditioned upon the fisher's own quota holdings and individual circumstances regarding fishing effort. The tools we apply are econometric in nature and allow us to obtain shadow values indirectly through observed fisher choices. In anticipation of the empirical results, we show that for the fishery investigated here smaller vessels less than 45 feet are not at a disadvantage in competition for quota.

The Scotia-Fundy mobile gear groundfishery, which takes place off the east coast of Canada, is the fishery under review for this chapter. In section 6.2, we briefly present the fundamentals for the fishery and a summary of the data available for analysis. The data are collected as a random sample of groundfishery vessels for each of the three years 1988, 1990 and 1991. The vessels range in size from 25 to 65 feet in length and harvest using otter trawls. In section 6.3, the restricted-profit function is presented and used to define shadow prices for quota-regulated fish. In the fishery examined here, fishers are constrained to quota levels in three fish species (cod, haddock and pollock) but are allowed access to non-quota fish species. The profit model must account for the interaction between quota and non-quota landings in calculating individual shadow values. Econometric results are reported in section 6.4, whilst section 6.5 concludes.

6.2 THE SCOTIA-FUNDY MOBILE GEAR GROUNDFISHERY

The Scotia-Fundy mobile gear groundfishery is an inshore fishery off the east coast of Canada, encompassing the Scotian Shelf and the Bay of Fundy.[1] Historically, fishers have caught predominantly cod, haddock, and pollock but, in more recent times, other groundfish and non-groundfish species have increased in harvest (Department of Fisheries and Oceans (DFO) Web Statistics). The fleet consists of vessels ranging from 25 to 65 feet in length that use otter trawls.

Individual transferable vessel quotas were introduced for the 1991 fishing season for the three main species: cod, haddock and pollock. The introduction of individual harvesting rights coincided with a subsequent reduction in the size of the principal fish stocks and harvests. Under the quota programme, vessel owner's individual quota allocations were based on the average of the best two years harvest during the 1986-1989 fishing seasons (Barbara *et al.*, 1995). Only temporary transfers of quota were officially permitted in 1991 and 1992. In 1993 the government removed the restriction on permanent transfers. Information from the first year of the programme (1991) reveals an active market with only about 20% of the fleet choosing not to participate – 68 licensees out of 325 had no net-trade activity. By contrast, 118 licensees were net buyers and a further 132 were net sellers. Of these net sellers, 41 sold more than 90% of their quota. By the third year of the programme (1993) only 58 licensees exhibited no net-trading activity while 112 were net buyers and 138 were net sellers, including 67 sellers of more than 90% of their quota. Of these 67, 14 licensees sold out of the fishery completely by selling 100% of their quota permanently.

While 325 vessels were given allocations of ITQs allocations, only 268 fished actively. In order to encourage quota trading and to discourage discarding arising from bycatches, managers put into place measures such as: unlimited trading, post-trip trading and even trading of small fractions of quota (Brander *et al.*, 1995). In order to encourage further rationalisation, the DFO subsequently (in 1996) required minimum species holdings and full payment of license fees prior to the approval of quota transfers. These fees were also made dependent upon the value of the fisher's quota holdings (Burke and Brander, 1999). In recent years the fleet has more or less

---

[1] For a more detailed overview of the Scotia-Fundy Mobile Gear Groundfishery, see Dupont *et al.* (2005).

stabilised at a level that is less than one half the active vessels that participated in the first year of the quota programme: 131 vessels in 2001. It is clear that the presence of the ITQ, which permitted trading activity, has given less efficient vessels the option to leave the fishery in a more orderly fashion that would have been the case in the absence of the ITQ programme. Moreover, given the low stock returns during the early 1990s, which necessitated the setting of successively lower total allowable catches for major species, it is fortunate that the quota programme was already in place. Burke and Brander (1999) argue that the groundfishery fleet is converging on what may be the most efficient size of vessel.

The data available for analysis are vessel-level observations on prices, landings and vessel size for a sample of groundfishery vessels for each of the three years 1988, 1990 and 1991. The data come from two sources: a costs and earnings survey conducted on a random sample of fishers by the Canadian Department of Fisheries and Oceans[2], supplemented by annual sales-slip data that include vessel level information on landings by species, value of landings by species, vessel characteristics (size), variable costs and usage (fuel and labour). From these data we construct annual ex-vessel species' prices and the variables of interest to the profit specification.

Table 6.1 presents summary statistics for landings of the three quota species and an aggregate sum of all other non-quota species harvested by our sample of vessels. The data are reported by year and by length of vessel; less than 45 feet and greater than or equal to 45 feet but less than 65 feet (hereafter, referred to as greater than 45 feet). (Vessels have been regulated according to length in this way for many years.) The most striking feature of the table is that, with the exception of pollock harvests for greater than 45 foot vessels, all categories of landings have decreased over the period. Haddock landings for vessels less than 45 feet show a 58% decline and for vessels greater than 45 feet they show a 47% decline in landings over the period. The most severe decline in landings is represented by a 70% drop in pollock landings by vessels less than 45 feet. The data dramatically reflects the decline in groundfish stocks off Canada's east coast during this period. For cod and haddock, in particular, the biomass fell sharply in the late 1980s, with a concomitant decline in harvests. According to the Canadian Stock Assessment Secretariat Science Sector (1998), a number of factors contributed to stock declines. These factors included: high quotas and over-optimistic assessments, high effort and wasteful fishing practices, inadequate enforcement, harsh environmental conditions, and seal predation.

Table 6.2 reports the average real price for each of the four fish categories for each of the data years and vessel size. The data are in real 1991 dollar terms. The interesting point of this table is the substantial price increase from 1990 to 1991 for the quota fish species.[3] Although the quota price effects observed for the fishery in 1991 are consistent with the predicted impacts associated with a shift to an ITQ regulatory scheme, it must be noted that output price increases can be a combination of both demand shocks and supply effects. Supplementary evidence available on price changes, however, supports the hypothesis that the ITQ programme brought about changes in prices for the ITQ species. For the Scotia-Fundy fishery the resulting

---

[2] DFO discontinued the survey after 1991.

[3] Dupont and Grafton (2000) compare prices for the same species for vessels in the ITQ fishery with fixed gear vessels not in an ITQ programme. Traditionally the fixed gear had resulted in higher prices (typically due to better handling) but the authors found that prices in the ITQ fishery started to climb above non-ITQ prices after the introduction of the ITQ programme.

adjustments in catching and landing meant both better handling of fish (leading to a higher quality output) and a greater dispersion in the pattern of deliveries, thereby reducing the landings gluts that might depress prices. As a result, even as early as the first year of ITQ operations, the average dockside price of ITQ species increased relative to previous years' prices.[4]

Table 6.1 Average harvest cod, haddock, pollock, and other fish (kg); vessel length and year.

		1988	1990	1991
Cod	<45 feet	124,311	98,129	51,484
	≥45 feet	183,013	166,054	156,917
Haddock	<45 feet	68,169	36,704	28,674
	≥45 feet	92,953	38,689	49,439
Pollock	<45 feet	58,921	26,330	17,456
	≥45 feet	70,246	50,400	81,752
Non-quota fish	<45 feet	57,750	44,272	31,266
	≥45 feet	61,559	45,349	49,255

Table 6.2 Average real price cod, haddock, pollock, and other fish; vessel length and year ($1991).

		1988	1990	1991
Cod	<45 feet	0.7703	0.8664	1.2511
	≥45 feet	0.7265	0.8016	1.1549
Haddock	<45 feet	1.2165	1.3491	1.5713
	≥45 feet	1.2069	1.3016	1.5693
Pollock	<45 feet	0.3867	0.6127	0.7528
	≥45 feet	0.3728	0.5979	0.7187
Non-quota fish	<45 feet	0.9258	1.0434	1.2631
	≥45 feet	0.7715	1.0776	1.2597

---

[4] Barbara *et al.* (1995) estimate prices for cod and haddock had the ITQ programme not been in place. They argue that cod prices would have been 12% lower and haddock prices would have been 15% lower in 1992.

Table 6.3 presents summary statistics on a number of variables of interest for the three year period. The table shows that the largest average catches of cod, haddock and pollock occurred in 1988 and that these are higher than the average catches observed in 1991, the year in which ITQs were introduced. On the other hand, the largest average revenues for both cod and pollock are observed in 1991. There are seven non-quota species: silver hake, redfish, flounder, halibut, cusk, catfish and an unspecified or remainder group of species. Fisher price and quantity indices are used to calculate an aggregate price and quantity index for all non-quota landings.

Table 6.3a   Summary statistics of sample vessels (1988).

Variable	Mean	Minimum	Maximum	Std. dev.
Observations	35.00	0	0	0.00
Gross registered tons	55.00	15	110	27.18
Average crew	2.34	1	5	0.97
Fuel expenditures ($)	26,209.23	3,618	60,000	13,573.23
Qty cod (kg)	139,305.20	0	546,600	130,609.20
Qty haddock (kg)	74,655.66	0	237,736	60,936.44
Qty pollock (kg)	57,364.40	0	184,950	49,832.54
Qty non-quota (kg)	59,716.20	5,294	245,000	59,535.00
Value cod ($)	88,543.37	0	327,961	74,863.11
Value haddock ($)	82,266.49	0	244,868	67,449.62
Value pollock ($)	19,241.09	0	66,582	16,279.30
Value non-quota ($)	47,159.20	3,376	175,000	44,639.00

Table 6.3b   Summary statistics of sample vessels (1990).

Variable	Mean	Minimum	Maximum	Std. dev.
Observations	35.00	0	0	0.00
Gross registered tons	61.71	14	110	26.91
Average crew	2.27	1	6	1.09
Fuel expenditures ($)	30,384.89	5,400	74,000	16,520.55
Qty cod (kg)	138,755.00	17,554	642,721	121,925.90
Qty haddock (kg)	37,559.14	1,800	106,527	29,546.41
Qty pollock (kg)	40,708.03	0	148,189	31,861.75
Qty non-quota (kg)	44,969.10	3,624	121,000	32,184.00
Value cod ($)	103,156.10	10,268	345,021	66,511.11
Value haddock ($)	47,562.91	2,541	134,543	36,861.53
Value pollock ($)	23,521.49	0	88,884	18,381.50
Value non-quota ($)	48,974.30	1,298	143,000	39,138.00

6.3   A RESTRICTED PROFIT MODEL

In modelling the Scotia-Fundy groundfishery, economic assumptions are imposed on the behaviour of fishing vessels. Specifically, the vessels are assumed to minimise the cost of harvesting the quota regulated species; cod, haddock and pollock, and are assumed to maximise profits over the unregulated non-quota landings. Because our interest is in measuring shadow values for the quota regulated species, the input side of harvesting is simplified to one variable input and one fixed or constrained input.

Table 6.3c Summary statistics of sample vessels (1991).

Variable	Mean	Minimum	Maximum	Std. dev.
Observations	38.00	0	0	0.00
Gross registered tons	51.89	13	110	27.12
Average crew	1.95	1	6	0.90
Fuel expenditures ($)	26,349.71	3,000	70,000	15,050.89
Qty cod (g)	98,430.97	8,245	819,753	128,608.60
Qty haddock (kg)	36,475.21	0	89,284	25,562.59
Qty pollock (kg)	45,302.32	0	189,276	45,389.06
Qty non-quota (kg)	39,761.10	2,797	149,000	34,851.00
Value cod ($)	104,306.10	11,404	543,552	89,358.06
Value haddock ($)	57,231.03	0	129,457	39,427.06
Value pollock ($)	33,643.34	0	92,683	27,435.13
Value non-quota ($)	48,953.30	3,180	153,000	41,068.00

The variable input is modelled using a Fisher price index of fuel consumption and labour expenditures. The fixed input factor is defined as the overall length of the vessel. These economic assumptions can be written out as a restricted profit function (π^R):

$$\pi^R = f(P_O, P_I; X_K, X_C, X_H, X_P) \tag{6.1}$$

where prices (P) with subscript O indicate the aggregate output price index of the non-quota landings and subscript I the aggregate variable input price, X specifies a restricted factor, with subscript K referring to the overall length of the vessel, C the quota landings of cod, H the quota landings of haddock and p the quota landings of pollock.

Equation (6.1) is defined as a dual profit function based on the behavioural assumptions imposed on fishing vessels in the groundfishery and characterises the underlying harvesting structure.[5] Lau (1976) provides a complete characterisation of the restricted profit function. However, it is important to note that, for equation (6.1) to be a meaningful and proper representation of the underlying harvesting structure, it must satisfy a number of regularity conditions. Specifically, the restricted profit function must be non-decreasing in P_O and non-increasing in P_I, homogeneous of degree one and convex in prices, and concave in restricted factors. Because of our interest in shadow values of the quota restricted landings it is essential that equation (6.1) satisfies the curvature properties of convexity and concavity. In estimation we test and, when necessary, impose correct curvature properties on the model.

---

[5] There are numerous applications of the dual restricted profit function in the fisheries economics literature, for example see Squires (1988), Dupont (1991) and Bjørndal and Gordon (1993).

The normalised quadratic functional form seems ideal as the specific choice of a flexible function form to represent restricted profit for the purposes at hand because of the ease of imposing curvature properties on the model (Diewert and Wales, 1987 and Kholi, 1993). For the variables defined in equation (6.1), our normalised quadratic restricted profit function is written as;

$$\pi^R = 0.5 \frac{\beta^T x}{P_I} \left[a_{OO} P_O^2 + a_{II} P_I^2 + 2 a_{OI} P_O P_I \right]$$

$$+ 0.5 \frac{\alpha^T p}{X_K} \left[b_{KK} X_K^2 + b_{CC} X_C^2 + b_{HH} X_H^2 + b_{PP} X_P^2 \right]$$

$$+ \frac{\alpha^T p}{X_K} \left[b_{KC} X_K X_C + b_{KH} X_K X_H + b_{KP} X_K X_P + b_{CH} X_C X_H + b_{CP} X_C X_P + b_{HP} X_H X_P \right] \quad (6.2)$$

$$+ c_{KO} X_K P_O + c_{CO} X_C P_O + c_{HO} X_H P_O + c_{PO} X_P P_O$$

$$+ c_{KI} X_K P_I + c_{CI} X_C P_I + c_{HI} X_H P_I + c_{PI} X_P P_I$$

Following Kohli (1993), $\beta^T x$ is defined as a Fisher Quantity Index over the restricted factors, X_j, $j = K, C, H, P$ and $\alpha^T p$ is defined as a Fisher Price Index defined over the variable output and input prices, P_i, $I = O, I$. The normalisation is carried through using the input price variable P_I and restricted input variable X_K.

In order that equation (6.2) represents a proper restricted profit function it must be convex in prices and concave in restricted factors. These regularity conditions can be assessed directly from the estimated coefficients in equation (6.2), where the price coefficients must be positive semi-definite and the restricted factor coefficients must be negative semi-definite (Diewert and Wales, 1987). Linear homogeneity is imposed on the quadratic function by the following restrictions: $b_{Kj} = 0$, $j = K, C, H, P$ and $a_{iI} = 0$, $i = O$ and I.

Diewert and Wales (1987) show that the normalised quadratic is particularly well suited for testing and imposing curvature properties, using a technique described by Wiley, Schmidt and Bramble (1973). It is important to note that imposing curvature does not destroy the flexibility properties of the functional form. Dupont (1991) describes how to impose the Wiley, Schmidt and Bramble curvature properties. Writing the coefficients for the restricted factors in matrix B, define a lower triangular matrix D such that
$B = -DD^T$ or

$$B = \begin{bmatrix} b_{KK} & b_{KC} & b_{KH} & b_{KP} \\ b_{KC} & b_{CC} & b_{CH} & b_{CP} \\ b_{KH} & b_{CH} & b_{HH} & b_{HP} \\ b_{KP} & b_{CP} & b_{HP} & b_{PP} \end{bmatrix} = neg \begin{bmatrix} 0 & 0 & 0 & 0 \\ d_1 & 0 & 0 & 0 \\ d_2 & d_3 & 0 & 0 \\ d_4 & d_5 & d_6 & 0 \end{bmatrix} * \begin{bmatrix} 0 & d_1 & d_2 & d_4 \\ 0 & 0 & d_3 & d_5 \\ 0 & 0 & 0 & d_6 \\ 0 & 0 & 0 & 0 \end{bmatrix}$$

the estimation requires a nonlinear maximum likelihood estimator to recover the parameters of the D matrix, from which the B matrix can then be recovered by simple matrix manipulation.

Imposing symmetry and linear homogeneity on equation (6.2), Hotelling's lemma is used to derive the output supply equation for the unrestricted non-quota landings and the input demand equation for the variable aggregate input factor. The output supply equation is written as:

$$Y_O = \frac{\beta^T x}{P_I}[a_{OO} P_O] + 0.5 \frac{\alpha_O}{X_K}\left[b_{CC} X_C^2 + b_{HH} X_H^2 + b_{PP} X_P^2\right]$$

$$+ \frac{\alpha_O}{X_K}\left[b_{CH} X_C X_H + b_{CP} X_C X_P + b_{HP} X_H X_P\right] \qquad (6.3)$$

$$+ c_{KO} X_K + c_{CO} X_C + c_{HO} X_H + c_{PO} X_P$$

And the input demand equation is written as:

$$-Y_I = -0.5 \frac{\beta^T x}{P_I^2}\left[a_{OO} P_O^2\right] + 0.5 \frac{\alpha_I}{X_K}\left[b_{CC} X_C^2 + b_{HH} X_H^2 + b_{PP} X_P^2\right]$$

$$+ \frac{\alpha_I}{X_K}\left[b_{CH} X_C X_H + b_{CP} X_C X_P + b_{HP} X_H X_P\right] \qquad (6.4)$$

$$+ c_{KI} X_K + c_{CI} X_C + c_{HI} X_H + c_{PI} X_P$$

Estimation of equations (6.3) and (6.4) recovers all parameters defined in equation (6.2).

After estimation we are interested in two fundamental characteristics of the production structure; the elasticity of intensity of unrestricted non-quota output with respect to quota output and, most importantly, the shadow value of quota, for each vessel and for each species. To be clear, the elasticity of intensity is a measure of the change in non-quota landings caused by a one percentage change in quota holdings for a specific species (Dupont, 1991). The elasticity of intensity of non-quota holdings associated with each quota restricted factor is defined as:

Elasticity of intensity of non-quota species with respect to cod quota:

$$\eta_{y_O x_C} = \{\frac{\beta_C}{P_I}[a_{OO} P_O] + \frac{\alpha_O}{X_K}[b_{CC} X_C]$$

$$+ \frac{\alpha_O}{X_K}[b_{CH} X_H + b_{CP} X_P] + c_{CO}\} * \left[\frac{X_C}{Y_O}\right] \qquad (6.5)$$

Elasticity of intensity of non-quota species with respect to haddock quota:

$$\eta_{y_O x_H} = \{\frac{\beta_H}{P_I}[a_{OO} P_O] + \frac{\alpha_O}{X_K}[b_{HH} X_H]$$

$$+ \frac{\alpha_O}{X_K}[b_{CH} X_C + b_{HP} X_P] + C_{HO}\} * \left[\frac{X_H}{Y_O}\right] \qquad (6.6)$$

Elasticity of intensity of non-quota species with respect to pollock quota:

$$\eta_{y_O x P} = \{\frac{\beta_P}{P_I}[a_{OO} P_O] + \frac{\alpha_O \beta_P}{X_K}[b_{PP} X_P]$$

$$+ \frac{\alpha_O}{X_K}[b_{CP} X_C + b_{HP} X_H] + C_{PO}\} * \left[\frac{X_P}{Y_O}\right] \qquad (6.7)$$

A negative elasticity of intensity implies a one percent increase in a quota species causes a decline in the harvest of the non-quota landings, whereas a positive elasticity of intensity implies a one percent increase in quota causes an increase in the harvest of the non-quota landings. It is possible in a multi-output fishery that we may well observe both substitute and complementary relationships between non-quota and quota landings.

The shadow value of quota landings is measured as the value to the vessel of a one unit increase in quota holdings. Within our restricted profit function framework a two-step approach is used in calculation. In the first step, we measure the change in restricted profit associated with non-quota catch from a one unit change in quota holdings. The change in restricted profit is caused by two-factors: i) An increase in marginal cost and subsequent decline in non-quota related restricted profit resulting from a one unit increase in quota landings. Marginal cost increases because of an increase in the variable input factor necessary to land the additional quota and this effect is unambiguously negative. ii) A change in marginal restricted profit caused by a change in non-quota landings. This secondary change is due to the substitute/complementary relationship between non-quota and quota landings. If a substitute relationship exists, then marginal restricted profit from non-quota landings will decline as landings of quota species increase. A complementary relationship will increase marginal restricted profit as landings of non-quota species increase. In the second step, we define a price received for the quota fish. This is a vessel specific price and reflects the quality and size characteristics etc. of quota landings. Finally, the shadow value to the vessel is price of quota fish plus change in restricted profit.

The change in restricted profit of non-quota landings associated with a one unit change in each of the quota species is defined as:

Cod

$$\frac{\partial \pi^R}{\partial X_C} = 0.5 \frac{\beta_C}{P_I}\left[a_{OO} P_O^2\right] + \frac{\alpha^T p}{X_K}\left[b_{CC} X_C\right]$$
$$+ \frac{\alpha^T p}{X_K}\left[b_{CH} X_H + b_{CP} X_P\right] + C_{CO} P_O + C_{CI} P_I \qquad (6.8)$$

Haddock

$$\frac{\partial \pi^R}{\partial X_H} = 0.5 \frac{\beta_H}{P_I}\left[a_{OO} P_O^2\right] + \frac{\alpha^T p}{X_K}\left[b_{HH} X_H\right]$$
$$+ \frac{\alpha^T p}{X_K}\left[b_{CH} X_C + b_{HP} X_P\right] + C_{HO} P_O + C_{HI} P_I \qquad (6.9)$$

Pollock

$$\frac{\partial \pi^R}{\partial X_P} = 0.5 \frac{\beta_P}{P_I}\left[a_{OO} P_O^2\right] + \frac{\alpha^T p}{X_K}\left[b_{PP} X_P\right]$$
$$+ \frac{\alpha^T p}{X_K}\left[b_{CP} X_C + b_{HP} X_H\right] + c_{PO} P_O + c_{PI} P_I \qquad (6.10)$$

Note that the change in restricted profit in non-quota landings for each individual vessel is conditioned on vessel characteristics and other quota holdings.

For completeness, the shadow value (SV) of the *jth* vessel, for the *ith* quota species, (*i*= cod, haddock and pollock) is written as:

$$SV_{ji} = P_{ji} + \frac{\partial \pi_j^R}{\partial X_i}. \qquad (6.11)$$

It is also of some interest to separate out the two factors that impact marginal restricted profit. We do this by calculating a marginal shadow value (MSV) that reflects only the shadow value to the vessel of an increase in one unit of quota but holding constant the restricted profit change due to changes in non-quota landings. In other words, the MSV focuses only on the decline in restricted profits resulting from the increase in marginal cost of landing an additional unit of quota.

The MSV is calculated by transforming the elasticity of intensity to measure a change in non-quota landings caused by a one unit increase in quota landings and this is multiplied by the price on non-quota landings. Finally, this value is subtracted from the shadow value to obtain the MSV. Our results will report both the shadow value and MSV for each vessel.

6.4 ECONOMETRIC SHADOW VALUES

A maximum likelihood estimator is used to estimate the parameters in the supply and demand equations defined in (6.3) and (6.4). The equations are non-linear in the parameters and estimated as a system. To initiate the optimising routine, starting parameters values are generated by forcing least squares estimates on the equations and allowing the routine to converge to optimal values. A number of different starting values were used to test the robustness of the optimising routine.[6] We are particularly interested that the estimated parameters are consistent with the regularity conditions necessary to ensure a proper restricted profit function. Specifically, the curvature properties of convexity in prices and concavity in restricted factor inputs must be satisfied. In our estimations, convexity in prices was not an issue as it is satisfied in the data set. On the other hand, eigenvalues for the Hessian used in testing concavity in restricted factor inputs show that this regularity condition is not satisfied in the data. Using the procedure described by Dupont (1991), we impose the proper curvature properties on the estimating equation. It is important to keep in mind that imposing curvature on the generalised quadratic function does not destroy the second order flexibility properties of the functional form (Diewert and Wales, 1987).

Table 6.4 reports the estimated parameters and standard errors for the final estimating equations satisfying all regularity conditions for our restricted profit function. Nine of the 15 parameters are statistically important at a conventional 10% level. The *B* matrix that contains the parameters of interest for the restricted factor inputs is calculated using the estimated vector *d* parameters in Table 6.4, as described above. It must be noted that only 3 of the 6 parameters in the *d* vector are statistically significant. This may be the consequence of modelling the harvest of cod, haddock and pollock prior to 1991 as restricted factors[7]. Nevertheless, we continue and

---

[6] Of course, we are not sure that we have obtained a global maximum but the convergence procedure is robust to the many different starting values used.

[7] A referee has pointed out that that modelling harvest prior to 1991 as restricted may introduce bias into the estimation. The extent of the bias will be a function of the correlation of the assumed restricted variable and the error term. We believe the size of the potential bias to be

examine the robustness of the estimates by reporting, in Table 6.5, the own-price elasticities for supply of non-quota landings and demand for the aggregate variable factor input by year and vessel size. Both supply and demand elasticities for the different vessel sizes and over the three-year period are inelastic. Vessels less than 45 feet show little variation in values over the three year period. On the other hand, vessels greater than 45 feet are measured to be more inelastic in response to supply price changes of non-quota landings over the period but more elastic in aggregate input price response.

Table 6.4   Restricted profit function.

Coefficient	Estimated value	Standard error	Coefficient	Estimated value	Standard error
a_{oo}	2.92E-02*	7.88E-03	C_{KO}	-498.16	536.7
d_1	-5.73E-03*	1.10E-03	C_{CO}	-0.13306*	7.17E-02
d_2	-1.21E-02*	3.08E-03	C_{HO}	-0.10244	0.1778
d_3	3.02E-10	1.18E-02	C_{PO}	0.26335*	0.1252
d_4	6.12E-03*	4.37E-03	C_{KI}	23.974	93.878
d_5	1.10E-11	1.49E-02	C_{CI}	-0.11437*	1.66E-02
d_6	5.51E-10	8.46E-03	C_{HI}	-6.49E-02*	3.99E-02
			C_{PI}	-6.85E-02*	2.11E-02

N = 101
Log-likelihood function = -2255.659
* statistical significant at 10% level

Table 6.5   Price elasticities, by year and vessel length.

		1988	1990	1991
Vessel length less than 45 ft	η_{oo} *	0.227	0.233	0.248
	η_{ii} **	-0.329	-0.256	-0.315
Vessel length greater than 45 ft	η_{oo}	0.492	0.329	0.344
	η_{ii}	-0.307	-0.433	-0.374

* Own price elasticity of supply
** Own price elasticity of demand

virtually zero since the large declines in biomass during the late 1980s required intensive monitoring by fisheries managers to ensure that the overall TAC not be exceeded. We believe that this in-season hands-on management, combined with biomass levels that fell dramatically each year over the period, effectively restricted the ability of vessel owners to freely choose their catch levels.

To add further to the characterisation of the structure of harvesting in the Scotia-Fundy groundfishery, Table 6.6 reports the elasticity of intensity between the variable non-quota landings and quota restricted landings by year and vessel size. The results here are interesting in that for the quota species, cod and haddock, we measure a substitute relationship with non-quota landings. This result is particularly strong for cod relative to haddock for each of the three years investigated. This reflects the fact that cod is by far the most important fish species in terms of landings and revenue for vessels in the groundfishery. It is also worth pointing out that for vessels greater than 45 feet in 1991 we measure a stronger output response to cod changes and a weaker output response to haddock changes compared to vessels less than 45 feet. In other words, it is more costly for larger vessels to acquire additional quota of cod and less costly to acquire additional quota of haddock than smaller vessels, in terms of loss in restricted profit because of a decline in non-quota landings. On the other hand, the quota species, pollock, shows a complementary relationship with non-quota landings. Recall that a complementary relationship implies that an increase in quota allocation causes an increase in non-quota landings; a substitute relationship implies the opposite. It is important to point out that the intensity results are robust over a wide spectrum of model specifications and estimations.

Table 6.6 Elasticity of intensity, by year and vessel length.

		1988	1990	1991
Vessel length less than 45 ft	η_{OC}[+]	-0.882	-0.566	-0.664
	η_{OH}	-0.324	-0.190	-0.231
	η_{OP}	0.655	0.454	0.434
Vessel length greater than 45 ft	η_{OC}	-0.779	-0.831	-0.749
	η_{OH}	-0.381	-0.174	-0.176
	η_{OP}	0.639	0.602	1.116

[+] η_{Oj} is elasticity of intensity of non-quota fish (O) with respect to quota fish j, j = Cod (C), Haddock (H) and Pollock (P).

These results accord with the physical characteristics of the Scotia-Fundy groundfishery. Cod and haddock are highly valued fish and the two most important fish species for the groundfishery in terms of revenue earned (Table 6.3). Moreover, the cod/haddock fisheries can be considered directed fisheries and thus an increase in quota allocation means more time spent fishing the quota species at the cost of less time spent fishing non-quota fish. Combine this with restrictions on vessel capacity, length of fishing seasons, weather, etc., and it is reasonable to observe such results. From a policy perspective, the results imply that if quota allocations for cod or haddock are reduced for, say, stock management purposes, our model predicts increased fishing effort directed at the non-quota species.

Of the three quota species, pollock is the least valued in terms of price and revenue. Pollock appears to hold a complementary relationship with non-quota species, reflected in the fact that pollock is part of a by-catch fishery with other non-quota species. In this case, an increase in pollock quota allocation indicates increased landings of both pollock and non-quota species. For management purposes a decline in pollock quota allocations implies reduced fishing effort directed at non-quota species. This is important for regulatory purposes as the setting of a TAC for pollock can be used as an instrument to control fishing effort directed at non-quota species.

The elasticity of intensity is important in calculating shadow values for the quota species. For cod and haddock the shadow value must include not only the increase in restricted profit from increased quota landings but also the decline in restricted profit from the decline in non-quota landings. What is interesting is that shadow value for pollock includes the increased value of restricted profit from increased quota landings plus increased restricted profit from increased non-quota landings.

We turn now to the focus of the chapter, the measurement of shadow values for the three quota species, cod, haddock and pollock for the Scotia-Fundy mobile gear groundfishery. To get an overall impression of the range and evolution of vessel level shadow prices, Figures 6.1 and 6.2 graph both the shadow value and the marginal shadow value by species and vessel size for the three year period 1988, 1990 and 1991. Recall, the marginal shadow value reports only the value to the vessel of changes in restricted profit from changes only in the marginal cost of landing one more unit of quota. For cod and haddock (Figures 6.1a,b and 6.2a,b), the marginal shadow value lies above the shadow value reflecting the substitute relationship between non-quota and quota species, and both series are trending upward over the period of study. The largest shadow values occur in 1991. For these fish species the shadow value falls substantially once we account for changes in restricted profit from reduced landings of non-quota species. There are two important points to emphasise from the figures. First, there is a substantial range in shadow values associated with different vessels. This range reflects vessel level characteristics and the different quota holdings by the individual vessels. This is important because it shows that different vessels value additional quota differently, and this is true for both small and large vessels. In other words, this is a necessary condition for an active trading market to develop in quota. Second, for cod in Figures 6.1a and 6.2a, for some vessels the shadow value is measured as negative. This indicates that for some vessels adding an additional unit of quota results in such a significant decline in restricted profit from the decrease in non-quota landings that the overall benefit of increased cod quota is negative. Cleary, for these vessels entering a quota market would be for the purposes of selling quota.

Figures 6.1c and 6.2c show the evolution of shadow prices for pollock for small and large vessels, restrictively. The noticeable point in these figures is that the marginal shadow value lies below the shadow value representing the complementary relationship between non-quota and pollock landings. In fact, notice that the trend in marginal shadow values is relatively flat compared to the trend in shadow values. Clearly, the major benefit of an additional unit of pollock quota is in the additional landings and profit obtained from non-quota species.

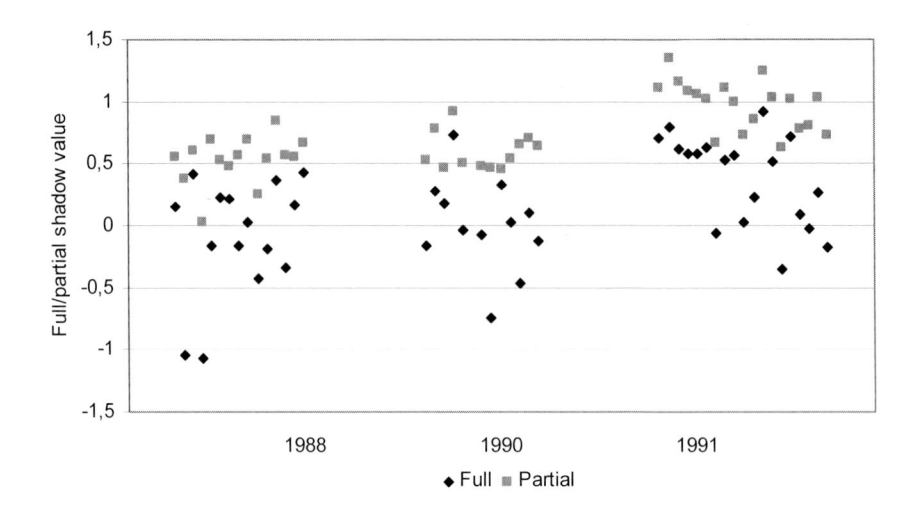

Figure 6.1a Shadow values cod, vessels less than 45 feet.

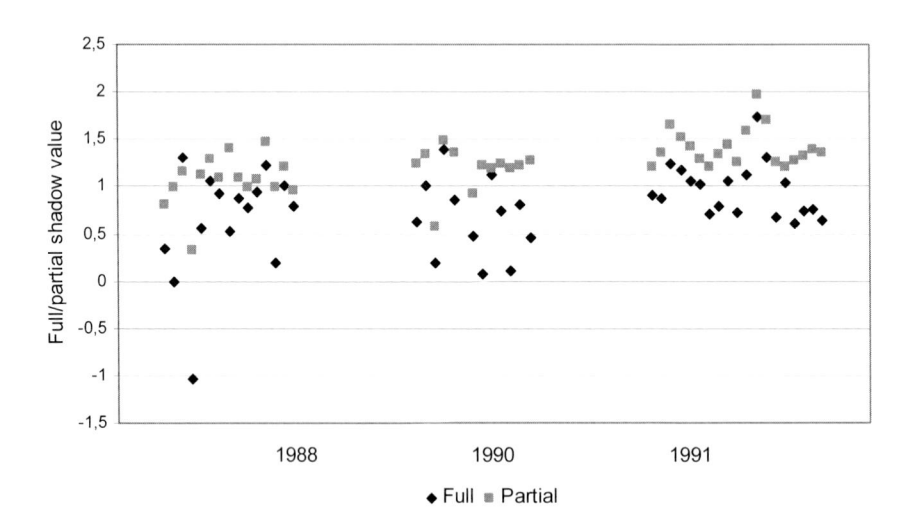

Figure 6.1b Shadow values haddock, vessels less than 45 feet.

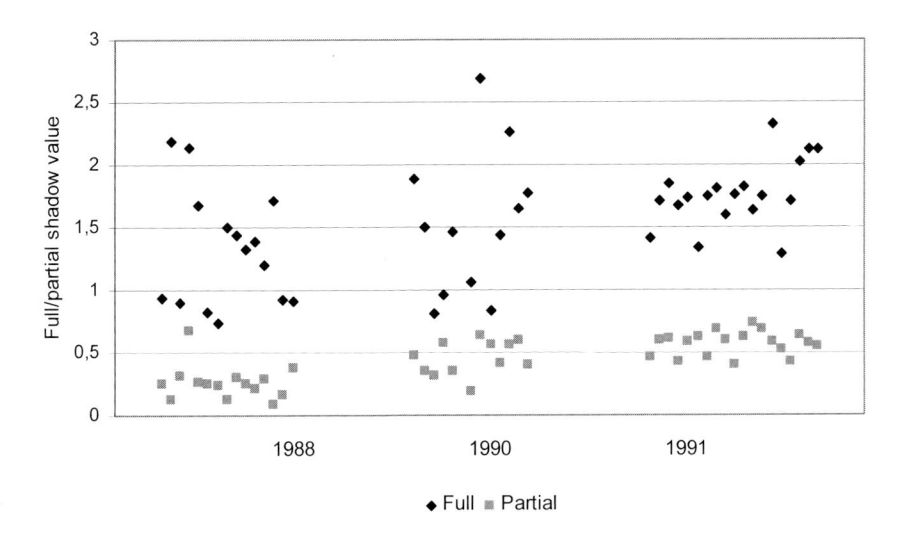

Figure 6.1c Shadow values pollack, vessels less than 45 feet.

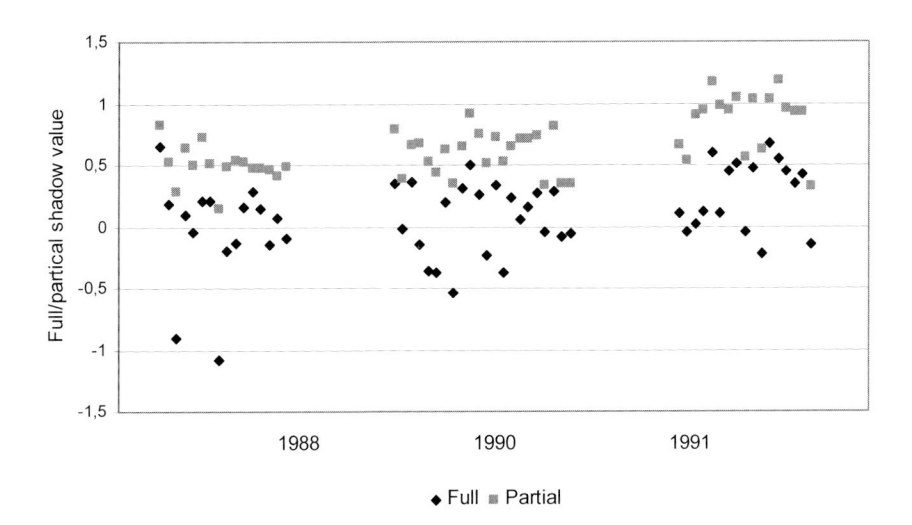

Figure 6.2a Shadow values cod, vessels more than 45 feet.

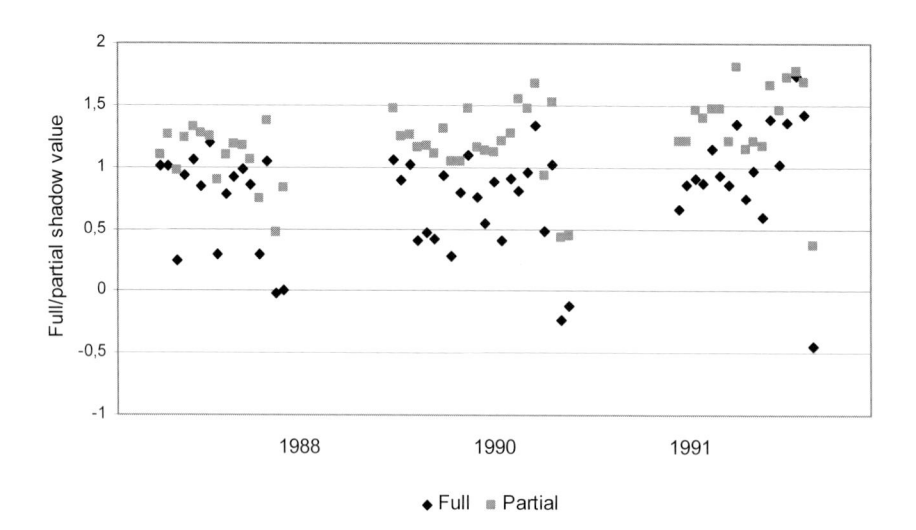

Figure 6.2b Shadow values haddock, vessels more than 45 feet.

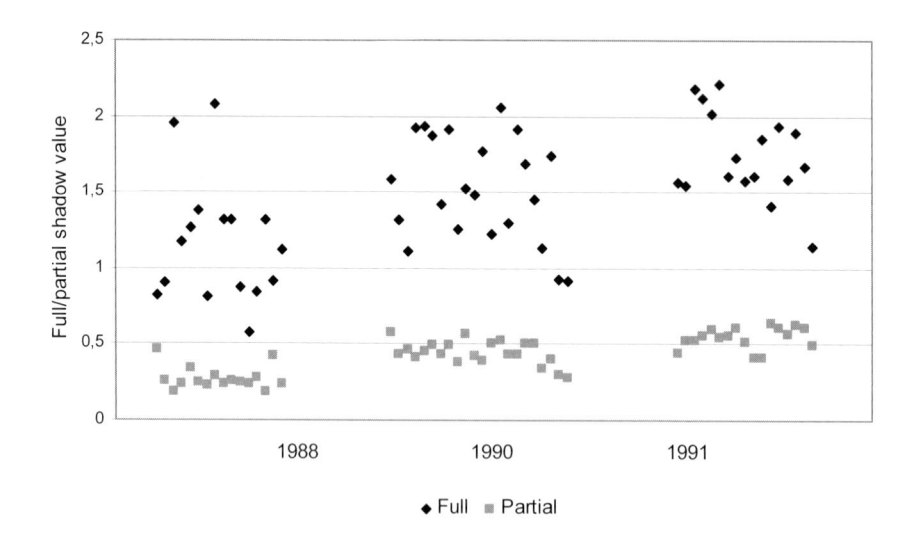

Figure 6.2c Shadow values pollack, vessels more than 45 feet.

Overall for both small and large vessels, there is substantial variation in shadow values that leads us to suspect that, in an active quota trading market, there will be a redistribution of quota across the different vessels. However, the figures do not make it clear as to the direction of the redistribution in terms of larger or small vessels. In Table 6.7 we report summary statistics that may be suggestive of redistribution in an active quota market.

It is important to stress that the measured shadow values are conditioned on the holdings of other quota species and the specific characteristics of production at the current level of landings. For instance, the harvest of all fish species in our data set declined over the period of study, but vessels less than 45 feet suffered this decline more seriously than the larger vessels. From Table 6.1, smaller vessels showed almost a 60% decline in cod and haddock landings and a 70% decline in pollock landings form 1988 to 1991. This is compared to a 14% decline in cod, a 47% decline in haddock and an increase of 16% in pollock landings for larger vessels over the same period. For smaller vessels the drastic change in landings forces disequilibrium in production and requires the vessel to operate on the downward slope of the marginal cost curve; far greater disequilibrium than experienced by the larger firms. Not surprisingly, if smaller firms can acquire additional quota the shadow values reflects the benefit obtained.

Table 6.7 Prices and shadow values (1991) by vessel length.

	Less than 45 feet	**Greater than 45 feet**
Price cod	**1.25**	1.16
Price haddock	**1.57**	1.56
Price pollock	**0.753**	0.719
MSV* cod	**0.969**	0.868
MSV haddock	**1.408**	1.379
MSV pollock	**0.570**	0.544
SV+ cod	**0.377**	0.256
SV haddock	0.956	**0.960**
SV pollock	**1.762**	1.743

*Marginal Shadow Value
+ Shadow Value
Values in bold represent the higher of the prices/shadow values estimated for the two groups.

6.5 CONCLUSION

Individual vessel shadow values can provide information on possible redistribution of quota in an active trading market. Those vessels that show a high shadow value can be expected to bid aggressively for additional quota. Based on the data set available for the Scotia-Fundy mobile groundfishery and the economic model used in this analysis, large vessels greater than 45 feet show no superior advantage over smaller vessels less than 45 feet in terms of measured shadow values for quota fish. In fact, a redistribution of quota by market trading would lead us to suspect that smaller vessels can bid aggressively with larger vessels for quota, especially in the cod and pollock markets. The results are conditioned on current quota holdings by vessels and the state of production disequilibrium particularly forced on smaller vessels due to the drastic decline in stock and harvest levels of all groundfish over this period.

Even in the early years of the Scotia-Fundy ITQ programme it is clear that the market for quota was an active one for the majority of vessels, as might be expected given the diversity in shadow values between different vessels (Figures 6.1 and 6.2). Information from the first year of the programme (1991) shows that 118 licensees were net buyers and a further 132 licensees were net sellers. Similar results were found for 1993. In addition, about half of the net sellers in 1993 sold off more than 90% of their quota. Of these, 14 licensees sold out of the fishery completely by selling 100% of their quota permanently. One prediction from the introduction of ITQs is the redistribution of quota from less efficient vessels (of both large and small sizes) towards more efficient vessels of both types. Liew (1999) shows how concentration of quota increased between 1991 and 1998. In 1991, 162 licensees held 80% of the quota. By 1998, 80% of the quota was held permanently by 109 licensees. Moreover, Burke and Brander (1999) argue that this led to a more efficient fleet composition.

Knowledge of individual vessel shadow values for quota is also useful for fisheries managers to access increasing pressure on fish stocks and to evaluate annual total allowable catch allocations. A high and increasing value placed on quota will indicate a desire by fishers to increase effort in the fishery and alert fisheries managers that changes may need to be made to programme details. For example, at the outset of the Scotia-Fundy ITQ programme the government permitted fishers to exchange excess catches of one species for another species at nominal, pre-determined terms of exchange. However, this provision was quickly dropped since managers felt it was being abused. By 1992 and 1993, for example, quota traded for between $1.00 and $1.50 per pound, Barbara *et al.*, 1995. These prices are very close to the values that we estimated for this fishery and imply increasing pressure upon available fish stocks.

Finally, changing regulatory polices will impact the value that fishers place on quota and allow regulators to identify types of transactions that might take place in a quota market. Forecasting the transactions that might occur will provide information on distributional implications of regulatory changes. Because shadow values are conditioned upon other quota holdings, fisheries managers can influence the value fishers place on quota by changes in other quota holdings. Simulating changes in shadow values will allow predictions on the distributional impact amongst vessels of different size. Such simulations will show also changing pressure on other quota fisheries, helping regulators design better adjustment mechanisms and in allocating departmental resources for monitoring and fisheries officers.

Acknowledgements

The authors acknowledge the comments and helpful suggestions by Scott Taylor, Frank Asche, Nils Arne Ekerhovd and two anonymous referees.

References

Barbara, R., L. Brander and D. Liew. 1995. *Scotia-Fundy Inshore Mobile Gear Groundfish ITQ Program.* Department of Fisheries and Oceans (DFO), Mimeo.

Burke, D.L. and G.L. Brander. 1999. Canadian Experience with Individual Transferable Quotas. In R. Shotton (ed.) *Use of Property Rights in Fisheries Management.* FAO Fisheries Technical Paper 404/1. FAO, Rome.

Bjørndal, T. and D.V. Gordon. 1993. The opportunity cost of capital and optimal vessel size in the Norwegian fishing fleet. *Land Economics* 69(1): 98-107.

Canadian Stock Assessment Secretariat, Science Sector 1998. *Stock Status of Atlantic Cod: Where are we now?* St. John's, February 1998. DFO. Accessible online http://www.dfompo.gc.ca/csas/csas/show/1998/index.htm

Department of Fisheries and Oceans 1998. Atlantic Coast Landed Quantities and Atlantic Coast Landed Values. DFO. Accessible online
http://www.ncr.dfo.ca/communic/statistics/

Department of Fisheries and Oceans, Maritimes Region 2003. *Groundfish Integrated Fisheries Management Plan.* Scotia-Fundy Fisheries, Maritimes Region, April 1, 2000-March 31, 2002. DFO, Accessible online
http://www.mar.dfo-mpo.gc.ca/fisheries/res/imp/2000grndfish.htm

Diewert, W.E. and T.J. Wales. 1987. Flexible functional forms and global curvature condition. *Econometrica* 55(1): 43-68.

Dupont, D.P. 1991. Testing for input substitution in a regulated fishery. *American Journal of Agricultural Economics* February: 155-64

Dupont, D.P. 1996. Limited entry fishing programs: Theory and Canadian practice. In D.V. Gordon and G.R. Munro (eds) *Fisheries and Uncertainty: A Precautionary Approach to Resource Management.* University of Calgary Press: 107-28.

Dupont, D.P. and R.Q. Grafton. 2000. Multi-species individual transferable quotas: The Scotia-Fundy mobile gear groundfishery. *Marine Resource Economics* 15(3): 205-20.

Dupont, D.P., K. Fox, D.V. Gordon and R.Q. Grafton. 2005. Profit and price effects of multi-species individual transferable quotas. *Journal of Agricultural Economics* 56(1): 31-57.

Grafton, R.Q., D. Squires and J. Kirkley. 1996. Private property rights and the crises in world fisheries: Turning the tide? *Contemporary Economic Policy* 14: 90-9.

Homans, F.R. and J.E. Wilen. 1997. A model of regulated open access resource use. *Journal of Environmental Economics and Management* 32: 1-21.

Kohli, U. 1993. A symmetric normalized quadratic GNP Function and the US demand for imports and supply of exports. *International Economic Review* 34(1): 243-55.

Lau, L.J. 1976. A characterization of the normalized restricted profit function. *Journal of Economic Theory* 12: 131-63.

Liew, D.S.K. 1999. Measurement of concentration in Canada's Scotia-Fundy inshore groundfish fishery. In R. Shotton (ed.) *Use of Property Rights in Fisheries Management.* FAO Fisheries Technical Paper 404/2. FAO, Rome.

Newell, R., J. Sanchirico and S. Kerr. 2005. Fishing quota markets. *Journal of Environmental Economics and Management* 49: 437-62.

Squires, D. 1987. Public regulation and the structure of production in multiproduct industries: An application to the New England otter trawl. *Rand Journal of Economics* 18: 232-47.

Squires, D. 1988. Production technology, costs, and multiproduct industry structure: An application of the long-run profit function to the New England fishing industry. *Canadian Journal of Economics* 21: 359-78.

Wilen, J. 1988. Limited entry licensing: a retrospective assessment. *Marine Resource Economics* 5: 313-24.

Wiley, D.E., W.H. Schmidt, and W.J. Bramble. 1973. Studies of a class of covariance structure models. *Journal of the American Statistical Association* 68: 317-23.

Section 2

Capital theory and natural resources

7

Rational Expectations and Fisheries Management

Colin W. Clark

7.1 INTRODUCTION

The two classic problems in commercial fisheries are widely perceived to be overfishing ('too few fish') and overcapacity ('too many boats'). Dealing effectively with these two problems has proven to be far more difficult than anyone would have expected (Caddy and Seijo, 2005).

For example, science-based management has often succeeded in estimating maximum sustainable harvest levels (usually implemented with some safety factor), leading to the standard TAC (Total Allowable Catch) system of management. Until fairly recently this was often considered to be adequate. The fishery was opened on some specified date, and the closed when the TAC had been taken, or when it was estimated that it had been taken. We all know the consequences of this system – the 'derby' fishery. Knowing that the managers will close the fishing season once the year's TAC has been taken, fishermen are forced to fish at maximum intensity while the season remains open. This typically results in a low quality product, often produced under exceedingly dangerous conditions at sea.

In addition, the TAC-based fishery often experiences extreme overcapacity of fishing vessels, a condition characterised by greatly shortened fishing seasons, sometimes lasting only a few days per year. A further consequence may be strong industry pressure for expanded quotas, which fishermen deem necessary in order to escape bankruptcy.

These descriptions apply in a nutshell to many of today's fisheries operating within, and in some cases beyond, 200-mile zones. Proposals to improve the situation have largely concentrated on limiting entry to the fishery, and if necessary removing excess capacity by means of buy-back programmes. Unfortunately, experience with existing buy-back programmes has not been encouraging. Three things that can, and do, go wrong are:

1. Though costly (typically in the hundreds of millions of dollars per fishery), most buy-back programmes have succeeded in reducing the fleet capacity by a relatively small amount (James, 2004).

2. After completion of the buy-back, fleet fishing capacity has tended to gradually increase again as a result of capital stuffing (Holland *et al.*, 1999; Weninger and McConnell, 2000).

3. In many cases, the prospect of a future buy-back programme has initiated a round of further anticipatory *expansions* of fleet capacity, in effect defeating the programme before it begins.

A series of theoretical papers by Gordon Munro and his collaborators has helped to explain the economic incentives underlying such behaviour. Here I briefly review the findings and implications of this theory.

7.2 INVESTMENT AND NON-MALLEABLE CAPITAL

A capital stock $K(t)$ is called perfectly non-malleable if, once acquired, it can never subsequently be reduced (except by being abandoned):

$$\frac{dK}{dt} \geq 0 \tag{7.1}$$

In contrast a perfectly malleable capital stock can be increased or decreased at will, with equal cost for investment and disinvestment. Finally, a quasi-malleable capital stock satisfies:

$$\frac{dK}{dt} = I(t) - \gamma K(t) \tag{7.2}$$

where the investment rate $I(t) \geq 0$, and investment costs $c(I)$ are given by:

$$c(I) = \begin{cases} c_1 I & \text{if } I > 0 \\ c_2 I & \text{if } I < 0 \end{cases} \tag{7.3}$$

with $0 \leq c_2 < c_1$, i.e. the resale price of capital is less than the purchase price. Here γ denotes the rate of depreciation.

In the fishery setting $K(t)$ is equal to fleet capacity, in the sense that $K(t)$ is the maximum level of effort that the fleet can deploy:

$$0 \leq E(t) \leq K(t) \tag{7.4}$$

where $E(t)$ denotes fishing effort. The fish population biomass $x(t)$ is assumed to satisfy:

$$\frac{dx}{dt} = G(x) - qEx \tag{7.5}$$

i.e. the standard general production model used in fisheries. Here $G(x)$ is the natural growth rate of the population, and $h = qEx$ is the harvest or catch rate; q is a constant (catchability). The net operating revenue flow from the fishery is:

$$R(t) = pqx(t)E(t) - cE(t) = (pqx(t) - c)E(t) \tag{7.6}$$

To begin, we assume that the fishery is entirely open-access and unregulated. We wish to predict, from our model, the time path of investment $I(t)$ in fleet capacity, as well as the time paths of effort $E(t)$ and fish stock $x(t)$. This problem was first formulated and solved by McKelvey (1985). (An optimisation version of the model had been solved earlier by Clark *et al.*, 1979).

Notice that fishermen/vessel owners are faced with two decisions, whether to obtain a vessel and enter the fishery, and having entered, how intensively to fish.

Because of the open-access assumption, we will assume that:

$$E(t) = \begin{cases} K(t) & \text{if } pqx(t) - c > 0 \\ 0 & \text{if } pqx(t) - c < 0 \end{cases} \quad (7.7)$$

In other words, fishing at full capacity occurs provided that this is profitable in the short term. Define the stock level x_{OA} by:

$$x_{OA} = c/pq \quad (7.8)$$

Then equation (7.7) becomes:

$$E(t) = \begin{cases} K(t) & \text{if } x(t) > x_{OA} \\ 0 & \text{if } x(t) < x_{OA} \end{cases} \quad (7.9)$$

Thus x_{OA} is a stock equilibrium for the unregulated open-access fishery, provided that existing capacity $K(t)$ is sufficiently large, namely provided that:

$$K(t) \geq K_{OA} = G(x_{OA})/qx_{OA} \quad (7.10)$$

where $G(x_{OA})$ is the resource productivity at $x = x_{OA}$; see equation (7.5). In fact, x_{OA} is the short-term, operating-cost bionomic equilibrium encountered in the basic Gordon-Scott theory of fishing (Clark and Munro, 1975). Note, however, that the definition of x_{OA} in equation (7.8) does not include fixed costs of vessel capital.

We therefore define a second stock level x_I by:

$$x_I = \frac{c + (\delta + \gamma)c_1}{pq} \quad \sqrt{b^2 - 4ac} \quad (7.11)$$

where δ is the rate of interest, γ is the rate of vessel depreciation as in equation (7.2), and c_1 is the capital cost of one vessel. It makes sense to think of x_I (subscript I for Investment) as the long-term, full-cost bionomic equilibrium, and this is indeed correct. It can be seen that investment in vessel capital occurs only if $x(t) \geq x_I$.

To explain this, note that if $x(t) < x_I$ then the initial flow of operating revenue per vessel, $pqx(t) - c$, is less than interest plus depreciation on the vessel, $(\delta + \gamma)c_1$. The potential vessel purchaser would thus not recover his capital costs if $x(t) < x_I$, so that no new investment would occur in this situation.

On the other hand, if $x(t) > x_I$ investment could occur. To predict the extent of investment in this situation we need to hypothesise how potential vessel owners will come to a decision. The driving mechanism, of course, is the fact that a large initial fleet will rapidly deplete the fish population, thereby reducing future per-vessel revenue. We assume that potential investors take this into account, and invest in a vessel only if they expect to ultimately make a profit. If all investors behave in this way, under open access the actual per-vessel profits will be approximately zero.

In other words, behaviour based on rational expectations in an unregulated, open-access fishery will tend to dissipate all potential economic rents, by a combined process of overcapitalisation and overfishing. This prediction extends the purely static theory of bionomic equilibrium, as originally proposed by Gordon (1954). Our present theory has important implications for fishery management, and these implications differ significantly from those of the static theory.

Let $PV_0(K)$ denote the total present value of fleet operating profits if initial fleet capacity equals K_0. We have:

$$PV_0(K_0) = \int_0^\infty e^{-\delta t}(pqx(t)-c)E(t)\,dt \qquad (7.12)$$

where $x(t)$ and $E(t)$ are as described earlier. We can now put together the dynamic equations, to predict the time trajectories of the two state variables $x(t)$, $K(t)$; see McKelvey (1985), Clark (1990), or Clark *et al.* (2005) for diagrams. We assume here that $I(t) \geq 0$ (i.e. there is no removal of vessels, for example because the resale value c_2 is zero) and $\gamma > 0$. In words:

1. First, there is an initial investment $K(0) = K_0$ that depends on the initial stock $x(0) = x_0$. Larger x_0 ($\geq x_1$) implies larger K_0.

2. For $t > 0$ fishing with $E(t) = K(t)$ begins to reduce the fish population $x(t)$. Fleet capacity gradually declines via depreciation, $K(t) = e^{-\gamma t}K_0$.

3. Fishing ceases if $x(t)$ fails below x_{OA}, so a temporary stock equilibrium occurs at $x = x_{OA}$, $E = E_{OA} = G(x_{OA})/qx_{OA}$.

4. Eventually fleet capacity falls (through depreciation) below E_{OA}, and the fish stock begins to recover.

5. When $x(t)$ reaches x_1, a further investment takes place, causing $K(t)$ to increase to $K_1 = G(x_1)/qx_1$. Thereafter, long-term equilibrium is maintained at $x = x_1$, $K = K_1$. This is the full-cost bionomic equilibrium.

Thus, our model predicts a single 'boom-and-bust' cycle of overcapacity and stock depletion, winding up at a long-term bionomic equilibrium. To reiterate, in a new fishery, capacity initially develops in the expectation of future revenues, which will be especially high while the resource remains abundant. Fishing then gradually reduces the stock of fish, eventually reaching the level of short-term bionomic equilibrium, where net revenue flow becomes zero. For a while, fishing capacity (which is never withdrawn from the fishery) may only be needed part-time. Sooner or later however, depreciation takes its toll and fleet capacity declines to such a level that biological recovery can begin. (Smith, 1969 came up with a similar prediction of joint biomass-effort dynamics, but without explicitly considering capital investment *per se*.) Eventually the stock recovers to a level where vessel replacement takes place; here a long-term bionomic equilibrium occurs, balancing gross revenues against total (variable plus fixed) costs of fishing:

By the way, the McKelvey model applies also to fisheries with a history of past exploitation. The stock levels x_0, x_1 are as specified above, and capital investment occurs as already described. If exogenous changes, for example in prices or costs, occur, the stock levels x_0, x_1 change accordingly, but the subsequent dynamics are as described before. For example, a price increase could (but might not) cause an expansion of capacity, and induce a new cycle of overcapacity and overfishing. (We assume here that such exogenous changes are not anticipated by the fishermen. But if price increases, for example, were anticipated, the result would be greater initial capacity than otherwise.)

The initial fleet capacity K_0 is given by:

$$PV_0(K_0) = c_f K_0 \qquad (7.13)$$

Here we are assuming rapid, indeed instantaneous expansion of the fleet at time $t = 0$. This is an extreme assumption, perhaps, but rapid expansion of a newly developing fishery is not uncommon. The first entrants usually enjoy the greatest profits from the fishery.

How does this predicted equilibrium compare with the industry-wide profit maximising solution? The answer is that the unregulated fishery experiences overfishing and overcapacity relative to the profit-maximising solution. With the advent of TAC-based regulation, overfishing may be controlled, but overcapacity may be exacerbated, as we next explain.

7.3 REGULATED OPEN-ACCESS

We next consider the effects of TAC-based management, designed to prevent overfishing. We first assume that no attempt is made to limit entry to the fishery. Suppose that the target stock level is x^*, with:

$$x^* > x_1 \qquad (7.14)$$

Maintaining the stock at x^* will result in larger catches (and net revenues) per unit effort than would otherwise be achieved. Hence it is likely that (with evident notation):

$$PV_{0,reg}(K_0) > PV_0(K_0)$$

By equation (7.13), we conclude that fishing capacity will be greater under regulated than under unregulated open access. It is this phenomenon of management-generated excess capacity that is the source of the current widespread difficulties in fishery management. Scientifically sophisticated management programmes have often had very little effect on the economic performance of major fisheries, potential profitability having been largely dissipated through overcapacity.

To reiterate: overcapacity (but presumably not overfishing) can actually be exacerbated by TAC-based management regimes. Empirical studies identifying this effect would be most welcome; there is no shortage of cases to study.

7.4 LIMITED-ENTRY AND BUY-BACK PROGRAMMES

Overcapacity of fishing fleets has prompted the introduction of licence-limitation programmes. The initial problem in such programmes is to reduce the existing level of overcapacity. As noted earlier, buy-back programmes introduced for this purpose have often been unsuccessful. Our model helps to explain why this is the case (here I follow a paper of Clark *et al.*, 2005).

We consider a TAC-regulated fishery, in which the TAC maintains a constant target stock level x. The annual catch H is:

$$H = qxE \leq H_{max} \quad (7.15)$$

where E denotes annual effort and H_{max} is the TAC. Let K denote fleet capacity, and assume that the full fleet is deployed during the regulated fishing season, which lasts for a portion D of the year, $0 < D \leq 1$. Then:

$$E = KD, \text{ so that } H = qxKD \quad (7.16)$$

The optimal fleet size K_{opt} is assumed to be the smallest fleet that can take the TAC over the full year ($D = 1$):

$$K_{opt} = H_{max}/qx \quad (7.17)$$

(Note that qx is the annual catch for a single vessel that fishes for the full season.) If $K > K_{opt}$ the managers are forced to shorten the fishing season to D, where:

$$D = H_{max}/qxK \quad (\text{if } K > K_{opt}) \quad (7.18)$$

Net annual operating revenue for the fleet is:

$$R = pH - cE = (pqx-c)KD \quad (7.19)$$

This expression is increasing in K for $K < K_{opt}$, and constant for $K \geq K_{opt}$, by equation (7.18). We have:

$$R = \begin{cases} (pqx-c)K & \text{if } K < K_{opt} \\ (p-c/qx)H_{max} & \text{if } K \geq K_{opt} \end{cases} \quad (7.20)$$

For this static equilibrium model (see Clark *et al.*, 2005 for a dynamic version) we have:

$$PV_0(K) = R/\delta \quad (7.21)$$

and the open-access fishing capacity K_{OA} satisfies:

$$PV_0(K_{OA}) = c_1 K_{OA} \quad (7.22)$$

If $K_{OA} > K_{opt}$ this implies that:

$$K_{OA} = (p - \frac{c}{qx}) \frac{H_{max}}{\delta c_1} \qquad (7.23)$$

As illustration we consider a numerical example with:

H_{max} = 10,000 tonnes
qx = 500 tonnes
p = $2,000 per tonne
c = $500,000 per vessel year
c_f = $2,000,000 per vessel
δ = 0.10

Thus K_{opt} = 20 vessels. For $K \geq K_{opt}$ we have:

$$R = \$10 \text{ million}$$
$$PV_0(K) = \$100 \text{ million}$$

Open-access capacity is therefore:

$$K_{OA} = 50 \text{ vessels}$$

What buy-back price p_b would be needed to induce the 30 excess vessels to leave the fishery? The post-buy-back value of a licence will be $PV_0(K_{opt})/K_{opt}$ = 5 million. Hence 30 vessel owners will withdraw if $p_b > \$5$ million. In other words, vessel owners who paid $2 million for their vessels will demand $5 million to withdraw them, if they know that 20 vessels will remain.

Why is this? By announcing its intent to operate the fishery with 20 vessels, the government has suddenly increased the value of a licence to $5 million. Vessel owners will wish to stay in the fishery unless they are compensated accordingly. This example, which is entirely typical, explains why many quite costly buy-back programmes have not succeeded in causing much reduction in fleet capacity. Note in the example that the cost of the buy-back programme would be $150 million, considerably larger than the present value of the fishery. We will consider alternate schemes in a moment.

First, however, let us note that in practice, the economic effect of a buy-back programme could actually be much worse than this. Suppose that potential fishermen get wind of an impending buy-back programme to be introduced sometime in the future. Anyone with a history of participation in the fishery will be eligible for a buy-back payment. With such prospects, new vessel owners enter the fishery in large numbers. An anticipated buy-back programme, or even the rumour of one, may entice additional overcapacity, at least if it is understood that the ultimate licensees will get to share the full rents from the fishery.

One alternative to a voluntary buy-back programme is a self-financed one, in which future profits from the reduced-capacity fishery are used to finance the buy-back payments. The necessary funds would be raised by a levy on future catches, which would reduce the value of a licence by a certain amount, say w. If the buy-back payment per vessel equals y, we have:

$$\text{total present value of future levies} = K_{opt} w$$
$$\text{total buy-back payments} = (K_{OA} - K_{opt}) y$$

For self-financing:

$$K_{opt} w = (K_{OA} - K_{opt}) y \qquad (7.24)$$

Also:

$$\text{licence value net levy} = \frac{R}{\delta K_{opt}} - w$$

This is the same as the buy-back payment:

$$\frac{R}{\delta K_{opt}} - w = y$$

Solving for w and y, we obtain:

$$w = \frac{R}{\delta K_{opt}} \frac{K_{OA} - K_{opt}}{K_{OA}}$$

$$y = \frac{R}{\delta K_{opt}} \frac{K_{opt}}{K_{OA}}$$

Finally, from equation (7.22) and the fact that $PV_0(K_{OA}) = R/\delta$, we obtain:

$$w = \frac{K_{OA} - K_{opt}}{K_{opt}} c_1$$

$$y = c_1$$

For the example, the buy-back payments are $2 million per vessel, and these are financed by catch levies that reduce the value of a licence from $5 million to $2 million. This amounts to a catch levy of $600 per tonne.

But will the new system, with K_{opt} (7.20) vessels competing for the annual TAC (10,000 tonnes) and paying a levy ($600/tonne) really work? I see two problems. First, can the catch levies actually be amortised – who will provide the lump-sum buy-back payments? Second, since net operating revenues will be positive ($200,000 per annum) in the limited-entry fishery, what will prevent the licensed vessel owners from practicing capital stuffing in an attempt to increase their individual shares of the catch (Weninger and McConnell, 2000)?

For the first problem, perhaps the government will agree to buy back the excess capacity, and then simply let the full revenues go to the licensed fleet. However this will not work! The reason has already been explained. For the second problem, there is no solution other than taxing away the entire net revenue (in which case no buy-back payments would be needed), or else instituting a system of IVQs (individual vessel quotas). It is possible that the licensed fishermen would themselves decide to set up an IVQ system, as has recently occurred for example in the British Columbia sablefish fishery.

7.5 CONCLUSION

Overcapitalisation of fishing fleets is often flagged as a major obstacle to the shift to a sustainable, profitable fishery. The presumed cure is usually taken to be the use of public funds to decommission a portion of the fleet, by paying vessel owners to give up their licences and withdraw from the fishery. Experience with existing buy-back schemes (on which billions of US dollars have been spent, and more billions proposed) has not been promising, however.

Three practical difficulties are recognised: (a) these expenditures have often had at best a marginal effect on total fleet capacity; (b) remaining vessels tend to be upgraded, pushing fleet capacity upwards once again; (c) whenever fishermen have gotten wind of an upcoming buy-back programme, there has been a sudden upsurge of fleet capacity – even though existing capacity is known to be excessive.

A dynamic model of investment under competitive access conditions (even with strict control of annual catches) explains all these phenomena, which are simply the predictable result of rational expectations concerning future revenue after the buy-backs have been completed. In essence, vessel buy-backs are a form of subsidy, with all the recognised undesirable effects of other forms of subsidisation. The oft-repeated mantra that buy-back subsidies are 'good subsidies' is not supported by theory or by empirical observation.

Acknowledgments

The helpful suggestions of an unknown reviewer are hereby gratefully acknowledged.

References

Caddy, J.F. and J.C. Seijo. 2005. This is more difficult than we thought! The responsibility of scientists, managers and stakeholders to mitigate the unsustainability of marine fisheries. *Philosophical Transactions of the Royal Society B*: 59-75.
Clark, C.W. 1990. *Mathematical Bioeconomics: The Optimal Management of Renewable Resources*, Second Edition. Wiley-Interscience, New York.
Clark, C.W. and G.R. Munro. 1975. The economics of fishing and modern capital theory: a simplified approach. *Journal of Environmental Economics and Management* 2: 92-106.
Clark, C.W., F.H. Clarke and G.R. Munro. 1979. The optimal exploitation of renewable resource stocks: problems of irreversible investment. *Econometrica* 47: 25-41.

Clark, C.W., G.R. Munro and U.R. Sumaila. 2005. Subsidies, buybacks and sustainable fisheries. *Journal of Environmental Economics and Management* 50: 47-58.

Gordon, H.S. 1954. The economic theory of a common property resource: The fishery. *Journal of Political Economy* 62: 124-42.

Holland, D., E. Gudmundsson and J. Gates. 1999. Do fishing vessel buyback programmes work? A survey of the evidence. *Marine Policy* 23: 47-69.

James, M. 2004. The British Columbia salmon fishery 'buyback' programme – a case study in capacity reduction. IIFET Japan Conference Proceedings.

McKelvey, R. 1985. Decentralized regulation of a common property resource industry with irreversible investment. *Journal of Environmental Economics and Management* 12: 287-307.

Smith, V.L. 1969. On models of commercial fishing. *Journal of Political Economy* 77: 181-98.

Weninger, Q. and K.E. McConnell. 2000. Buyback programmes in commercial fisheries: Efficiency versus transfers. *Canadian Journal of Economics* 33: 394-412.

8

Linking Natural Capital and Physical Capital: A Review of Renewable Resource Investment Models

Anthony Charles

8.1 INTRODUCTION

Research to determine optimal levels of physical capacity and capital investment has a lengthy history. Investment models are well-established in economics as in other fields, such as operations research. Some of these are behavioural models – focusing on predicting the investment dynamics of economic actors – while others are optimisation studies. Of the latter, some focus on determining investment in conjunction with other inputs to maximise profits, rents or another utility measure, while others emphasise minimising the costs of investment required to meet a certain demand. This is the so-called capacity expansion problem (Manne, 1967; Freidenfelds, 1981; Luss, 1982), dealing with the optimal extent and timing of capacity additions in order to meet changing demand – for example, in a factory production system or a telephone utility's network.

Natural resource industries have just as much a need as any part of the economy for such capacity investment studies. For example, fishery planning involves determining an optimal number of fishing vessels, while forestry operations will need to know the capital investment required in harvesting machinery and lumber mills. However, a unique consideration arises with such investment problems in renewable resource sectors – they cannot be addressed properly without an appreciation that decisions about investment in physical capital must be made simultaneously with, and integrally linked with, decisions on harvest of the natural resource itself.

This leads to one of the key contributions arising out of a partnership between the two prominent researchers, Gordon Munro and Colin Clark. Before the concept of 'natural capital' had risen to the stature it enjoys today, Clark and Munro made a conceptual breakthrough in treating the natural resource itself as a capital stock, with conservation and sustainable use seen as an investment in the resource (or as it might be put today, investment in natural capital). This represented a natural-resource-focused analogue to the traditional study of investments in physical capital such as roads, factories, and fishing boats. Indeed, just as an investment in physical capital involves up-front capital costs incurred to generate future (uncertain) returns, so too an investment in the resource capital (e.g. by reforestation programmes or reduced fish harvests) typically necessitates short-term costs for potential (uncertain) future gains.

But how much investment in the natural capital is optimal? Focusing on renewable resources, Clark and Munro developed a systematic bio-economic framework to model and thereby determine the 'optimal' investment in natural capital (e.g. Clark and Munro, 1975) – the level of the renewable resource stock desirable to maintain, in parallel to the physical capital. Determination of an optimal natural capital stock at any point in time clearly must take into account both the desired output, related to demand, and the available renewable resource inputs. This capital theoretic approach to resource management has come to dominate in recent decades.

This 'investment in natural capital' focus inevitably led Munro and Clark to explore how the process of investing in the natural resource (using input or output measures to manage the resource biomass left for the future) interacts with the conventional investment problem of optimising physical capital (such as the number of fishing boats) that in turn generates the harvesting inputs. A key goal has been to determine 'optimal investment' policies and the resulting joint dynamics of both the physical capital and the natural capital, through dynamic optimisation of resource investment together with physical capital investment.

The seminal paper accomplishing this involved a collaboration of Clark and Munro with their colleague Frank Clarke. The paper of Clark, Clarke and Munro (1979) combined the bio-economic approach to investment in natural capital described above, with Arrow's (1968) emphasis on the irreversibility of investment in physical capital – a reality, to varying degrees, in many resource industries, where the capital involved in specialised harvesting technology typically has limited alternative uses. The analysis they achieved is described in some detail herein.

Indeed, this chapter uses the Clark, Clarke and Munro analysis as a key element around which an array of related work on investment in renewable resource sectors is reviewed. Emphasis is placed on models that contribute to the theory of investments in renewable resources, but some models developed for specific case studies are also discussed. I focus on research in which investment in physical capital is a key element, alongside investment in the resource (natural capital). Section 8.2 of the chapter takes a step back to briefly summarise three 'predecessors' of Clark, Clarke and Munro (CCM), while section 8.3 focuses on that particular paper. Section 8.4 then reviews a series of 'descendants' of the CCM work – research studies building on the CCM analysis – and in contrast, sections 8.5 and 8.6 briefly examine some alternate research directions that use different approaches to explore investment in natural resource industries. The chapter concludes in section 8.7 with a discussion of the current state of research on renewable resource investment models, and the impact such research has had, and continues to have, on our understanding of resource use and management.

8.2 PREDECESSORS

This section reviews three frequently-referenced early works focusing on the economic modelling of investments, by Arrow (1968), Smith (1968) and Burt and Cummings (1970). The work of Arrow has set the tone for much subsequent research in developing the economic theory of optimal investment, while the papers of Smith and of Burt and Cummings present models on investment and physical capital, applicable to a variety of natural resources.

8.2.1 *Arrow*

A key predecessor to the Clark *et al.* (1979) paper, and indeed to much of the research that has been done on dynamic investment modelling, was the work of Kenneth Arrow. In particular, the analysis in Arrow (1968) focused on optimisation of net social benefits accruing from economic production over time. The model is noteworthy for its combination of dynamic investment processes, and for the realisation that investment decisions are typically at least partially irreversible. Specifically, if the physical capital stock is denoted by *K(t)* and the gross investment level by *I(t)*, then the Arrow model begins with the dynamic equation:

$$dK/dt = I - \gamma K \tag{8.1}$$

where γ is a constant depreciation rate in the capital stock, and *dK/dt* is the time rate of change of capital. As Arrow points out, 'the assumption of irreversible investment means that gross investment must be non-negative', producing the constraint:

$$I(t) \geq 0 \tag{8.2}$$

The optimal investment problem can then be stated as:

$$\text{Maximise} \quad \int \alpha(t) \{P(K,t) - \kappa I\} \, dt \tag{8.3}$$

where *P(K,t)* is the measure of instantaneous net benefits (apart from investment costs), κ is the unit cost of capital, and $\alpha(t)$ is the discount factor, as a function of time *t*. Thus, the present value of the instantaneous benefits at time *t* is given by multiplying the level of those benefits by the discount factor $\alpha(t)$. Note that the returns to capital are assumed to satisfy the inequalities $P_K > 0$, $P_{KK} < 0$.

Given the initial capital stock *K(0)*, the investment policy *I(·)* determines the evolution of the capital level over time. Arrow dealt with this optimisation problem using methods of optimal control theory, in a relatively early application of the Pontryagin maximum principle (Pontryagin *et al.*, 1962). The optimal solution Arrow obtained is one in which 'free intervals' and 'blocked intervals' alternate over time, the latter being ones in which the non-negativity constraint on investment is binding, i.e. within which we would prefer to be able to dis-invest.[1] This work of Arrow has a parallel in the realm of natural resource investments, to be discussed later.

---

[1] Arrow applied the same 'free' and 'blocked' interval approach to an optimal growth model, also with irreversible investment, in Arrow and Kurz (1970), and attempted to incorporate uncertainty into the analysis of optimal investments in Arrow and Lind (1970).

8.2.2 Smith

Around the same time as Arrow was pioneering work on optimal investment modelling, others were looking at problems arising specifically within natural resource sectors, and developing models of these sectors that incorporated investment dynamics. Some of these involved normative optimisation studies of how the resource systems *should* be operated (see below), but others focused on modelling the *natural* behaviour of such systems. In particular, the seminal paper of Smith (1968) modelled investment behaviour from a dynamic non-optimising point of view, using qualitative analysis of differential equations. The model of Smith's involves a system of two autonomous first-order differential equations in the resource stock x and the capital stock K:

$$dx/dt = F(x,K) \tag{8.4}$$

$$dK/dt = I(x,K) \tag{8.5}$$

This system can be written explicitly if the following terminology and assumptions are specified:

(i) Given K units of capital (e.g. lumber mills or fishing vessels) each producing at a rate y, the total production rate is $K \cdot y$, generating a revenue flow $R(K \cdot y)$. Hence the revenue per unit is $R(K \cdot y)/K$ and the implicit average price is $R(K \cdot y)/K \cdot y$.

(ii) There is an optimisation element, in that, given a cost function of $c(y,x,K)$ for each firm, profits for the firm are maximised by equating the given price to marginal cost:

$$R(K \cdot y) / K \cdot y = \partial c(y,x,K)/\partial y \tag{8.6}$$

Hence $y = y(x,K)$ defines the individual unit rate of production implicitly. The firm's profit function:

$$\pi(x, K) = R(K \cdot y) / K \cdot y - c(y, x, K) \tag{8.7}$$

is assumed to drive entry and exit decisions. Building on these assumptions the dynamic system can now be written:

$$dx/dt = f(x) - Ky \tag{8.8}$$

$$dK/dt = \delta\pi \tag{8.9}$$

where $y = y(x,K)$ and $\pi = \pi(x,K)$, and δ is a response rate parameter. Note that, in contrast to the industry-level or 'sole owner' model of Arrow, this behavioural model allows for either an open access situation, in which anyone can enter (or exit) the resource sector, driven by relative profit levels, or a limited entry scenario, in which the set of participants is limited – typically to those with permits, licences or harvesting capacity.

Using this system, Smith undertook a variety of phase-plane analyses to explore the dynamic evolution of fishery and forestry, as well as mineral resources. Particularly interesting for purposes of this chapter are his comparison of unregulated and optimal dynamics, and the extension of the investment dynamics to the case of nonmalleable (immobile) capital:

$$dK/dt = \delta'\pi \text{ if } \pi > 0 \quad (8.10)$$

$$\delta''\pi \text{ if } \pi < 0 \quad (8.11)$$

where $\delta' > \delta''$ (implying entry into the resource sector occurs more rapidly than exit) and $\delta'' = 0$ in the case of totally irreversible investment (no exiting). Smith's analysis of these situations has direct parallels with research to be described below.

8.2.3 Burt and Cummings

A third key historical paper relating to the theme of investment modelling is that of Burt and Cummings (1970) – focused specifically on natural resource industries (as with Smith) but analysing a dynamic optimisation model using optimal control theory (as did Arrow). Investment and production decisions are treated simultaneously within the Burt and Cummings model, which is formulated in discrete time, with the period-by-period natural resource stock *(X)* and capital stock *(K)* following the dynamics:

$$X_{t+1} = X_t + F(q_t, X_t) \quad (8.12)$$

$$K_{t+1} = K_t - D(q_t, I_t, X_t) \quad (8.13)$$

over the time horizon $t = 0, 1, ..., T-1$. Here q is the rate of resource use at time t (extraction rate, harvest rate or effort applied in the process), and is assumed to be constrained by the availability of resource and capital stocks, so $0 \leq q_t \leq h(X_t, K_t)$. Furthermore, the growth in the resource stock is given by the function $F(q, x)$ and the net depreciation of physical capital is $D(q, I, K)$ where I is the investment input. Burt and Cummings (1970) assumed that society wishes to maximise the discounted sum of social benefits accruing over time:

$$\text{Maximise } \sum \alpha^t B_t(q_t, I_t, X_t, K_t)$$

where the benefits B depend on the state *(x, K)* together with the controls *(q, K)*, and α is an appropriate discount factor. Using Lagrange multipliers and a Kuhn-Tucker analysis, Burt and Cummings derived conditions for optimal investment and production. Essentially, they showed that the solution involves a level of investment satisfying the economically intuitive equation:

Marginal social cost of investment = $\sum$ (Discounted Marginal Values of future capital increments due to a unit of current investment)

The theoretical framework developed by Burt and Cummings, as with those of Arrow and Smith, had considerable impact – it has been much referenced in subsequent work, and has been subject to a number of extensions (e.g. Rausser, 1974).

8.3 CLARK, CLARKE AND MUNRO

As noted earlier, the Clark *et al.* (1979) paper reflects a logical union of two elements. First is the focus on 'natural capital', with resource management seen as a matter of investment in natural capital, arising in the context of a bioeconomic modelling framework. Second is the recognition that in a resource industry context, investing in physical capital is essentially an irreversible decision, raising the variety of issues noted by Arrow (1968). Thus CCM undertook a dynamic optimisation analysis, at an industry-wide (or sole owner) scale, based on a continuous-time deterministic model. The model contains two state variables – natural capital (the resource) and physical capital, together with two decision variables – the harvesting effort (related to disinvestment in the resource), and the capital investment. Irreversibility of investment in specialised and/or immobile capital is incorporated in the analysis.

The CCM model builds directly on standard bioeconomic models, in which a renewable resource is of size $x(t)$ and has population dynamics based on a reproduction function (F) and instantaneous harvest level $h(t) = qE(t)x(t)$, where q is a constant and $E(t)$ is the fishing effort at time t:

$$dx/dt = F(x) - h(t) = F(x) - qEx \qquad (8.14)$$

The amount of harvesting is limited by the fishing effort, which in turn is limited by the available physical capital stock (or 'fleet capacity') $K(t)$ with $0 \le E(t) \le K(t)$. In turn, the capital stock dynamics are determined by the investment $I(t)$ and the capital depreciation, at a rate given by a constant fraction of the capital stock, i.e. γK, so that:

$$dK/dt = I - \gamma K \qquad (8.15)$$

where it is assumed that $I(t) \ge 0$ for all t, i.e. no disinvestment is possible.

The objective of management is assumed to be to maximise the discounted sum of rents *[ph-cE-πI]* over time (where p is the unit price, c is the unit cost of effort, and π is the unit cost of capital).

The structure of the model – namely, the fact that the range of values for the control variable (E) depends on a state variable (i.e. $E \le K$) – required shifting from standard optimisation methods to apply a solution verification method known as 'the royal road of Caratheodory' (Clark *et al.*, 1979). In this way, they were able to produce a complete optimal feedback control solution in each of four specific cases: (a) the relatively simple situation of reversible investment, i.e. perfectly malleable capital (with no sign restrictions on I), (b) totally irreversible investment, i.e. perfectly non-malleable capital *($\gamma = 0$)*, (c) irreversible investment with depreciation *($\gamma > 0$)*, which is the major case discussed below, and (d) the situation of a 'market for scrap', allowing for disinvestment but at a loss (scrap value less than initial capital cost).

The full solution in case (c) – with depreciation – is a fascinating one, in which there are basically three possible scenarios in which the fishery can be found, defined in terms of combinations of the resource stock (x) and the capital stock (K):

1. The first scenario is one in which there is a relatively large resource stock and a relatively small physical capital stock. This situation suggests that a higher level of harvesting than currently possible would be optimal and indeed, the solution calls for achieving this through an immediate investment up to a certain level. That desired investment is a function of the current resource stock (x), so that investment takes place until the physical capital is increased to the level $K=\sigma_2(x)$, referred to as a switching curve.

2. The second scenario follows from the first, being a case of sufficient physical capital and a substantial resource stock. The optimal solution now calls for harvesting at full capacity, causing a decline in the resource. This is accompanied by a decline in the physical capital through depreciation, so that at a certain point, this physical capital (and the corresponding fishing effort) have shrunk to a point incapable of keeping the resource at a low level. The resource begins to increase, and when x reaches a particular level, a second round of investment is called for, to bring the resource-capital combination to a final equilibrium (x^*, K^*) at which continual harvesting and investment occurs thereafter to maintain the equilibrium.

3. The third scenario is one of a very low resource stock, not a desired situation, but one which could have been caused by prior over-harvesting. If the fishery is in such a situation, the optimal solution is to cease harvesting, allow the resource to recover, and only re-commence harvesting once the resource stock x has risen sufficiently. A 'sufficient' stock in fact depends on how much harvesting capacity is in place – i.e. there is no harvesting if x lies below a second switching curve $x=\sigma_1(K)$, but if x is above this level, there is full use of the capital stock to harvest back down to this level.

Through this process, the fishery approaches the final equilibrium noted above. These three essential features of the optimal investment strategy are maintained under a range of circumstances – e.g. if the model itself is seasonal (discrete-time) rather than continuous (see Charles, 1983a). The CCM analysis has produced a wide range of useful insights that have been examined and indeed extended subsequently by many authors. Some of these advances are discussed below.

8.4 DESCENDANTS OF CCM: SOME SUBSEQUENT MODELS

8.4.1 Case studies

An important initiative to compare investment theory and practice was that of Clark and Lamberson (1982), who applied the results of the CCM analysis to explore the evolution of whaling fleets, and how well the actual history of those fleets matched the economically optimal temporal path. In a similar vein, McKelvey (1987) developed a dynamic investment model, using control theory, to connect open access dynamics and irreversible investment behaviour for the cases of the whaling fleet and the Pacific fur seal. Further examples of empirically-oriented investment models building on the CCM analysis are included in the various sections below.

8.4.2 Investment under uncertainty

As highlighted by a wide range of authors (e.g. Arrow and Fisher, 1974; Brennan and Schwartz, 1985), investment decision making and uncertainty are inextricably linked. Thus an obvious direction to expand on the CCM work has been to explore how uncertainty affects the joint optimisation of investments in physical and natural capital within a renewable resource context. This theme has been particularly important in fisheries, where stochastic fluctuations and structural uncertainty in the resource are especially strong. Indeed, a variety of papers have dealt with investment models for a stochastic fishery that build on the CCM approach – e.g. Charles (1983b, 1985), Charles and Munro (1985), Flam (1986, 1990), and Hannesson (1987, 1993, 1994).

The early work on this theme can be illustrated by the stochastic analogue of the CCM model examined by Charles (1983b) and Charles and Munro (1985). This discrete-time (year by year) model involves two state variables (the renewable resource stock R and the capital stock K at the start of a fishing year) and two control variables (the harvest H taken during that year, and the investment level I to become available in the following year). The dynamics are such that the end-of-season fish population S (given by $R-H$, the initial stock R minus the harvest) is assumed to reproduce and thus generate next year's (average) resource stock R', while the investment I made one year produces a new capital stock K' next year:

$$R' = F(S) \cdot Z \quad and \quad K' = (1-\gamma)K + I \qquad (8.16)$$

where Z is an independent log-normal random variable with mean 1, $F(\cdot)$ is the stock-recruitment function, and γ is the capital depreciation factor. The decision variable S is constrained by fishing-effort capacity, while I is constrained to be non-negative $(I \geq 0)$ reflecting the irreversibility of investment. The objective of the fishery is assumed to be the maximisation of the expected present value of discounted annual rents $\pi(R,K,S,I)$ summed over time.

Qualitatively, the optimal investment strategy in this model turns out to be independent of the level of uncertainty (random fluctuations) in the biomass, the solution being the same for both deterministic and stochastic cases. The strategy is in fact qualitatively similar to that of Clark *et al.* (1979). In any given year, the target end-of-year resource stock is determined by a switching curve as a function of K, i.e. $S^* = s(K)$, while the capital stock for the following year (and thus the investment I^*) should be picked so that the harvest capacity K' (for next season) is as close as possible to the switching curve target $K^*=h(S)$, depending on the biomass level S.

While qualitatively the presence of uncertainty does not affect the optimum, there can be substantial quantitative differences between the deterministic and stochastic optima, given realistic levels of the variance in the random fluctuations. The optimal level of investment under uncertainty relative to the deterministic analogue reflects a balancing of upside and downside risks (the risk of having too much capital versus the risk of having too little). For example, a fast-growing fish stock and a low unit cost of capital tend to produce a higher optimal level of investment in the stochastic case, while a slow growth rate and expensive capital produce the opposite result.

In renewable resource industries, the uncertain investment environment involves not only random effects, but problems of parameter and model uncertainty as well. Charles (1992) and Clark *et al.* (1985) incorporated Bayesian methods into an optimisation model to examine the case in which model parameters, particularly biological ones, are known only approximately. Viewed as an adaptive control problem, it may be optimal, for example, to have investments in fishing capital exceed what would otherwise appear to be optimal levels, in order to increase harvest levels and thereby hasten the acquisition of valuable information about the resource stock.

8.4.3 Investment behaviour and regulation

The CCM paper determined an industry-level optimisation, but how does this sole-owner situation compare with an open access resource sector, in which entry and exit are not restricted? McKelvey (1985) addressed this question by modelling investment behaviour in a manner compatible with the theory of open-access fisheries, and comparable with corresponding optimisation models. Assuming that entry to the fishery (investment) occurs up to the level ($I^\#$) at which average cost of capital matches the unit returns, based on 'rational expectations' about the future, McKelvey derived a dynamic system describing the evolution of an open-access fishery:

$$dx/dt = F(x) - h \quad ; \quad h \leq q\, xK \qquad (8.17)$$

$$dK/dt = I^\# - \gamma K \quad ; \quad I^\# \geq 0 \qquad (8.18)$$

$$d\mu/dt = (\delta + \gamma)\mu - [pq\, x - c]^+ \qquad (8.19)$$

involving the fish stock (x), capital stock (K), harvest rate (h), investment rate ($I^\#$) and the current value of a unit investment at time $t(\mu)$. Here the discount rate (δ), depreciation rate (γ), catchability coefficient (q), price (p) and unit cost of effort (c) are model parameters. McKelvey used this model to address the optimal levels of regulation required to convert an open-access fishery to an optimal one.

A number of other avenues have been explored in relation to investment behaviour. For example, Lane (1988) used probabilistic dynamic programming to develop a predictive model for investment decision making by fishermen, and to compare actual with predicted results in an empirical case study. Jensen (1998) explored the impact of tax policy on investment in a fishery, using a behavioural model of capital dynamics. Squires *et al.* (1994) examined the investment decision making of fishing firms in a US sablefish fishery. In a more theoretical vein, Mchich *et al.* (2000) developed a model building on the Smith (1968) tradition – a set of differential equations describing the evolution of a fishery with two fish stocks moving between two fishing zones, two fleets harvesting them, and effort dynamics driven by the profit levels of the fleets.

8.4.4 Game theoretic and multi-player problems

Another significant area of work building on the CCM approach has been that of modelling natural resource investment problems involving interaction between two or more players, such as competing harvesting groups (e.g. multiple fishing fleets) or government decision makers versus a resource industry (e.g. a foreign fishing fleet). Research using game theory has included that of Kivijarvi and Soismaa (1992) and Sumaila (1995), for a forestry and a fishery case respectively. The first of these modelled a forest management problem involving investment, harvest and taxation decisions, using a differential game approach to analyse forest tax policies, assuming irreversible investment. The second paper considered a non-co-operative game in which two fishing fleets harvest a single fish stock – a key goal of the analysis was to compare the cases of malleable and non-malleable capital (the latter implying irreversible investment).

A related approach is that of principal-agent analysis – Novak *et al.* (1995) developed a model to explore the investment practices of an agent to whom the fishery owner has leased the rights to the resource, finding that the agent may optimally choose periodic capital investments. A related form of multi-player problem arises between a coastal state and distant water fishing fleets. The dynamic optimisation of the balance between domestic and foreign harvesting capacity is explored by Charles (1986) and Charles and Yang (1990, 1991), providing guidance for coastal states deciding between the development of their own domestic fishing fleets or reliance on the collection of royalties from foreign fleets harvesting within their jurisdiction. Similar decisions were explored by Meuriot and Gates (1983), using mathematical programming to determine the optimal balance.

8.4.5 Financial dynamics

Jorgensen and Kort (1997) expanded on the CCM model to incorporate financial dynamics. Specifically, their model envisions a sole-owner fishery in which the firm must decide on harvesting effort levels and investments in physical capital, but the latter are determined via financial management decisions concerning (a) financing investments through retained profits, and (b) borrowing to obtain additional funds. Thus, in addition to resource stock *(S)* and capital stock *(K)* variables, they add a variable *B* reflecting the level of debt or lending by the firm, and $K = X + B$ (where X is the accumulated equity). This added complexity comes at the cost of allowing completely reversible investment, a fundamental change from the CCM model. However, the authors argue that this should have little effect on the results, since disinvestment occurs as only a minor aspect of their solution.

8.4.6 Nonlinearities

An insightful paper by Boyce (1995) examined the implications arising from relaxing two key linearity assumptions in the CCM paper – linearity in investment costs and in profits (with respect to harvest, implying a constant price). In particular, making both these functions nonlinear leads to a more gradual optimal capital accumulation path than the 'bang-bang' investment policy in CCM. Boyce noted that this result meshes more closely with the rather gradual capacity expansion in the whaling fleet, as discussed by Clark and Lamberson (1982).

8.4.7 Investment in resource protection

Investment may take place not to increase harvesting capacity, as usually assumed, but for some other objective – such as to increase protection of a natural resource from catastrophic destruction. As Reed (1989) noted, such destruction can arise from fire (in forests or rangeland), pests (in forests or agricultural systems), wind or environmental change (such as the El Nino phenomenon). Reed modelled the optimal investment in such protection with CCM-like investment assumptions but utilising a hazard function approach based on a survivor function $S_p(x)$, the probability that the resource survives to age x, and applying the Pontryagin maximum principle.

8.5 PARALLEL THREADS

Most of the research described above has focused on deriving conceptual and theoretical insights from dynamic optimisation models. This section outlines two of the most prevalent alternate threads in research on renewable resource investment models. The first of these has been particularly prevalent in fishery studies – shifting away from dynamic modelling, this approach involves static, empirically-oriented optimisation of harvesting capacity. The second approach has had a significant presence in forestry studies and to some extent in rangeland management – these models are oriented to providing theoretical insights, but in this case through optimisation of dual investments not in natural and physical capital, but rather in natural capital and resource productivity.

8.5.1 Static optimisation models

In contrast to the dynamic bioeconomic models described above, an alternative approach to renewable resource investment problems has been through optimisation of static, linear models. Such models are typically analysed using linear programming, are structurally more detailed than their counterpart dynamic models, and tend to focus on empirical or policy applications (rather than on obtaining theoretical insights, the goal in most dynamic bioeconomic models). A number of such analyses are contained in Haley (1981), typically dealing with determining optimal capacity requirements for specific fisheries, seeking a match between fleet capacity and available catch allocation levels.

An example of this approach is that of Flam and Storoy (1982) who developed a steady-state optimisation model of the Norwegian industrial fishery. The analysis generated the maximum net economic benefits, by optimising fleet sizes, and allowed a comparison of the current harvesting capacity with the optimal level. The model focuses on the operation of the fishery during a single year, sub-divided into suitable time periods, and can be stated as:

$$\text{Maximise } \sum c_{tvsaf} x_{tvsaf} \text{ subject to: } \sum a_{tvsaf} x_{tvsaf} \leq b_i; \quad i = 1, ..., m \quad (8.20)$$

where x_{tvsaf} is defined as the number of trips in time period t completed by vessel group v bringing species s, caught at the fishing area a, to factory f. A particular trip, *tvsaf*, is assumed to generate net profits c_{tvsaf} and to require resources a_{tvsaf}. These physical or biological resource inputs are constrained by a set of limits on fleet capacities, factory capacities and/or harvests, represented by the terms $\{b(i)\}$. Constraints are also imposed on catch levels, based on the total allowable catch level for species s, and the percentage of the allowable catch that is to be caught in each area during each period of time.

Detailed mathematical programming models such as these, while obviously missing the dynamic element, benefit from a higher degree of complexity that can be incorporated, compared with most bioeconomic models. Squires *et al.* (1994), for example, used linear programming models of investment decision making to examine the performance of a fishery quota management system, generating significant empirical results.

8.5.2 Investment in resource productivity

The literature on renewable resource investment analysis – particularly in forest and rangeland management – includes a component that, unlike the research described earlier, does not include physical capital (harvesting capacity) as a variable. Instead, investments are oriented toward increasing resource productivity, i.e. the capability of the renewable resource stock itself to generate harvests over time. This might be achieved, for example, through decisions to invest in forest silviculture practices (such as reforestation) or rangeland development. This investment in increasing the resource stock differs from the investing-in-natural-capital approach highlighted earlier, in that this is deliberate monetary investment in measures that produce greater resource growth, rather than indirect investment in resource growth achieved through limitations on the harvesting activity.

This approach is particularly common in forestry modelling. Examples include the Adams *et al.* (1982) simulation model of simultaneous short-term harvesting and long-term investments in forest management, and optimisation models involving silviculture as a form of investment in the resource, by Brodie and Haight (1985), Reyner *et al.* (1996) and Siry *et al.* (2004). Related forest management models include those of MacMillan and Chalmers (1992), and Yin and Newman (1996). It might be noted that many analyses of investment in forests take what might be considered a financial-model approach, focusing on risk and rate of return from investments in forests versus other economic activities (e.g. Lonnstedt and Svensson, 2000).

With respect to optimal investment in pasture and rangeland improvements, as a form of resource productivity, a classic work is Burt (1971), with subsequent research in this area including that of Lambert and Harris (1990). An interesting variation on rangeland management, with a focus on wildlife, is the paper of Skonhoft (1998), who uses a bioeconomic model to explore the balance by African pastoralists between rangeland cattle herding and wildlife harvesting – which compete over available grazing land.

8.6 NEW DIRECTIONS

Certainly, research on the interaction of investment modelling and natural resources will continue into the future along many lines. It is only possible in the space here to briefly note two such directions, ones that seem particularly prevalent in the new literature on analysis of renewable resource investments. One of these focuses on assessing (rather than optimising) the level of harvesting capacity utilisation, by applying the method of data envelopment analysis, while the second focuses on addressing uncertainty in resource investment through models based on options theory. Both approaches are noticeably distinct from the types of models described earlier in this chapter.

8.6.1   Capacity utilisation and data envelopment analysis

The dynamic analyses of Clark, Clarke and Munro and subsequent studies provided a rich understanding of the dual investment problem involving a renewable resource stock and a corresponding physical capital. Yet most such models involved a very simple conception of the physical capital, often as a single variable K. In recent years, responding to a concern about over-capitalisation and over-capacity in fisheries, a new research thrust has focused on empirically measuring fishery harvesting capacity and the degree of capacity utilisation.

This work has drawn in particular on the application of a relatively new analytic approach called data envelopment analysis (DEA), found to be useful in a fishery context (Kirkley and Squires, 1999). The DEA methodology measures the technical efficiency and degree of utilisation of existing capacity within a given economic activity. As Pascoe *et al.* (2003) note, 'it estimates a frontier level of production, and measures inefficiency and capacity utilisation as deviations from the frontier'. Key advantages of DEA lie in its non-parametric nature, and its formulation as a fairly straightforward mathematical programming problem.

Various authors in the collection of Pascoe and Greboval (2003) apply DEA methodology to examine fishing capacity utilisation in such locations as Denmark, Canada, the United States and Malaysia. The approach has also been applied to the English Channel fishery of the United Kingdom (Tingley *et al.*, 2003), to a multi-product and multi-fleet fishery in Canada (Dupont *et al.*, 2002) and to tuna fisheries in the Pacific Ocean (Reid *et al.*, 2003), among other examples. Such studies complement the dynamic optimisation described earlier, providing detailed, empirical, cross-sectional results as a base-line, with respect to current harvesting capacity, from which dynamic investment analyses can better proceed.

8.6.2   Investment, uncertainty and options

The interaction of investment, uncertainty and risk in a resource industry continues to attract attention (e.g. Bohn and Deacon, 1997), and a notable new direction in the research literature has been the application of options theory. So-called real-options studies typically involve continuous-time models and the use of stochastic calculus to address investment under uncertainty – building on classic works such as Brennan and Schwartz (1985) and Dixit and Pindyck (1994). There have been several applications of this approach to forestry – see Duku-Kaakyire and Nanang (2004), and the references therein – although the method has been rather more frequently applied to exhaustible resource industries. For example, option theory was used by Harchaoui and Lasserre (1996) to address the problem of capacity choice in the mining sector, by Cortazar and Casassus (1998) in a stochastic study of optimal mine investment timing, as well as by Cortazar and Casassus (1999), Lumley and Zervos (2001) and others.

8.7 CONCLUSIONS

This chapter has highlighted the development and evolution of models to help understand the time-dependent and multi-dimensional aspects of investment problems that arise in renewable resource sectors. Such an understanding has progressed considerably over time, from the early works of Arrow, Smith, and Burt and Cummings, through the seminal paper of Clark, Clarke and Munro, to subsequent work branching out in a wide range of directions. Attention has focused here on three key conceptual considerations that have played a particularly major role in research on the subject over the years: (i) the dual processes of investment in the physical capital and investment in the resource stock (natural capital), (ii) irreversibility of investment in specialised and/or immobile capital, and (iii) the complexities arising from resource stock structure and inherent uncertainties.

Indeed, the review here suggests that the extent of the research undertaken in the more than a quarter century since the Clark, Clarke and Munro (1979) paper appeared has led to a maturing of the theoretical framework developed by CCM and extended by many authors. New research contributions in that vein have now become less frequent, but as seen above, there continues to be strong new directions of theoretical research in the broad area of resource investment modelling, utilising tools such as options theory. Further, there have been significant contributions to empirical work on resource investment and harvesting capacity, notably through data envelopment analysis (DEA).

It is also notable that there have been enduring policy impacts and insights arising from the principal research approaches described in this chapter. In particular, the focus that CCM and others placed on the dual-investment concept, on the irreversibility of physical capital investment in natural resource industries, and on the dynamic complexities in the evolution of a renewable resource sector, have together formed an integral part of our understanding in resource analysis. These insights continue to be referenced in a wide range of research literature. Consider but one example, using an innovative combination of bioeconomic modelling and anthropological analysis, a recent paper by Janssen and Scheffer (2004) addresses the role of sunk costs (non-malleable capital) in resource over-exploitation by ancient societies (such as Easter Island, Mesopotamia and the Norse in Greenland). The authors draw on the results of the CCM paper as a strong conceptual illustration of the impacts of sunk costs in renewable resource sectors.

Clearly, from theoretical, empirical and policy-oriented perspectives, the modelling of dual investments in physical and natural capital will remain an important area of research within the renewable resource sector, and beyond, well into the future.

Acknowledgments

My exploration of the material described in this chapter began rather long ago, while a student of Gordon Munro, Colin Clark and others at the University of British Columbia, and I am very grateful for the strong support I received in those early days. I am also grateful to several anonymous referees for comments that greatly improved the chapter. Finally, I would like to acknowledge the financial support received from the Natural Sciences and Engineering Research Council of Canada, the Social

Sciences and Humanities Research Council of Canada, and the Pew Fellows Program in Marine Conservation.

References

Adams, D., R.W. Haynes, G.F. Dutrow, R.L. Barber and J.M. Vasievich. 1982. Private investment in forest management and the long-term supply of timber. *American Journal of Agricultural Economics* 64: 232-41.

Arrow, K.J. 1968. Optimal capital policy with irreversible investment. In J.N. Wolfe (ed.) *Value, Capital and Growth: Papers in Honour of Sir John Hicks*. Edinburgh University Press, Chicago.

Arrow, K.J. and A.C. Fisher. 1974. Environmental preservation, uncertainty, and irreversibility. *Quarterly Journal of Economics* 88: 312-19.

Arrow, K.J. and R.C. Lind. 1970. Uncertainty and the evaluation of public investment decisions. *American Economic Review* 60: 364-78.

Arrow, K.J. and M. Kurz. 1970. Optimal growth with irreversible investment in a Ramsey model. *Econometrica* 38: 331-44.

Bohn, H. and R.T. Deacon. 1997. Ownership risk, investment and the use of natural resources. Discussion Paper 97-20. *Resources for the Future*. Washington D.C.

Boyce, J.R. 1995. Optimal capital accumulation in a fishery: A nonlinear irreversible investment model. *Journal of Environmental Economics and Management* 28: 324-39.

Brennan, M.J. and E.S. Schwartz. 1985. Evaluating natural resource investments. *Journal of Business* 58: 135-57.

Brodie, J.D. and R.G. Haight. 1985. Optimization of silvicultural investment for several types of stand projection systems. *Canadian Journal of Forest Research* 15: 188-91.

Burt, O.R. 1971. A dynamic model of pasture and range investments. *American Journal of Agricultural Economics* 53: 197-205.

Burt, O.R. and R.G. Cummings. 1970. Production and investment in natural resource industries. *American Economic Review* 60: 576-90.

Charles, A.T. 1983a. Optimal fisheries investment: Comparative dynamics for a deterministic seasonal fishery. *Canadian Journal of Fisheries and Aquatic Sciences* 40: 2069-79.

Charles, A.T. 1983b. Optimal fisheries investment under uncertainty. *Canadian Journal of Fisheries and Aquatic Sciences* 40: 2080-91.

Charles, A.T. 1985. Nonlinear costs and optimal fleet capacity in deterministic and stochastic fisheries. *Mathematical Biosciences* 73: 271-99.

Charles, A.T. 1986. Coastal state fishery development: Foreign fleets and optimal investment dynamics. *Journal of Development Economics* 24: 331-58.

Charles, A.T. 1992. Uncertainty and information in fishery management models: A Bayesian updating algorithm. *American Journal of Mathematical and Management Sciences* 12: 191-225.

Charles, A.T. and G.R. Munro. 1985. Irreversible investment and optimal fisheries management: A stochastic analysis. *Marine Resources Economics* 1: 247-64.

Charles, A.T. and C. Yang. 1990. A decision support model for coastal fishery planning: Optimal capacity expansion and harvest management. In A.J.M. Guimaraes Rodrigues (ed.) *Operations Research and Management in Fishing.* Kluwer Academic Publishers, Dordrecht.

Charles, A.T. and C. Yang. 1991. A strategic planning model for fisheries development. *Fisheries Research* 10: 287-307.

Clark, C.W. and G.R. Munro. 1975. The economics of fishing and modern capital theory: A simplified approach. *Journal of Environmental Economics and Management* 2: 92-106.

Clark, C.W., F.H. Clarke and G.R. Munro. 1979. The optimal exploitation of renewable resource stocks: Problems of irreversible investment. *Econometrica* 47: 25-47.

Clark, C.W. and R. Lamberson. 1982. An economic history and analysis of pelagic whaling. *Marine Policy* 6: 103-20.

Clark, C.W., A.T. Charles, J.R. Beddington and M. Mangel. 1985. Optimal capacity decisions in a developing fishery. *Marine Resource Economics* 2: 25-53.

Cortazar, G. and J. Casassus. 1998. Optimal timing of a mine expansion: Implementing a real options model. *Quarterly Review of Economics and Finance* 38: 755-69.

Cortazar, G. and J. Casassus. 1999. A compound option model for evaluating multistage natural resource investments. In M.J. Brennan and L. Trigeorgis (eds) *Project Flexibility, Agency, and Product Market Competition: New Developments in the Theory and Application of Real Options Analysis.* Oxford University Press.

Dixit, A. and R.S. Pindyck. 1994. *Investment Under Uncertainty.* Princeton University Press, NJ.

Duku-Kaakyire, A. and D.M. Nanang. 2004. Application of real options theory to forestry investment analysis. *Forest Policy and Economics* 6: 539-52.

Dupont, D.P., R.Q. Grafton, J. Kirkley and D. Squires. 2002. Capacity utilization measures and excess capacity in multi-product privatized fisheries. *Resource and Energy Economics* 24: 193-210.

Flam, S.D. and S. Storoy. 1982. Capacity reduction on Norwegian industrial fisheries. *Canadian Journal of Fisheries and Aquatic Sciences* 39: 1314-17.

Flam, S.D. 1986. Variable quotas, irreversible investment and optimal capacity in the fisheries. *Modeling, Identification and Control* 7(2): 93-105.

Freidenfelds, J. 1981. *Capacity Expansion: Analysis of Simple Models with Applications.* North Holland, New York.

Haley, K.B. 1981. *Applied Operations Research in Fishing.* Plenum Press, New York.

Hannesson, R. 1987. Optimal catch capacity and fishing effort in deterministic and stochastic fishery models. *Fisheries Research* 5: 1-21.

Hannesson, R. 1993 Fishing capacity and harvest rules. *Marine Resource Economics* 8: 133-43.

Hannesson, R. 1994. Optimum fishing capacity and international transfer of excess allowable catches. *Land Economics* 70: 330-44.

Harchaoui, T.M. and P. Lasserre. 1996. Capacity choice as the exercise of a financial call option. *Canadian Journal of Economics* 29: 271-88.

Janssen, M.A. and M. Scheffer. 2004. Overexploitation of renewable resources by ancient societies and the role of sunk-cost effects. *Ecology and Society* 9(1): 6.

Jensen, C.L. 1998. Investment behaviour and tax policy. *Marine Resource Economics* 13: 185-96.

Jorgensen, S. and P.M. Kort. 1997. Optimal investment and financing in renewable resource harvesting. *Journal of Economic Dynamics and Control* 21: 603-30.

Kirkley, J.E. and D. Squires. 1999. Measuring capacity and capacity utilization in fisheries. In D. Greboval (ed.) *Managing Fishing Capacity: Selected Papers on Underlying Concepts and Issues*. FAO Fisheries Technical Paper. No.386. Food and Agriculture Organization of the United Nations, Rome.

Kivijarvi, H. and M. Soismaa. 1992 Investment and harvest strategies of the Finnish forest sector under different forest-tax policies: A differential game approach with computer-based decision aid. *European Journal of Operational Research* 56: 192-209.

Lambert, D.K. and T.R. Harris. 1990. Stochastic dynamic optimization and rangeland investment decisions. *Western Journal of Agricultural Economics* 15: 186-95.

Lane, D.E. 1988. Investment decision making by fishermen. *Canadian Journal of Fisheries and Aquatic Sciences* 45: 782-96.

Lonnstedt, L. and J. Svensson. 2000. Return and risk in timberland and other investment alternatives for NIPF owners. *Scandinavian Journal of Forest Research* 15: 661-69.

Lumley, R. and M. Zervos. 2001. A model for investments in the natural resource industry with switching costs. *Mathematics of Operations Research* 26: 637-53.

Luss, H. 1982. Operations research and capacity expansion problems: A survey. *Operations Research* 30: 907-47.

MacMillan, D.C. and N.A. Chalmers. 1992. An investment model for commercial afforestation in Scotland. *Forestry* 65: 171-88.

Manne, A.S. 1967 *Investments for Capacity Expansion: Size, Location and Time-Phasing*. MIT Press, Cambridge, MA.

McKelvey, R. 1985 Decentralized regulation of a common property renewable resource industry with irreversible investment. *Journal of Environmental Economics and Management* 12: 287-304.

McKelvey, R. 1987. Fur seal and blue whale: The bioeconomics of extinction. In Y. Cohen (ed.) *Applications of Control Theory in Ecology*. Lecture Notes in Biomathematics 73, Springer-Verlag.

Meuriot, E. and J.M. Gates. 1983. Fishing allocations and optimal fees: A single- and multilevel programming analysis. *American Journal of Agricultural Economics* 65: 711-21.

Mchich, R., P. Auger and N. Raissi. 2000. The dynamics of a fish stock exploited in two fishing zones. *Acta Biotheoretica* 48: 207-18.

Novak, A., V. Kaitala and G. Feichtinger. 1995. Resource leasing and optimal periodic capital investments. *Mathematical Methods of Operations Research* 42: 47-67.

Pascoe, S. and D. Greboval. 2003 *Measuring Capacity in Fisheries*. FAO Fisheries Technical Paper. No.445. Food and Agriculture Organization of the United Nations, Rome.

Pascoe, S., J.E. Kirkley, D. Greboval and C.J. Morrison-Paul. 2003. *Measuring and Assessing Capacity in Fisheries. 2. Issues and Methods*. FAO Fisheries Technical Paper. No.433/2. FAO, Rome.

Pontryagin, L.S., V.S. Boltyanskii, R.V. Gamkrelidze and E.F. Mishchenko. 1962. *The Mathematical Theory of Optimal Processes*. Wiley-Interscience, New York.

Rausser, G.C. 1974. Technological change, production and investment in natural resource industries. *American Economic Review* 64: 1049-59.

Reed, W.J. 1989. Optimal investment in the protection of a vulnerable biological resource. *Natural Resource Modeling* 3: 463-80.

Reid, C., D. Squires, Y. Jeon, L. Rodwell and R. Clarke. 2003. An analysis of fishing capacity in the western and central Pacific Ocean tuna fishery and management implications. *Marine Policy* 27: 449-69.

Reyner, K.M., W.A. Leuschner and J. Sullivan. 1996. A silviculture investment model for industrial forests. *Forest Products Journal* 46: 25-30.

Siry, J.P., D.J. Robison and F.W. Cubbage. 2004. Economic returns model for silvicultural investments in young hardwood stands. *Southern Journal of Applied Forestry* 28: 179-84.

Skonhoft, A. 1998. Investing in wildlife: Can wildlife pay its way? *Journal of African Economies* 7: 237-62.

Smith, V.L. 1968. Economics of production from natural resources. *American Economic Review* 58: 409-31.

Squires, D., M. Alauddin and J. Kirkley. 1994. Individual transferable quota markets and investment decisions in the fixed gear sablefish industry. *Journal of Environmental Economics and Management* 27: 185-204.

Sumaila, U.R. 1995. Irreversible capital investment in a two-stage bimatrix fishery game model. *Marine Resource Economics* 10: 263-83.

Tingley, D., S. Pascoe and S. Mardle. 2003. Estimating capacity utilisation in multi-purpose, multi-metier fisheries. *Fisheries Research* 63: 121-34.

Yin, R. and D.H. Newman. 1996. The effect of catastrophic risk on forest investment decisions. *Journal of Environmental Economics and Management* 31: 186-97.

9

Fisheries Management with Stock Uncertainty and Costly Capital Adjustment

Matthew Doyle
Rajesh Singh
Quinn Weninger

9.1 INTRODUCTION

Stock growth uncertainty and costly adjustment of fishing capital are fundamental features of the fisheries management problem (Pindyck, 1984). Analytical complexity in solving stochastic dynamic optimisation models however has forced researchers to address these key features in isolation.[1] This chapter jointly incorporates uncertainty regarding the growth of the fish stock, and capital adjustment costs in a model of fisheries management. Numerical techniques are used to identify the optimal, or resource rent maximising management policy. We use the model to study the impacts of key economic and biological parameters on catch rates, stock levels, and the size of the fishing fleet.

It is natural that the physical and human capital that is embodied in a fishing fleet of fixed size will exhibit diminishing marginal productivity. As more fish are harvested by a fleet of fixed size, marginal harvest costs will eventually rise and consequently, the marginal net benefit from harvesting more fish will decline. In this environment, the returns from harvesting fish will decline under harvest policies that allow for a widely fluctuating catch. Including costly adjustment of capital into the calculation of the net benefits from harvesting fish thus introduces an incentive to smooth the catch over time.

The catch smoothing incentive has been overlooked in the analysis of fishery management under stochastic stock growth, in part because of difficulties that arise in solving stochastic dynamic optimisation models. A classic paper by Reed (1979) derives an optimal constant escapement policy in a fishery with stochastic stock growth, and *linear* net harvest benefits. The linearity assumption facilitates an analytical solution to the model and is crucial for the optimality of the constant escapement policy. Under a constant escapement policy, the per-period catch is chosen

---

[1] See Smith (1968, 1969), Clark, Clarke and Munro (1979), Reed (1979), Berck and Perloff, (1984), Boyce (1995).

to maintain the unharvested stock, i.e. the escapement, at a target and constant level.[2] The corresponding per period harvest fluctuates widely; if a large unanticipated increase in the stock is observed, the per period harvest must also be large to return the stock to the target escapement level. If poor environmental conditions cause stock growth to fall below the target escapement, the harvest is zero, in other words, the fishery is closed. With a linear payoff from harvesting fish there are no *costs* that arise due to large fluctuations in the harvest over time.

Consider the temporal harvest decision under costly capital adjustment and diminishing returns to the current period catch. Suppose again that favourable environmental conditions have resulted in large, unanticipated growth of the fish stock. The manager can harvest the entire surplus in the current period or alternatively can *bank* some of the excess fish for future harvest. In our model, the optimal harvest decision depends critically on the number of boats that are available to harvest the surplus growth. This is because banking a portion of the excess growth will allow time to invest in additional harvesting capital.[3] The surplus growth can then be harvested by a larger fishing fleet at lower average harvesting costs. Similar catch smoothing incentives are present in the event that unanticipated stock growth is low. In general, and in contrast to the results in Reed (1979), we find that the optimal (present value maximising) harvest policy in a fishery with stochastic stock growth and costly capital adjustment involves considerable smoothing of the catch over time.[4]

Our model and numerical solution technique represent a powerful tool to improve fisheries management. Fisheries management has recently received sharp criticism for failing to meet conservation goals and for failing to protect the fishing communities whose livelihood depends on healthy fisheries resources.[5] These criticisms generally focus on biological or economic aspects in isolation. In practice, however, managers must balance biological, economic and social goals under uncertain environmental conditions. Because our model can incorporate the complexity that fisheries managers actually face it provides the requisite analytical framework to balance key management trade-offs.

In addition to identifying the rent maximising policy, the numerical methods that we employ characterise the optimal policy over a range of initial conditions. This feature enables us to analyse 'out of steady state' dynamics, which may be crucial

---

[2] The optimal escapement satisfies a modified golden rule where the rate of return from harvesting one more fish is just equal to the expected rate of return from leaving the fish in the sea.

[3] Diminishing marginal consumer benefits, which will result under a downward sloping demand for fish, will provide an similar incentive to bank a portion of the surplus stock for future consumption.

[4] Clark and Kirkwood (1986) modify the Reed (1979) model to consider uncertainty over stock measurement. Recent work by Costello, Polasky and Solow (2001), and Sethi *et al.* (2005) consider variations of the fisheries management problem under stochastic stock growth. These papers assume that harvest net benefits are linear in the catch. Charles (1983, 1985) analyses capital investment in a fishery with stochastic growth. In Charles (1983) price and marginal harvesting costs are assumed linear in the catch, however, vessels face a maximum feasible harvest capacity. These assumptions lead to quite different investment incentives, and stock dynamics than are featured in our model. Charles (1985) relaxes the assumption of linear harvesting costs but must assume temporally independent stock growth to solve his model.

[5] Munro (1998), FAO (1999), Federal Fisheries Investment Task Force (1999), Eagle *et al.* (2003), National Marine Fisheries Service (2003), US Commission on Ocean Policy (2004).

given the current state of many fisheries. For example, oversized fishing fleets have been identified as a significant source of management problems in many fisheries throughout the world (FAO, 1999). Policies to reduce the size of fishing fleets such as vessel buyback programmes are a popular solution. In implementing these programmes, managers face the difficult and unresolved problem of determining the number of vessels to remove from the fishery, and the rate at which vessels are decommissioned. Both the rate and magnitude of fleet reductions will have significant economic implications. Importantly, the optimal rate of vessel decommissioning must be jointly chosen with the optimal harvest policy. We are able to identify the least cost fleet reduction policy in a fishery that is initially overcapitalised and thus provide valuable direction for the design of fleet reduction programmes.

Finally, fisheries managers may face the difficult problem of rebuilding depleted fish stocks. Stock rebuilding requires reductions in current harvest rates, but excessive reductions in the current catch can lead to idled fishing fleets with severe economic consequences for fishermen and fishing communities. Our model identifies catch and capital investment, or divestment, policies that determine the least-cost approach to rebuilding overfished stocks and thus provides valuable guidance for fisheries managers.

The remainder of the chapter is organised as follows. The next section introduces a model to formally investigate intertemporal catch smoothing incentives that arise under costly capital adjustment. Section 9.3 briefly discusses the empirical work performed to calibrate the model. Section 9.4 characterises the optimal policy and discusses the implications of various parameter changes. Section 9.5 examines dynamic responses to a number of scenarios that have received attention in the literature. Section 9.6 presents concluding remarks.

9.2 THE MODEL

We consider a planner who jointly chooses both the capital stock and the harvest (or, alternately, the escapement) with the goal of maximising the net present value of a fishery. Let $t = 1, 2, ...$ index a particular fishing period. Each period is subdivided into a harvest season, and a period that is closed to harvesting. We assume that all stock growth occurs during the time the fishery is closed to harvesting. The exploitable biomass in period t is denoted x_t. We assume that x_t is observed with certainty at the time that harvest quantity h_t is chosen.[6] The fish stock that is left following the harvesting, i.e. period $t + 1$ escapement is $s_{t+1} = x_t - h_t$, where $h_t \in [0, x_t]$ is the period t catch. Escapement is non-negative and cannot exceed the available biomass; $s_{t+1} \in [0, x_t]$.

---

[6] While common in the literature (Reed, 1979), the assumption that is observed without error does not describe real world fisheries management. An analysis of the effects of stock mismeasurement is reserved for future work. The assumption that harvest is selected after observing the realisation of random stock growth is somewhat representative of the actual timing of events in the halibut fishery, where biologists at the International Pacific Halibut Commission derive a Bayesian updated estimate of x_t prior to the selection of h_t.

Fish stock growth is density dependent and is influenced by random growth conditions in the ocean environment. The exploitable biomass is assumed to follow a Markovian process with transitions governed by

$$x_t = z_t G(s_t) \qquad (9.1)$$

where z_t is a random variable that represents an environmental shock, and $G(s_t)$ is the deterministic growth function which satisfies $G'(\cdot) > 0$, $G''(\cdot) < 0$.[7] The shock z_t is a mean 1 random variable with finite support $[\underline{z}, \overline{z}]$, where $0 < \underline{z} < \overline{z} < \infty$. Shocks follows a Markov process with known transition probabilities. We allow for the possibility that environmental shocks are serially correlated. Note that under serially correlated shocks, the distribution of the period $t+1$ shock depends on z_t, and z_t is a state variable in the model. The special case of independently distributed shocks is investigated in section 9.4.

This model of fish stock growth implies an intermediate level of abundance at which expected per-period growth attains a maximum. With a density-dependent growth function, a cost is incurred, in the form of reduced expected yields, when the variation in the per-period harvest is reduced. While, in principle, the planner chooses the escapement, the escapement and harvest decisions are constrained by the fact that $s_{t+1} = x_t - h_t$. This implies that a reduction in the variation of escapement can only be achieved by allowing wider fluctuations in the harvest. For example, the planner could choose to set escapement constant, in which case all of the variation in growth would be absorbed by the harvest. Alternately, the planner could choose a constant harvest over time in which case escapement will vary in response to changing stock growth.

The benefit of harvest quantity h_t includes the consumer surplus plus industry revenues and is denoted $B(h_t)$, where $B(\cdot)$ is a concave function of the total catch. If the consumer demand for fish is less than perfectly elastic $B(\cdot)$ will be strictly concave in the harvest.

The harvest technology utilises a single capital input which takes the value k_t in period t. For concreteness k_t will represent the number of fishing vessels in the harvest fleet. Capital is an essential and normal input in the harvesting process. Individual vessel harvesting costs are assumed to be increasing and strictly convex in harvest quantity, and non-increasing in the stock abundance. Fleet level harvesting costs are denoted $C(h_t, k_t, x_t)$. These assumptions imply that for fixed k_t, fleet harvesting costs are strictly convex in h_t, and non-increasing in the stock abundance, x_t. Furthermore, the period t net harvest benefits, $B(h_t) - C(h_t, k_t, x_t)$, are a strictly concave function of the catch.

The vessel capital (i.e. boats) can be moved in and out of the fishery but capital adjustment is costly. We assume a one period delay is required before capital investment is operational. The productive capital stock in period $t + 1$ will be equal to the current period capital stock plus period t investment, which we denote i_t.

---

[7] Following Reed (1979) and others, we assume that growth shocks enter multiplicatively. Alternative specifications are plausible and could lead to different results. An analysis of alternate stochastic growth models is reserved for future work.

Capital is assumed to depreciate at a constant rate δ, where 0 < δ < 1. Period *t+1* capital is:

$$k_{t+1} = (1-\delta)k_t + i_t$$

We assume that fishing vessels cannot be costlessly reallocated to uses outside of the fishery.[8] For example, capture gear is often specific to a particular species of fish, and the skills of the captain and crew are developed to operate a particular gear type, target a particular species of fish and operate within specific geographical boundaries.[9] Specificity of physical and human fishing capital is appropriately modelled as non-convex capital adjustment costs. We allow for a wedge between the capital *purchase* and *resale* price, and denote them by p_k^+ and p_k^-, respectively, with $p_k^+ > p_k^-$. The difference between the purchase and resale price introduces the capital adjustment cost.

Subject to the constraints described above, the social planner chooses both the annual harvest and the future capital stock to maximise the expected present discounted value of the fishery:

$$\max_{\{h_t, i_t\}} E_0 \sum_{t=0}^{\infty} \beta^t \left[B(h_t) - C(h_t, k_t, x_t) - p_k i_t \right], \quad (9.2)$$

where E_0 is the expectations operator conditional on currently available information, β is the discount factor[10] and p_k denotes the price of capital:

$$p_k = \begin{cases} p_k^+, \text{ if } k_{t+1} > (1-\delta)k_t \\ p_k^-, \text{ if } k_{t+1} \leq (1-\delta)k_t \end{cases}.$$

It is instructive to review the timing of the harvest and investment decisions and the information available when decisions are made. At date *t*, the manager observes the exploitable biomass x_t and the number of boats k_t in the fishing fleet. We assume that the Markov process that governs the biomass transition, Equation 9.1, is known. Since past escapement choice is known, the current period shock is also known.[11] Summarising, period *t* choice variables are (i_t, h_t) and period *t* state variables are (k_t, s_t, z_t), where z_t is an exogenously determined state and k_t and s_t are (predetermined) endogenous states. Notice that harvest choice in period *t*

---

[8] Clark *et al.* (1979), Matulich *et al.* (1996), and Weninger and McConnell (2000) have emphasised the importance of non-malleability of fishing capital.
[9] Knowledge of the location of fish across space and time is essential for a successful fishing operation. This knowledge may take years to acquire and likely involves costly investments in information, i.e. costly search which generates information but not necessarily a saleable catch. While some skills may be transferable to other fisheries, knowledge about the location fish within the geographical boundary of the fishery is likely to have discretely lower, possibly zero, value elsewhere.
[10] The discount factor is equal to the reciprocal of 1 plus the rate of interest. The results that follow assume $\beta = 0.96$.
[11] In the Pacific halibut fishery the growth of the stock is affected by surface water temperatures that follow decadal oscillations (Clark *et al.*, 1999). Water temperatures are easily measured and thus the assumption of observable shocks is not unreasonable.

determines the period $t + 1$ escapement, and the investment can be expressed as a choice of period $t + 1$ capital. The choice variables in the model can be expressed equivalently as (k_{t+1}, s_{t+1}).

The planner's problem can be formulated as a recursive stationary, dynamic programming problem. To ease notation, time indexes are dropped and a prime (') is used to distinguish one-period ahead state variables. Current period state variables are thus (k, s, z) which correspond to (k_t, s_t, z_t) and future states, current period choice variables are (k', s') (which correspond to (k_{t+1}, s_{t+1})). The Bellman equation for the planner's problem is given by:

$$V(k,s,z) = \max_{k',s'} \{B(zG(s)-s') - C(zG(s)-s',k,zG(s)) - p_k(k'-(1-\delta)k) + \beta EV(k',s',z')\} \quad (9.3)$$

subject to $0 \leq s' \leq zG(s)$.

The value function $V(k, s, z)$ *is* the discounted present value of the fishery given the initial state (k, s, z). This value consists of two components, the current period net benefits and the expected discounted present value of the fishery given next period's state (k', s', z'). E denotes the expectations operator conditional on current period information, so that $EV(k', s', z')$ represents the optimised expected value of the fishery in the next period (this value is not known with certainty because the realisation of z' is not known in the current period).

If V were known, the solution to the fisheries management problem would be a pair of policy functions $k' = K(k, s, z)$ and $s' = S(k, s, z)$ describing the optimal choices of k' and s' for all possible initial states (k, s, z). We solve the problem numerically by using value function iteration (see Judd, 1998 for a detailed description of the methodology). This technique entails discretising the state space and then iterating on the Bellman equation, starting with an initial guess for the value function. Essentially, we insert our starting guess for V(·) into the Bellman equation, solve for the optimal policy functions and then insert these solutions back into the Bellman equation. Since our initial guess for V(·) will not be correct, the result of this substitution is a new value function different from the starting guess. We then repeat this procedure using the new value function as our guess. It can be shown (see Stokey and Lucas, 1989) that these iterations converge to the true value function, for every possible starting guess. Subsequently, we use information on the distribution of z along with these policy functions to compute the invariant joint distribution of (k', s'), which is then used to calculate descriptive statistics for the model choice variables, and for obtaining forecast functions under various scenarios.

9.3 CALIBRATION

We calibrate the model to the Alaskan Pacific halibut fishery. Due to space limitations, an overview of the empirical analysis and results is presented here. A detailed discussion of the model calibration is available in Doyle *et al.* (2005).

An estimate of the own-price halibut demand elasticity is obtained from a recent study by Herrmann and Criddle (2005). Remaining model components are estimated using data from: a survey of halibut fishing costs conducted by the Institute of Social and Economic Research at the University of Alaska; landings data from the National Marine Fisheries Service, Restricted Access Management Division; stock abundance data maintained by the International Pacific Halibut Commission (IPHC). These data were supplemented with information gathered from a survey of vessel captains, and information from industry members and managers at the IPHC.

Harvest cost data are available for Alaskan halibut fishermen only. While Canadian boats use similar capture techniques, extrapolation of US fleet cost estimates is problematic. For example, the distance between the vessel's port and the halibut fishing grounds, an important factor in variable harvesting costs, varies across regions. Extrapolating US-based harvest costs to the Canadian fleet could bias the results.

Stock abundance data are not available for all of the management regions of the halibut fishery. Given these data limitations we focus our calibration to management unit 3A, which represents the Gulf of Alaska region. Management Unit 3A produces roughly 55% of all commercial harvests (US and Canada), is well-represented in the 1997 harvesting cost data, and has the most complete stock abundance information within the US segment of the halibut fishery.

9.3.1 Harvest benefit function

Herrmann and Criddle (2005) estimate the own-price halibut demand flexibility for the period 1976-2002 in the US wholesale market (the main market for Pacific halibut) at -0.29. Following Herrmann and Criddle we assume a linear inverse demand for halibut, $P(h) = b_1 - b_2 h$. Using consumer surplus as a measure of consumer welfare, the total (current period) benefit from the halibut harvest is the sum of consumer surplus plus fishing industry revenues:

$$B(h) = b_1 h - \tfrac{1}{2} b_2 h^2, \qquad (9.4)$$

where h is current period harvest, measured in tonnes, and benefits $B(h)$ are denoted in 1997 US dollars (all subsequent values are also in US$ 1997). Parameter estimates for b_1 and b_2 are summarised below in Table 9.1.

9.3.2 Harvesting costs

Halibut fishing involves steaming from port to a chosen fishing site where a heavy long line is lowered to the sea bottom. Smaller lines with baited hooks are attached to the long line at roughly 18 foot intervals. The long line is soaked and then recovered using a hydraulic winch. Hooked fish are retrieved, eviscerated and placed on ice. The catch is then returned to port and sold primarily to fish brokers who distribute the halibut to consumer retail markets and restaurants. A typical fishing trip will last 4-6 days.

Table 9.1. Model calibration summary.

Component	Functional form	Base case parameters
Harvest benefits[a]	$B(h) = b_1 h - \frac{1}{2} b_2 h^2$	$b_1 = 1.354, \; b_2 = 1.633e^{-5}$
Fleet costs	$C(h,k,x) = k \cdot c(q\|x), \; q = h/k$	$FC = 177.951; \; c_1 = 0.059$
Vessel costs	$c(q\|x) = FC + \sum_{j=1}^{3} c_j q^j + c_x x$	$c_2 = -1.755e^{-4}; \; c_3 = 3.402e^{-7}$ $c_x = -2.195e^{-4}$
Capital prices	$p_k = \begin{cases} p_k^+, \text{if } k' > (1-\delta)k \\ p_k^-, \text{if } k' \leq (1-\delta)k \end{cases}$	$p_k^+ = \$236,500, \; p_k^- = \$160,000$ $\delta = 0.1$
Stock growth[b]	$G(s) = s + \alpha s(1 - s/x^c)$	$\alpha = 0.128; \; x^c = 201,119$

[a] Benefits and costs are denoted in millions of 1997 US dollars.
[b] Stock quantity units are tonnes.

The main variable operating expenses are from fuel, bait and ice, food and supplies for the captain and crew, and lost gear. The cost data are used to estimate a cubic functional form to allow for a flexible relationship between costs and catch;

$$c(q|x) = FC + c_1 q + c_2 q^2 + c_3 q^3 + c_x x. \quad (9.5)$$

where q is the vessel harvest in tonnes, FC denotes annual fixed costs which include expenses for vessel mooring and storage, permits and licence fees, and fees for accountants, lawyers, office support, and routine maintenance and repairs.[12] We include the labour services of the captain and crew as a fixed cost component. While labour is often treated as a variable input, crew services are not easily adjusted in the short run.[13]

---

[12] The cubic functional form is simple and can be easily restricted to satisfy the theoretical curvature properties of cost functions. Doyle et al. (2005) discuss the estimation of the vessel-level cost function in detail.
[13] Fishing vessels are designed to accommodate a particular crew size. On occasion crew size may be increased or decreased but adjustments tend to be infrequent.

We assume that the total harvest in the fishery is distributed equally among active vessels. The fleet harvesting cost for a fleet consisting of k vessels is thus $C(h,k,x) = k \cdot c(q \mid x)$, where $q = h/k$.

9.3.3 Capital adjustment costs

Many vessels in the halibut fishery spend only a portion of the year fishing halibut (vessels operate in other fisheries for the remainder of the year). We surveyed halibut captains to determine the average costs of refitting a vessel to fish in a fishery other than the halibut fishery. We found that refit costs depend largely on which fishery the vessel is moved to or from. If a vessel is moved from a fixed gear fishery (into the halibut fishery), refit costs are considerably less than if the vessel is moved from a trawl gear fishery. Modelling the set of fisheries in which vessels participate is beyond the scope of this study. Instead we use our survey data to generate plausible ranges for refit costs.

Halibut fishermen inform us that the cost of refitting a boat which already uses fixed gear requires a relatively modest refit at a cost of approximately $27,000. If the boat is switched from a trawl gear fishery, the refit costs increase to $85,000. It should be noted that the refit cost estimates do not include human capital adjustment costs, e.g. the costs to retrain the captain and crew to fish for a different species, using different gear.

9.3.4 Stock growth model

The International Pacific Halibut Commission (IPHC) has developed a region and sex specific, age structure model of halibut stock abundance. The model tracks the number of fish, and average weight at age by region, sex and age. Changes in the survival, growth and recruitment of young fish into the exploitable population is tracked using a computer-based model which incorporates harvest selectivity (the likelihood that longline gear will intercept halibut of a give size and age), fecundity at age, recruitment of young into the commercial fishery, weight at age, among other factors (see Sullivan *et al.*, 1999). Each year new data are collected from commercial and survey sources and the number of fish and average weight at age by sex and age is re-estimated using a Bayesian updating procedure.

The computational time required for value function iteration increases exponentially with the number of state variables. Adopting the sex and age specific IPHC stock model directly would increase the number of state variables to well over 50 and thus was not practical. Our approach is to fit a simpler parametric model to characterise the halibut stock growth. For this purpose we aggregate across sex and age classes to obtain estimates of the pre-harvest exploitable biomass, catch, and escapement for management unit 3A from 1974-2003.

The logistic stock growth function, $G(s) = s + \alpha s(1 - s/x^c)$, is fitted to the 1974-2003 data using an iterative feasible generalised least squares procedure. The model provided a good fit to the data, although because the data set is itself generated from a model, formal tests of alternate empirical specifications are difficult to interpret. In addition to the logistic growth model, we estimated a Ricker growth function and found no differences in model fit. The state space for the random shock Z and the Markov transition matrix are calculated following Judd (1998: 85-88).

Additional details for the estimation are provided in Doyle et al. (2005). Table 9.1 summarises the calibration results.

9.4 RESULTS

In this section we discuss the optimal management policy for the baseline calibration. We emphasise the joint determination of biological and economic variables and describe the factors affecting the determination of each. We also illustrate some of the main mechanisms operating within the model by examining the effects of a variety of parameter changes on model choice variables.

First, note that the optimal s' and k' are jointly determined; factors that affect the evolution of the biological stock have implications for the choice of capital stock, and vice versa. A key advantage of this modelling approach, in relation to models that focus on one component of a fishery in isolation, is that it allows us to analyse the interactions between these two important variables.

Figure 9.1 plots the optimal escapement policy as a function of s (measured in thousands of tonnes) and k (measured in number of boats)[14] for the case of $z = 1$. Recall that different escapement and investment policies can be drawn for each value of z. Observe that the optimal escapement is increasing in s and decreasing in k. When past escapement s is large, the current stock size is also relatively large for $z=1$. This large stock of fish could be harvested immediately. However, due to the diminishing marginal net benefits in the current period harvest, it is best to increase escapement and bank some of the excess growth for future harvest. Distributing the excess catch over multiple periods keeps average harvesting costs from getting too high and, if warranted, allows time to increase the size of the fishing fleet.

With a large fleet, the average harvesting costs rise at a slower rate as catch increases than with a small fleet. In other words, the rate at which current period marginal net benefits decline is smaller when k is large. With a larger fishing fleet, more of the surplus stock is harvested in the current period than when the fleet is small, and thus with larger capital, optimal escapement declines.

The policy function for the capital stock is more complicated. Essentially, the optimal policy is to add more capital if the expected (value) marginal product of capital exceeds the price of new capital, p_k^+, and to divest if the expected marginal product of capital falls below the capital resale price, p_k^-. The gap between p_k^+ and p_k^- means that for particular combinations of (k, s, z), it is optimal to maintain the vessel capital stock at it current level, i.e. no investment in new boats and no divestment of the current vessel capital stock occurs. Because higher levels of escapement imply higher expected harvest in the future, the expected marginal product of capital is increasing in escapement. Therefore, the optimal capital for next period, k', is non-decreasing in s.

---

[14] Given the numerical approach taken, the capital stock is necessarily a discrete variable. This is natural under our interpretation that the capital variable corresponds to the number of vessels in the fishery.

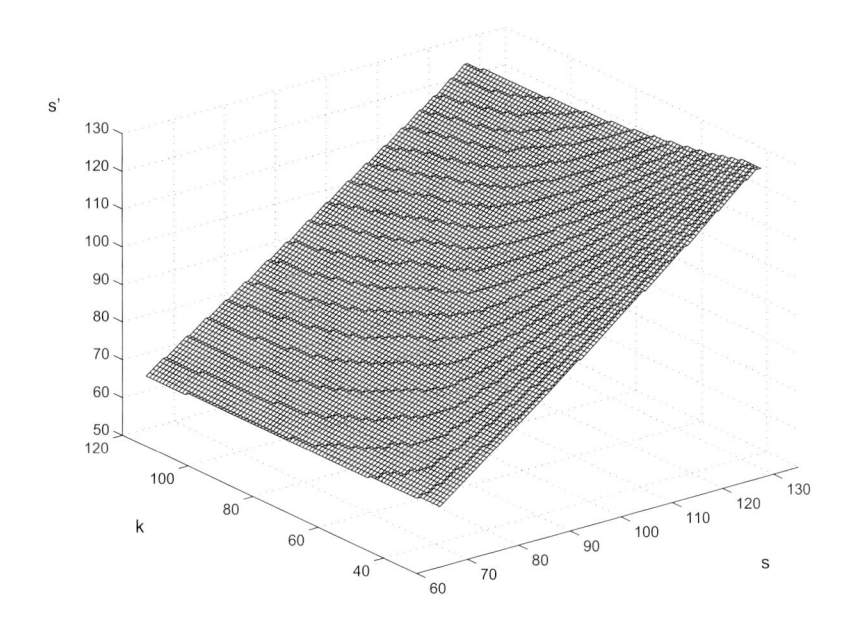

Figure 9.1 Optimal escapement.

The fact that z follows a random process implies that capital and escapement can be characterised by examining their invariant distributions.[15] Table 9.2 reports mean values, standard deviations and 99% confidence intervals for boats, escapement, harvest, catch per boat, and the harvest rate, i.e. the proportion of the total biomass that is harvested in each period. The results for the baseline calibration are presented in row 1 of Table 9.2. Rows 2-5 of Table 9.2 report results under alternate parameterisations of the model.

For the baseline calibration,[16] the mean number of boats is 61.8, and 99% of the time the number of boats in the fleet lies between 37 and 92. Mean escapement is near 92.5 thousand tonnes of fish, and escapement lies between 71 and 119 thousand tonnes 99% of the time. Though not reported in the table, escapement and capital are positively serially correlated; the correlation of s' and s is 0.8946, and the correlation of k' and k is 0.8879. Positively serially correlated escapement and capital investment illustrate the combined effects of serially correlated environmental shocks as well as the economic forces in the model which provide incentives to smooth the catch over

---

[15] The optimal policy functions imply a long run joint distribution of k' and s' which is independent of the initial state (k, s, z).
[16] The baseline calibration corresponds to the point estimates obtained from the estimation of data from the Pacific Halibut fishery outlined in the previous section.

time. The value of the optimally managed fishery under the baseline calibration and evaluated at the means of the state variables is roughly $1.645 billion.

Table 9.2   Descriptive statistics for invariant distributions.

	Scenario	Capital (boats)	Escapement ('000 tonnes)	Harvest ('000 tonnes)	Catch/boat (tonnes)	Harvest rate (%)
1	Baseline case	61.76[a] (10.97)[b] [37, 92][c]	92.49 (9.55) [70.8, 119.3]	14.06 (2.23) [8.7, 19.9]	229.3 (11.60) [197.5, 258.9]	13.2 (0.80) [10.7, 14.5]
2	i.i.d. shocks	61.45 (7.10) [44, 80]	92.47 (6.78) [72.6, 111.1]	14.07 (1.51) [10.1, 18.1]	229.3 (10.4) [201.3, 56.4]	13.2 (0.5) [11.7, 14.3]
3	Maximised industry profits	63.95 (9.47) [54, 103]	109.18 (17.43) [71.2, 156.9]	13.84 (1.22) [10.0, 16.2]	218.0 (14.4) [174.7, 251.1]	11.3 (0.8) [9.3, 12.8]
4	High capital adjustment cost	52.70 (8.68) [32, 74]	92.18 (9.96) [69.4, 119.8]	14.08 (2.21) [8.9, 19.7]	267.8 (11.3) [233.3, 294.6]	13.2 (0.7) [9.3, 12.8]
5	Constant escapement (maximum value)	67	88.45	14.02 (4.29) [1.8, 23.2]	209.22 (64.04) [26.3, 386.3]	13.5 (3.7) [2.0, 21.0]

a - denotes mean value, *b* - denotes standard deviation, *c* - denotes 99% confidence interval.

Row 2 of Table 9.2 presents the results for a case where shocks to the stock growth are intertemporally independent rather than serially correlated. The main effect of independent shocks versus serially correlated shocks is that the dispersion of model variables is noticeably reduced. This is an intuitive result. In the presence of positively serially correlated shocks, unusually high stock growth in the current period suggests that growth will also be high in the next period. The optimal policy function prescribes investment in new vessel capital in order to harvest the expected additional stock abundance. If shocks are independent, then the optimal investment response to unusually high stock growth is more muted, since the shock is not expected to persist. Both k and s remain highly serially correlated (the correlation coefficients between today and tomorrow's values are 0.897 and 0.803 for k and s respectively) even though the shocks are independently distributed.

Row 3 of Table 9.2 reports the results for the case where the fishery is managed in order to maximise the present discounted value of industry profits, rather than the present discounted total value (industry profits plus consumer surplus). The benefit function in this case is more concave than in the baseline calibration because marginal industry revenues decline more rapidly than marginal total benefits. Not surprisingly, the main effect on the results is an increase in the standard deviation of escapement and a reduction in the standard deviation of the harvest. The increased concavity in the current period benefit function induces greater catch smoothing at a cost of increased escapement volatility. Also, if the consumer surplus is not valued by the fisheries manager, the response is to leave more fish in the sea (i.e. mean escapement goes up and mean harvest goes down). This is because the costs of harvesting decline when there is greater stock abundance. When the benefits of catching fish are lower (when consumer surplus is ignored), the incentive to leave more fish in the sea to take advantage of the stock effect on harvesting costs becomes relatively more important.

Row 4 of Table 9.2 presents the results of an experiment in which the costs of adjusting capital are increased by widening the gap between the purchase and sale price of boats. Our survey of vessel captains indicated that the price of a new fully equipped fishing vessel is roughly $800,000 and that a vessel that is scrapped for metal and parts would fetch roughly $25,000. Column 2 reports the results for p_k^+ = $800,000 and p_k^- = $25,000 (the base case prices are p_k^+ = $236,500 and p_k^- = $160,000). The primary effects of this change, as might be expected, are a reduction in the volatility of the capital stock and a corresponding increase in the volatility of escapement. Essentially, as it becomes more costly to adjust to changes in the harvest by changing the number of boats, i.e. adjusting along the extensive margin, we see a greater tendency for movement along the average cost curves of individual boats (more of the adjustment occurs on the intensive margin). Since the increase in the costliness of capital reduces the overall economic flexibility, it is also optimal to reduce the volatility of the harvest to some degree, which implies that the volatility of escapement must rise.

9.4.1 Constant escapement policy

Finally, row 5 of Table 9.2 presents the results for the case where the biological resource is managed according to a constant escapement policy as opposed to the optimal policy. This policy is examined primarily for purposes of comparison, because constant escapement policies are popular in the literature on stochastic fisheries management, and because they are used in some fisheries. A constant escapement policy takes the form

$$s = \begin{cases} s^* & \text{if } x \geq s^* \\ x & \text{if } x < s^* \end{cases},$$

which implies that $h = \max(x - s^*, 0)$.

The preferred value of s^* will differ depending on the management objective, for example, the fish stock may be managed with the objective of maximising sustainable yield. Here we present the result where s^* is chosen to maximise the expected discounted present value of the fishery, evaluated at state variables equal to the means of their respective invariant distributions. We note that the two cases are qualitatively similar. Observe that with serially correlated shocks, the escapement that maximises either the sustainable yield or the value of a fishery will depend on z. In order to generate a fair comparison between a *constant* escapement policy and the optimal policy, we restrict attention in this section to the case where the shocks are serially independent. Under the timing conventions of our model, in which the shock is realised prior to the determination of the catch, and under independent shocks, the expectation of future shocks is always identical.[17] As a result, the appropriate comparison is between the constant escapement policy and the optimal policy under i.i.d. shocks (presented in row 2 in Table 9.2).

Notice first that under the constant escapement policy the distributions of escapement and capital are degenerate. As a consequence, there are no fish stock or vessel capital dynamics present. Essentially, the constant escapement policy eliminates any harvest smoothing. When escapement is set to a constant (around 88.45 thousand tonnes, according to this calibration) the harvest must absorb all of the volatility induced by changing environmental conditions. The results in the table show a marked increase in the volatility of the harvest relative to all of the other cases. When the biological resource is managed under a constant escapement policy, the optimal capital policy involves maintaining a constant fleet size of 67 boats in each period. This is an extreme example of how changes in the policy for managing the stock affect the optimal investment profile. Since with constant escapement future harvests are entirely unpredictable, there is never an opportunity to adjust the number of boats in the fleet. As a result, all economic adjustments come from individual boats moving along their average cost curves. These movements are much more extreme than in the case where the biological stock is optimally managed.

Finally, comparing the fishery value functions reveals that the costs associated with managing the stock with the optimal constant escapement harvest policy is on the order of 4% of the total value of the fishery.

9.5 COMPARATIVE DYNAMICS

This section uses the model to analyse the optimal dynamic response of escapement and capital to various initial conditions. Beginning with an initial state (k, s, z) we identify the optimal future states k' and s', under the assumption that the future realisations of the shock are equal to unity. The optimal values k', s' and $z' = 1$ become the one period ahead state variables and the process is repeated. Iterating this procedure identifies the evolution of the states under the optimal management policy and for future shocks equal to unity. A similar procedure could be used in practice, although the actual realisations of environmental shocks would replace the

---

[17] The fact that escapement is literally a constant is to some extent an artefact of the finite grid for z, which places a lower bound on per period stock growth. For the baseline calibration, it turns out that the optimal constant escapement level is sufficiently high that even for the worst possible outcome z, growth exceeds the escapement target (i.e. the probability that x < s* is zero). If z were to take on more extreme values, the stock would sometimes fall short of the target escapement level, in which case the fishery would be closed. This would also reintroduce dynamics into the capital investment policy, as the distribution of future harvests would differ depending on whether the fishery was closed or open.

assumed mean value shock. The experiments that follow are conducted using the baseline calibration reported in Table 9.1.

Panel (a) of Figure 9.2 illustrates the optimal response of the fishery to a high realisation of the shock ($z = 1.12$) when k and s are initially set equal to the means of their invariant distributions. For a given s, a high shock means high current period stock abundance. The left hand diagram of panel (a) shows that escapement increases in response to a large growth shock; rather than fish the excess stock back down immediately, which would involve costly increases in short run average harvest cost as well as a movement down the consumer demand curve, the optimal policy banks a portion of the excess. The right-hand diagram shows that the fleet size is increased concurrently; the optimal k' increases with the shock z. Over time the excess catch is harvested by the larger fleet until escapement falls back toward its starting value. Along the way, the capital stock is allowed to depreciate until it too approaches its initial value.

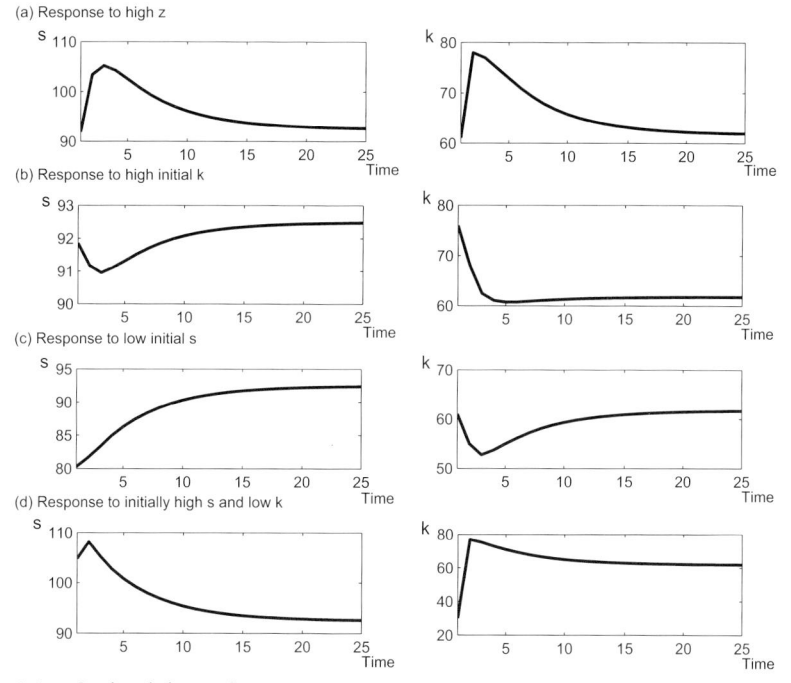

Figure 9.2 Optimal dynamic responses.

Panel (b) of Figure 9.2 plots the optimal dynamic response of the fishery to a high initial capital stock (where the fishery starts out overcapitalised, with $k = 75$ boats) and an initial escapement equal to the mean of its invariant distribution, roughly 92.5 thousand tonnes. The initial value of z is chosen to equal its mean value of 1. The results can be interpreted as the optimal expected response, since future shocks are equal to unity, to an initially overcapitalised fleet. The optimal response to a high capital stock is to sell off excess boats as soon as possible since the value of the boats in the salvage market exceeds their value in the fishery. The initial sale of the excess capital generates revenue p_k^- for every boat sold. Notice that capital divestment is delayed one period, due to the fact that current fishing fleet is determined by the previous period investment decision, and is taken as fixed in the current period. Since it takes time to reduce the capital stock, it is preferable to increase the current harvest, in other words, reduce escapement in the initial period rather than allow the excess capital to remain idle. As the capital stock is reduced to its desired long-term level, the planner gradually increases escapement. i.e. lowers the per period harvest.

Panel (c) of Figure 9.2 examines the optimal dynamic response to a depleted biological stock (inherited escapement is only 80 thousand tonnes) where the fleet size is equal to the mean of its invariant distribution (i.e. 61 boats). This scenario represents a case where stocks must be rebuilt, and the fleet size is larger than desired given current harvest levels, conditions which may describe many real world fisheries.

Notice that it is optimal to gradually rebuild the fish stock to the preferred level; optimal escapement is initially low and gradually rises to the long-term desired level. It is clear that the optimal escapement policy is not *a bang-bang* or most rapid approach path policy which would emerge in a model of costless capital adjustment. The incentive to smooth the catch across periods leads to more moderate stock adjustments. For instance, even though initial abundance is low the fishery does not close; it is optimal to maintain a positive catch in each period.[18] This is true even in the first period when stock size is at its lowest level. While it is true that positive harvest delays the stock rebuilding process, the immediate harvest value of the resource is large, particularly at the initial capital stock k.

Panel (d) of Figure 9.2 plots the dynamic response to an initially high escapement of 105 thousand tonnes of fish, and an initially small fleet size (set to 30 boats). These initial conditions emulate a previously unexploited fishery. Under the optimal policy, escapement is initially large; the fleet is too small to harvest significant quantities of the excess stocks. As the fleet is built up, escapement declines gradually, i.e. harvest increases. Turning to capital, the initially large increase in the fleet size is followed by a slow decline in the capital stock to the long term desired level.

9.6 CONCLUDING REMARKS

In this chapter we solved a model of a fishery incorporating uncertainty regarding the growth of the fish stock and capital adjustment costs. Value function iteration was used to solve for the resource rent maximising management policy. The model is applied to the Alaskan Pacific halibut fishery to illustrate key management insights.

Costly capital adjustment and diminishing marginal net harvest benefits imply that the policy that maximises the value of a fishery will involve smoothing the catch over time. The optimal amount of catch smoothing balances the benefits from a stable per period harvest against the yield losses that result under excessive fluctuations of

---

[18] Although not directly apparent from the figure, the invariant distribution of the harvest level does not have positive mass at zero harvest quantities.

the *in situ* fish stock. In addition, the optimal harvest policy is sensitive to the size of the fishing fleet that is available to carry out harvesting activities. When the fleet is large, the immediate harvest value of the fish stock is relatively large favouring higher current consumption of the resource. On the contrary, if the fishing fleet is small, banking some of the stock for future harvest (when more vessels will be available to harvest fish) may be preferred.

The model and numerical solution technique presented in this chapter represents a powerful tool to improve fisheries management. We show that incorporating two realistic features of real world fisheries, stochastic stock growth and costly capital adjustment, leads to policy prescriptions that differ sharply from previous literature. The catch smoothing incentives that are outlined are broadly applicable, and calibrating our model to other fisheries should yield similar insights (although precise management policies will differ under alternate model calibrations). In addition to identifying rent maximising management policies, the model and numerical methods can assist in the design of optimal (cost minimising) fleet reduction programmes, stock rebuilding programmes, or in efforts to tackle fleet reduction and stock rebuilding goals simultaneously.

The model provides a framework to assess the bioeconomic performance of actual fisheries management programmes under realistic conditions. Additional analysis of the Alaskan Pacific halibut fishery management programme undertaken in Doyle *et al.* (2004) finds that the harvest rule that is actually used by fisheries managers involves setting the annual harvest equal to 20% of the total exploitable halibut biomass. While this harvest rule departs from the optimal harvest policy, Doyle, Singh and Weninger find that it provides a reasonable approximation to the optimal harvest policy and causes negligible reductions in the value of the fishery. Their results show, however, that regulatory constraints that maintain a large and part-time fishing fleet are responsible for economic losses that range between 13% and 19% of the fishery value. These results inform ongoing debates over the underlying causes of management problems in an important US fishery. Application of the model to other fisheries could provide similar guidance for improving fisheries management.

Extensions of the methodology used in this chapter could provide additional insights for fisheries management. For instance, we have focused on uncertainty in stock growth, assuming throughout that true stock abundance is observed at the time harvest and investment decisions are made.[19] Other sources of uncertainty likely to be important in fisheries management include stock measurement error, uncertainty regarding the true stock growth function and the influence of random environmental shocks (e.g. additive versus multiplicative shocks), output price uncertainty and, from the perspective of the fisheries manager, uncertainty regarding the true cost of harvesting fish and the capital adjustment costs. Our model and the solution technique can be adopted to analyse the effects of these or other sources of uncertainty on the optimal harvest and investment policies.

---

[19] Clark and Kirkwood (1986) study a fishery management problem in which stock abundance is unobserved at the time the harvest decision is made and, as in our model, growth is influenced by multiplicative random shocks. Sethi *et al.* (2005) add a third source of uncertainty, mismeasurement of the actual catch of the fishing fleet. These papers do not consider fishing capital investments and assume a linear-in-catch payoff from harvesting fish.

Acknowledgements

The authors thank Steven Hare, Gunnar Knapp, Marcelo Oviedo, participants at the University of California, Davis, the University of British Columbia Conference in honour of Gordon Munro, staff at the National Marine Fisheries Service and the International Fisheries Management Council.

References

Berck, P. and J.M. Perloff. 1984. An open access fishery with rational expectations. *Econometrica* 2: 489-506.

Boyce, J.R. 1995. Optimal capital accumulation in a fishery: A nonlinear irreversible investment model. *Journal of Environmental Economics and Management* 28: 324-39.

Charles, A.T. 1983. Optimal fisheries investment under uncertainty. *Canadian Journal of Fisheries and Aquatic Sciences* 40: 2080-91.

Charles, A.T. 1985. Nonlinear costs and optimal fleet capacity in deterministic and stochastic fisheries. *Mathematical Biosciences* 73: 271-99.

Clark, C.W., F.H. Clarke, and G.R. Munro. 1979. The optimal exploitation of renewable resource stocks: Problems of irreversible investment. *Econometrica* 47: 25-47.

Clark, C.W. and G.P. Kirkwood. 1986. On uncertain renewable resource stocks: Optimal harvest policies and the value of stock surveys. *Journal of Environmental Economics and Management* 13: 235-44.

Clark, W.G., S.R. Hare, A.M. Parma, P.J. Sullivan and R.J. Trumble. 1999. Decadal changes in growth and recruitment of Pacific halibut *(Hippoglossus stenolepis)*. *Canadian Journal of Fisheries and Aquatic Sciences* 56: 242-52.

Costello, C., S. Polasky and A. Solow. 2001. Renewable resource management with environmental prediction. *Canadian Journal of Economics* 34: 196-211.

Doyle, M., R. Singh and Q. Weninger. 2005. Fisheries management with stock uncertainty and costly capital adjustment: Extended appendix. Iowa State University Economics Working Paper 05011.

Eagle, J., S. Newkirk and B.H. Thompson, Jr. 2003. *Taking Stock of the Regional Fisheries Management Councils*. Pew Science Series on Conservation and the Environment, Washington D.C.

Federal Fisheries Investment Task Force, Report to Congress, July 1999.

Food and Agricultural Organization. 1999. *Managing Fishing Capacity: Selected Papers on Underlying Concepts and Issues*. In D. Greboval (ed.) FAO Fisheries Technical Paper. No. 386. FAO, Rome.

Herrmann, M. and K. Criddle. 2005. An econometric market model for the Pacific halibut fishery. Unpublished manuscript.

Judd, K.L. 1998. *Numerical Methods in Economics*. MIT Press, Cambridge, MA.

Matulich, S.C., R. Mittelhammer and C. Reberte. 1996. Toward a more complete model of individual transferable quotas: Implications of incorporating the processing sector. *Journal of Environmental Economics and Management* 31: 112-28.

Munro, G.R. 1998. The Economics of Overcapitalization and Fishery Resource Management. Discussion Paper No. 98-21. Department of Economics, University of British Columbia.

National Marine Fisheries Service. 2003. *United States National Plan of Action for the Management of Fishing Capacity.* US Department of Commerce.

Pindyck, R.S. 1984. Uncertainty in the theory of renewable resource markets. *Review of Economic Studies* 51: 289-303.

Reed, W. J. 1979. Optimal escapement levels in stochastic and deterministic harvesting model. *Journal of Environmental Economics and Management* 6: 350-63.

Sethi, G., C. Costello, A. Fisher, M. Hanemann and L. Karp. 2005. Fishery management under multiple uncertainty. *Journal of Environmental Economics and Management* 50: 300-18.

Smith, V.L. 1968. Economics of production from natural resources. *American Economic Review* 58: 409-31.

Smith, V.L. 1969. On models of commercial fishing. *Journal of Political Economy* 77: 181-98.

Stokey, N and R.E. Lucas, Jr. 1989. *Recursive Methods in Economic Dynamics,* Harvard University Press, Cambridge, MA.

Sullivan, P.J., A.M. Parma and W.G. Clark. 1999. The Pacific Halibut Stock Assessment of 1997. International Pacific Halibut Commission Scientific Report No. 79.

US Commission on Ocean Policy. 2004. Preliminary Report of the U.S. Commission on Ocean Policy, Governors Draft.

Weninger, Q. and K.E. McConnell. 2000. Buyback programs in commercial fisheries: Efficiency versus transfers. *Canadian Journal of Economics* 33: 394-412.

Section 3

Game theory and international fisheries

10

The Incomplete Information Stochastic Split-Stream Model: An Overview

Robert McKelvey
Peter V. Golubtsov
Greg Cripe
Kathleen A. Miller

10.1 INTRODUCTION: COMPETITIVE HARVESTING IN A FLUCTUATING ENVIRONMENT

The economic and biological implications of competitive harvest of a reproducing biological stock are well-known in the fisheries economics literature. The common-property externality – implying that no competitor bears the full economic costs of his harvesting – leads to the well-known 'tragedy of the common' consequences of over-harvesting: both fish-stock depletion and the dissipation of economic rents.

Analyses of circumstances where there are many independently-acting fishermen go back to the beginnings of modern fisheries economics. In that circumstance, no individual fisherman can significantly impact the stock by a unilateral decision to forego harvesting. Innovative analyses, by Gordon (1954) and Scott (1955) first captured this phenomenon, initially through stylised static formulations.

The more complex situation, where the harvesters are grouped as members of two or more competing national fleets, was first analysed around 1980, using sophisticated dynamic bio-economic models (Clark, 1980; Levhari and Mirman, 1980). Initially it was assumed that the harvests occurred at a common location. Subsequent model variants have allowed for harvesting occurring at spatially distinct places and times. Most of these models are deterministic, though a few are stochastic.

These competitive-harvest models have proved to be useful in guiding practical management of fisheries. All tell similar stories, centring on the destructive effects of open access. Other factors, interacting with the common property externality, can further complicate the situation. For example, a substantial literature has developed around the implications of fleets employing different harvesting technologies and the consequences of fleet overcapitalisation and of capital immalleability.

Recently there is also increased awareness of the extent to which oceanic climatic shifts, affecting the marine environment, may be implicated – often with dramatic implications for the productivity and migratory behaviour of economically important fish stocks (McFarlane *et al.*, 2000; Hare and Mantua, 2000; Stenseth *et al.,* 2002;

Miller and Munro, 2004). Biologically-important climatic phenomena include El Niño events, the Pacific Decadal Oscillation, and the North Atlantic Oscillation. Furthermore, human-induced global climate change will lead to long-term changes in ocean temperature and circulation patterns, with implications that cannot yet be foreseen (Barnett *et al.*, 2001).

The modelling reported on in this chapter represents an extension of the classic bioeconomic harvest game analysis mentioned above, and not only incorporates the climatic shift phenomena but more importantly attempts to capture the implications of uncertainty and in particular the poor predictability of the onset of such climatic shifts. Technically, this modelling focuses on the implications of imperfect information concerning the timing and intensity of relevant environmental shifts.

A particular feature of our study is to examine the role of transparency of the known information in a harvesting game. It is well understood that in co-operative resource management transparency of information is a positive asset. As we shall see here, in a competitive harvesting game transparency may actually be corrosively destructive.

In earlier work, sometimes also with other colleagues, we have examined the ways in which poorly anticipated oceanic climate changes have adversely impacted the joint bi-national management by the US and Canada of the North American Pacific Salmon fishery (see for example, Miller *et al.*, 2001 and McKelvey *et al.*, 2003).

There are many other multilateral fisheries, in all of the world's oceans, for which one may anticipate similar disruptions as a result of shifting oceanic environments. For example, the particular susceptibility to over-harvesting of tightly-schooled small pelagic fishes, such as anchovies and sardines, is a classic story in the fisheries literature. This effect may be even more important where stocks migrate across jurisdictional boundaries in response to oceanic temperature shifts.

In the following we formulate our two basic versions of the split-stream model, with particular emphasis on characterisation of stochastic elements and alternative information states. We explore the models' implications, both analytically and numerically through simulation, displaying particularly striking features graphically. In particular, we systematically characterise how the specified information structure impacts the outcome of the competitive harvesting game, and then provide theoretical and intuitively plausible explanations of why this should be so.

10.2   THE COMPETITIVE SPLIT-STREAM HARVEST GAME

The Split Stream Harvest Model is a discrete-time dynamic model of the evolution of a marine fishery on a single non-overlapping-generation fish stock. It incorporates successive life-cycles of spawning, growth and harvest. A single such cycle may be represented schematically as follows:

$$S \to R = F(S, \phi) \begin{matrix} \nearrow & R_\alpha = \theta_\alpha R & \to S_\alpha & \searrow \\ & & & \\ \searrow & R_\beta = \theta_\beta R & \to S_\beta & \nearrow \end{matrix} S^+ = S_\alpha + S_\beta \to \cdots \quad (10.1)$$

The cycle shown begins at the spawning stage. Here S is the spawning stock biomass of the parental generation. Following the spawn, and over a period of time, the offspring-generation's biomass grows, while experiencing some mortality losses, ultimately maturing into the recruitment stock biomass:

$$R = F(S, \phi). \tag{10.2}$$

Here R is measured at the beginning of the harvesting season. In this stock-recruitment relation ϕ is a growth parameter. Depending on ϕ the stock-recruitment function F may be compensatory, depensatory or even may display critical depensation. (See Figure 10.1*a-d*.)

This recruited biomass then splits, with sub-stocks migrating along two separate streams, in which each is exposed to harvest by just one of two competing national fleets. For a symmetric presentation, let ν denote either α or β. The split-fraction θ_ν, with $0 \leq \theta_\nu \leq 1$ and $\theta_\alpha + \theta_\beta = 1$, determines the $\nu[0,1]$-substream recruitment biomass:

$$R_\nu = \theta_\nu R, \tag{10.3}$$

which is accessible to harvest by the ν–fleet.

The harvested sub-stream biomass is denoted H_ν, and the corresponding sub-stream escapement is denoted S_ν, with:

$$S_\nu = R_\nu - H_\nu. \tag{10.4}$$

Following the harvest, the two sub-stream escapements merge, to form the total escapement biomass:

$$S^+ = S_\alpha + S_\beta. \tag{10.5}$$

This escapement forms the brood stock for the next generation, and the cycle then repeats, through subsequent generations, out to the horizon $T \leq \infty$.

The main focus of our analysis is on the implications of environmental stochasticity for the outcome of the game – but especially on the profound, and perhaps surprising, implications when the fleet managers possess only imperfect information on the current state of the random dynamic game.

We shall assume in particular that the environmental parameters in the dynamical system, namely $\{\phi_\nu(t)\}$ and $\{\theta_\nu(t)\}$, form stochastic random processes which furthermore are observed imperfectly by the fleet managers. Here we provide an overview of our modeling approach, and report on some of our findings. A more detailed mathematical analysis of the incomplete-information dynamic split stream game can be found in McKelvey and Golubtsov (2002).

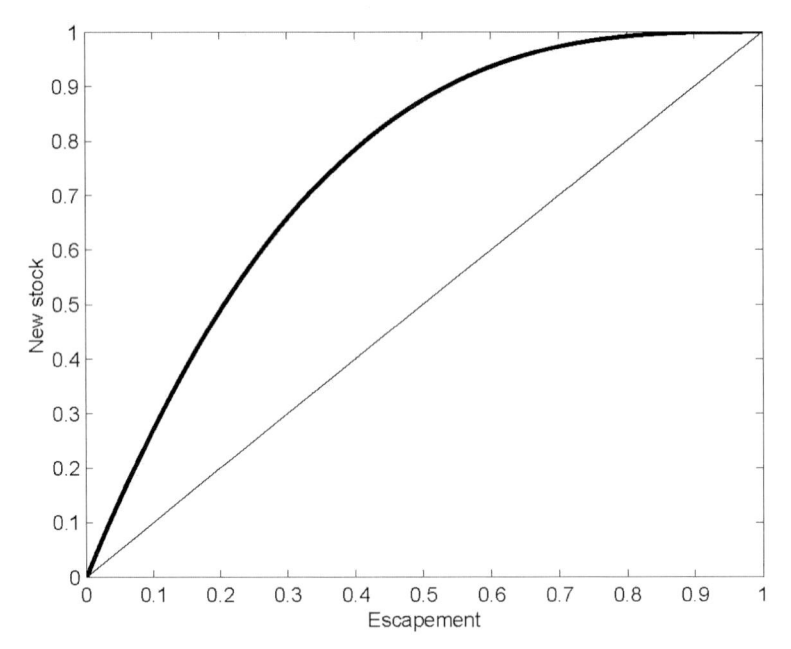

Figure 10.1a　　Growth function: High compensation.

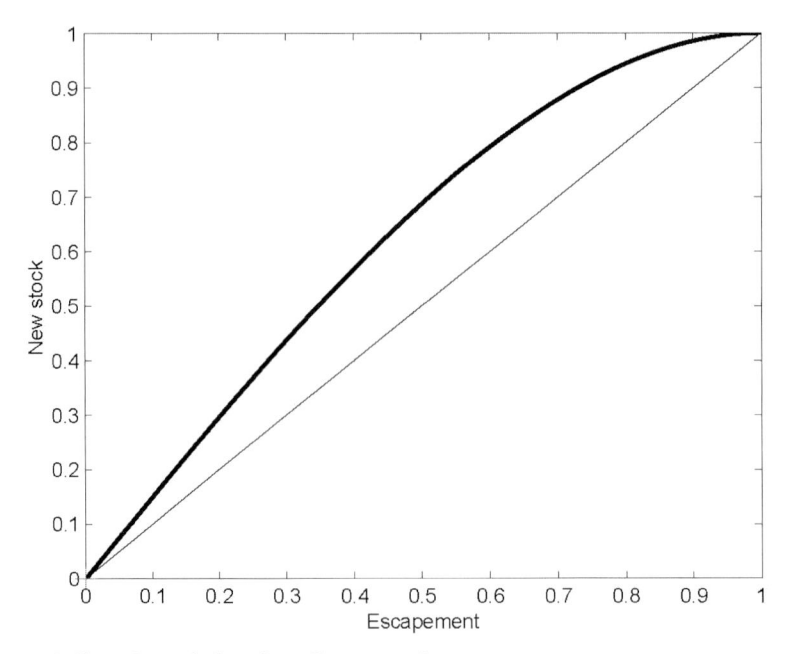

Figure 10.1b　　Growth function: Compensation.

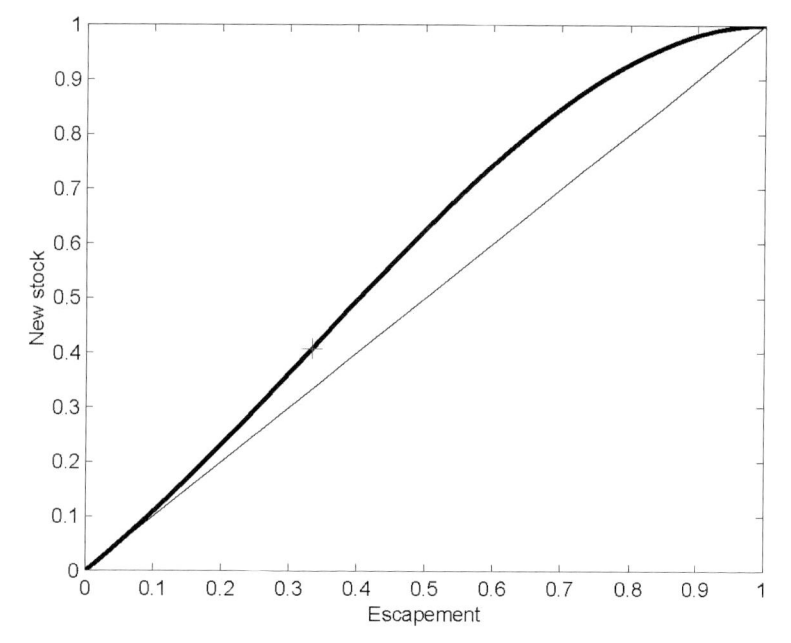

Figure 10.1c   Growth function: Depensation.

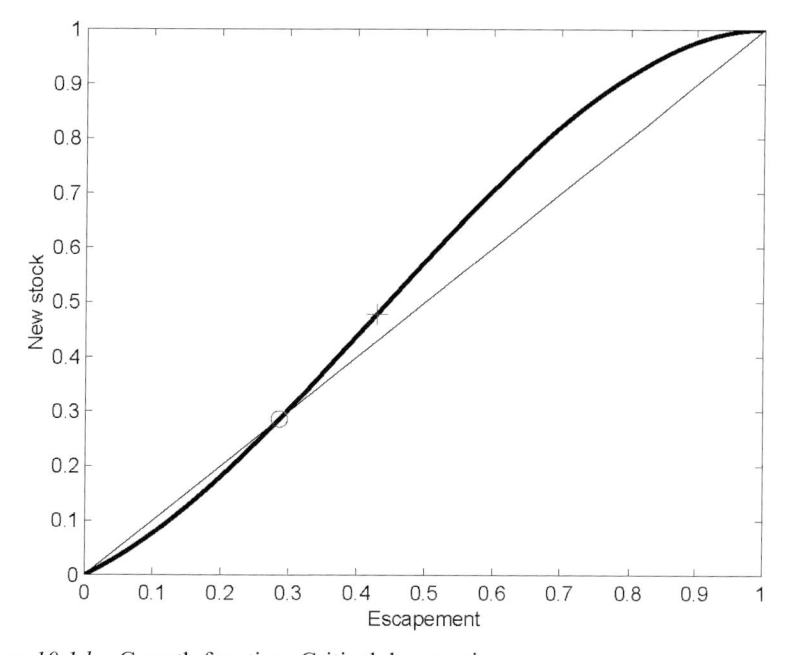

Figure 10.1d   Growth function: Critical depensation.

The examples described in this chapter all have a relatively simple information structure. First of all, the sequence $\{\phi(t)\}$ for $1 \leq t \leq T$, forms a two-valued stationary Markov chain with known initial probability distribution:

$$p_m(t=1) = prob[\phi(1) = \phi_m] \qquad (10.6)$$

for $m = 1, 2$, and known transition matrix:

$$p^+{}_{j|i} = prob[\phi^+ = \phi_j \mid \phi = \phi_i] \qquad (10.7)$$

for $i, j = 1, 2$.

Similarly, the stock-split parameter sequences θ_ν form i.i.d. chains with specified values $\theta_\nu(t) = \theta_1^{(\nu)}$ or $\theta_2^{(\nu)}$, where $\theta_i^\alpha + \theta_i^\beta = 1$ for $i = 1$ or 2, and stationary probability distribution:

$$q_i \stackrel{\Delta}{=} prob[\theta_\nu(t) = \theta_i^{(\nu)}], \qquad (10.8)$$

for $\nu = \alpha$ or β.

This probabilistic information about the environmental parameters is common knowledge shared by the fleet managers. Each fleet manager must adopt a long-term harvest policy for his fleet. We assume that, at the time when he must make his current harvest decision $H(t)$, he will know the immediately-prior brood-stock biomass S. However, the manager does not directly observe the environmental parameters ϕ and θ_ν. Instead he estimates them indirectly, by direct observation of measurement variables φ_ν and ϑ_ν.

The random variable φ_ν takes on the same two realized values as does ϕ, but with different frequencies. Specifically it registers the true-value correctly, with a known frequency:

$$prob[\varphi_\nu = \phi], \qquad (10.9)$$

which is ≥ 2. We define measurement precision r_ν as:

$$r_\nu = 2 \cdot prob[\varphi_\nu = \phi] - 1. \qquad (10.10)$$

Thus $r = 1$ means that ϕ_ν measures ϕ with complete accuracy, while $r = 0$ means that the measuring instrument cannot distinguish between the realisations of ϕ. In the latter case (of minimal information) the manager will assume only that the Markov chain of growth parameters $\{\phi_\nu\}$ has achieved its steady-state distribution.

Likewise, the random variable ϑ_ν takes on the same two realised values as does θ_ν, but with different frequencies. Specifically it registers the true value correctly, with the known frequency:

$$prob[\vartheta_\nu = \theta_\nu] \geq 1/2. \qquad (10.11)$$

We define measurement precision s_v as:

$$s_v = 2 \cdot \text{prob}[\vartheta_v = \theta_v] - 1. \tag{10.12}$$

By assumption, neither fleet manager is able to observe directly either recruitment R or substream recruitment R_v prior to setting his current harvest decision. Hence unless a fleet manager knows the growth parameter ϕ with precision, he cannot know the current recruitment R with precision. Likewise, unless he knows both ϕ and θ_v with precision, he cannot know current $R_v(n!/r!(n-r)!)$ with precision. In general his harvest choices will reflect this degree of ambiguity.

We will examine both the case where the realised measurement values φ_v and ϑ_v are private information, known only to the v-fleet's manager, and also the case where this information is transparent, i.e. $\varphi_\alpha, \vartheta_\alpha, \varphi_\beta,$ and ϑ_β are known to both fleet managers.

Finally, we shall examine potential co-operative bargaining solutions of the games. We shall assume that utility transfers through monetary side-payments are feasible, making possible the redistribution of the overall benefits of co-operation to conform to the relative competitive strengths of the fleets. For simplicity, we shall assume that the negotiations over co-operative harvest policies, and the *a priori* distribution of the competitive surplus, will lead to the classical Nash bargaining solution of our imperfect-information harvesting game.

10.3 RISK-NEUTRAL FLEET MANAGERS

Commonly one considers risk-neutral fleet managers. For that case we take the net seasonal payoff to harvest by the v-fleet to be of the form:

$$\pi_v = p(H)H_v - C_v(R_v, H_v). \tag{10.13}$$

Here the unit landing price $p(H)$ is a monotone decreasing function of the total seasonal harvest $H = H_\alpha + H_\beta$ and the seasonal harvest cost is:

$$C_v = \int_{S_v}^{R_v} \varsigma_v(x) dx \tag{10.14}$$

with unit harvest cost ς_v a monotone non-increasing function of the in-season stock-level x.

In our simulated examples we shall take:

$$P(H) = p - p_1 H \text{ and } \varsigma_v(x) = c_v/x. \tag{10.15}$$

Here p, p_1 and c_v, for $v = \alpha$ and β are given deterministic parameters. In this risk-neutral case, the v-fleet chooses its harvest policy:

$$\{H_v[R_v(t)]\}, \text{ for } t = 1 \text{ to } T \tag{10.16}$$

to optimise the expected long-run discounted-payoff, or objective function:

$$\Pi_v = E\{\sum_{t=1}^{T} \delta_v^{t-1} \pi_v(t)\}, \qquad (10.17)$$

given the anticipated policy of its competitor. The game solution is taken to be a standard dynamic Nash equilibrium – more technically, we seek a Markov-perfect equilibrium pair of policies, each policy being the optimal response to the other. With our specific model assumptions, this equilibrium solution-pair turns out to be unique.

There follow simulation studies of several examples of the risk-neutral model. No attempt has been made to set parameter values to mimic any particular fishery. Rather, our goal here is to explore the model's parameter space, to illustrate the kinds of phenomena that are possible in principle, and to develop intuition concerning the mechanisms for their occurrence.

Each fleet's choice of current harvest landings represents a trade-off at the margin between the expected net value of its immediate landings, and the expected contribution to future harvest returns of its sub-stream escapement. As always, in a competitive fishery each fleet's calculation of future returns ignores the positive external effect of each fleet's current escapement on the future harvest returns to the other. Thus, from a societal perspective, competitive management tends to induce over-harvesting as compared to co-operative management

On the other hand, incomplete information about the stochastic oceanic environment will lead each fleet to set a strategy that is, in some sense, a compromise between the differing strategies it might employ if it knew the current environmental data with precision. Both current harvest return and expected future returns will be estimated imperfectly, so that the balance struck between them may be skewed in either direction.

Our central goal will be to examine how the quality of each individual fleet's available oceanic environmental information will influence its strategic trade-off decisions, and the consequent biological and economic implications of the two fleets' interacting policy choices.

10.3.1 Interplay of price and degree of recruitment compensation

Our first simulations examine the role of information through examination of the interplay between the landings price $p(H)$ and the growth rate inherent in the stock-recruitment function. Harvest return-rate of course influences the value of current substream harvest landings H_v, while stock-recruitment growth influences the value of current sub-stream escapement S_v in determining expected future fleet landings values.

Since in this section we restrict attention to stock-recruitment functions which are deterministic, the stock-recruitment growth at given escapement is characterised through its compensatory or depensatory characteristics. We contrast, in Figures 10.2a-d, compensatory growth and critically-depensatory growth, over a range of values of p, for fixed p_1 and $c_1 = c_2$. The two fleet's situations are symmetric, except possibly for information precision. The figures show the results of pure competition but also the contrasting results from co-operation based on the Nash's bargaining process (which still takes into account their relative competitive strengths).

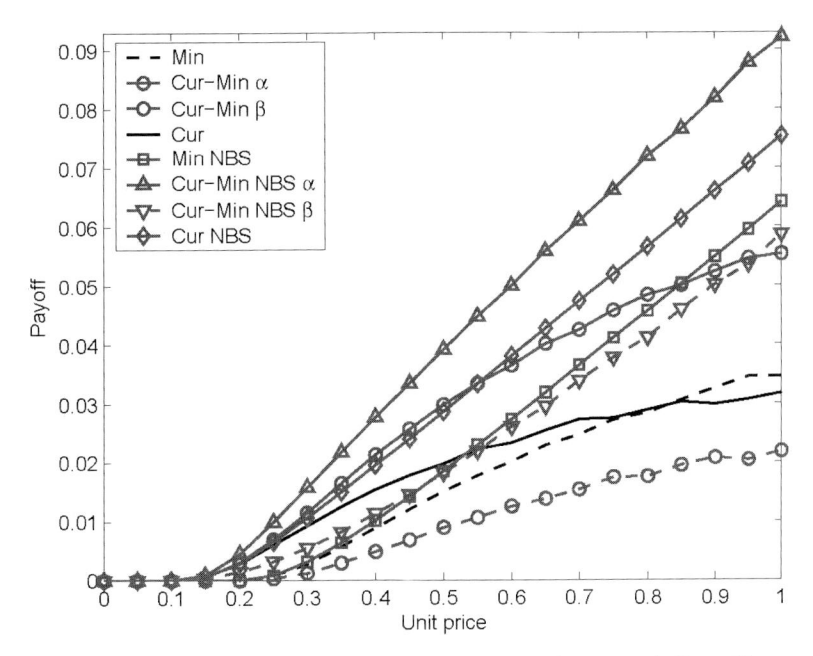

Figure 10.2a Influence of landings price for compensation growth: Payoff

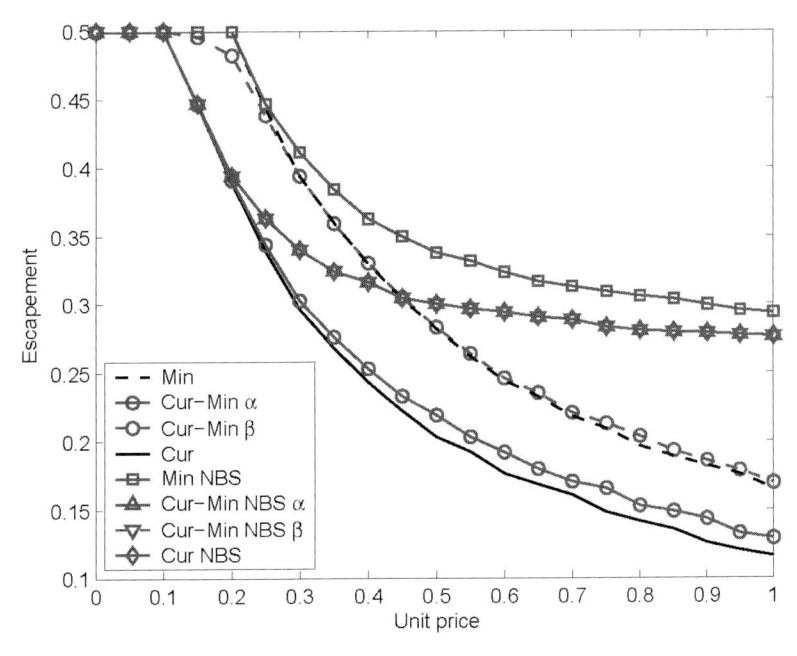

Figure 10.2b Influence of landings price for compensation growth: Escapement

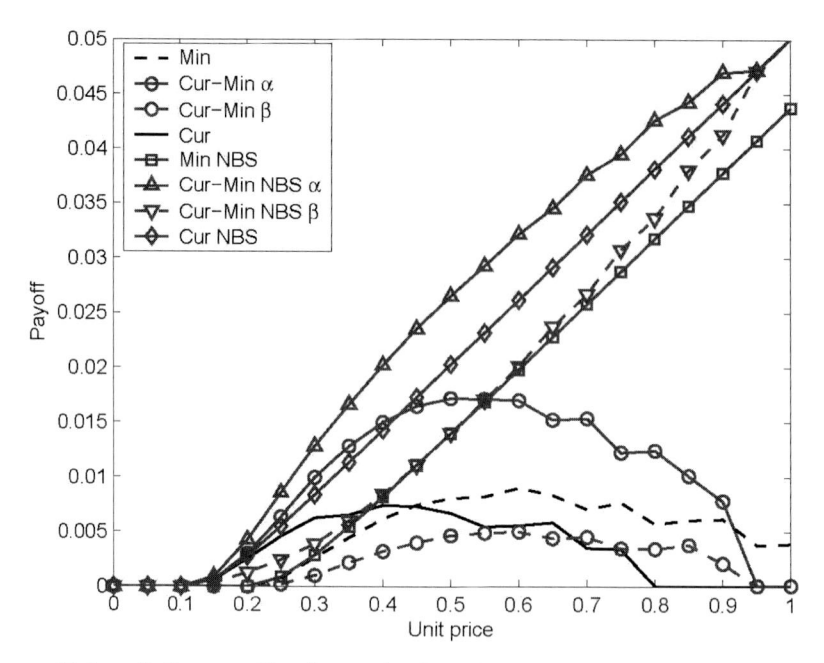

Figure 10.2c   Influence of landings price for critical depensation growth: Payoff.

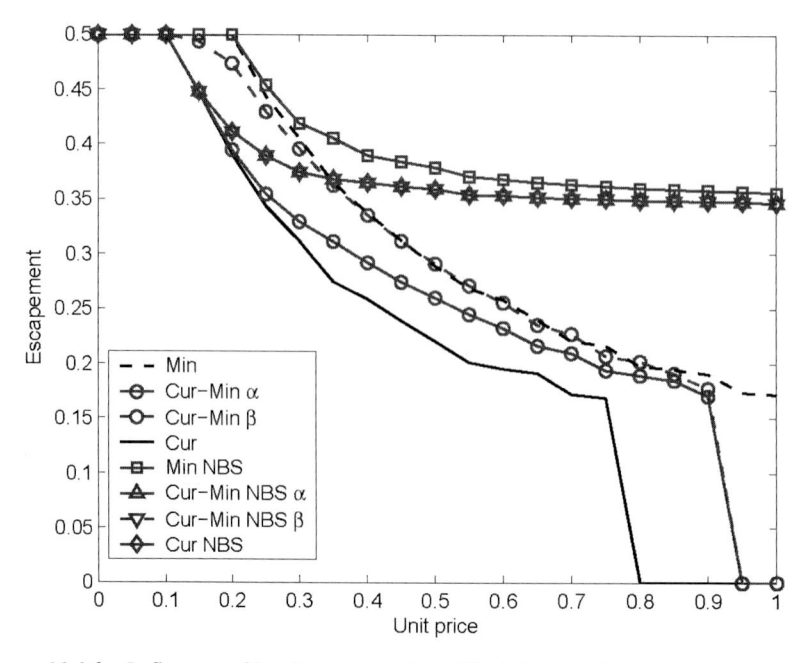

Figure 10.2d   Influence of landings price for critical depensation growth: Escapement.

Consider first those cases where the two fleets' knowledge levels coincide. From full symmetry, their expected long-run payoffs also will coincide. Furthermore, in the bargaining mode their long-run expected payoffs will grow with p and will be greater when they both possess full current knowledge (denoted 'cur') than when they both have only minimal ('min') knowledge (i.e. when they know only the steady-state probability distributions of the environmental parameters).

But when they compete, the payoffs begin to fall off as p increases. Indeed, in the case of critical depensation, eventually they drop to zero for sufficiently large p. This outcome reflects an extreme trade-off decision, favouring current over future returns – to the point of mining out the fish stock – as is evident in the escapement curves. Note too that the payoffs may be greater when the fleets possess only minimal knowledge than when they have full current knowledge. In effect, increased knowledge, combined with high landing prices, permits more intensive and hence more destructive competition.

The situation becomes more complex when the fleets' knowledge levels differ. Then the competitive advantage seems always to lie with the fleet with greater knowledge, and this is reflected also in the Nash bargaining solution (NBS). (The notation here is, for example, that 'cur-min alpha' refers to the payoff to the first fleet when it possesses current knowledge and its competitor possesses only minimal knowledge.)

In fact, with co-operation always entailing transparent utility (as we have been assuming), the Pareto Boundary will be symmetric whenever the positions of the fleets are symmetric other than in information, see Figure 10.3a-b. The NBS fleet payoff proportions then closely reflect those of the corresponding competitive payoffs, and thus will be asymmetric whenever private information held by the competitors is asymmetric, i.e. will be determined by the location of the 'threat point'.

10.3.2 Information of varying quality

In Figure 10.4a,b we look in another way at this phenomenon, now fixing p at an intermediate value, and allowing measurement precision of the stock split to vary continuously between the extremes of minimal information (on the left axis) and full information (on the right axis.)

A comparison of Figures 10.4a-d serves to emphasise that the phenomena observed in Figure 10.2 for critical depensation, also occurs under compensatory growth conditions, and is simply more intense when the growth is strongly depensatory. Of course this is no longer true when the players agree to a cooperative bargaining resolution of the game. In that case, making information transparent ensures a better outcome for both, while acknowledging the stronger bargaining position of the fleet which is asked to reveal its private information.

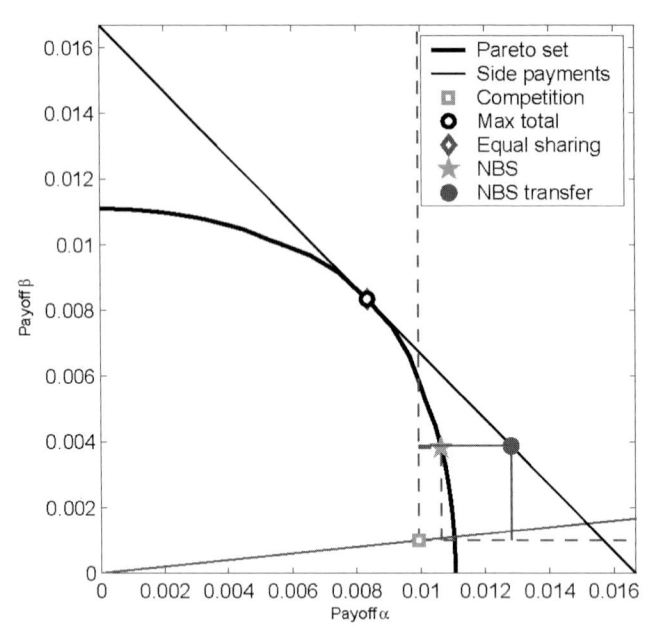

Figure 10.3a   Pareto set and NBS for critical depensation: Price=0.3.

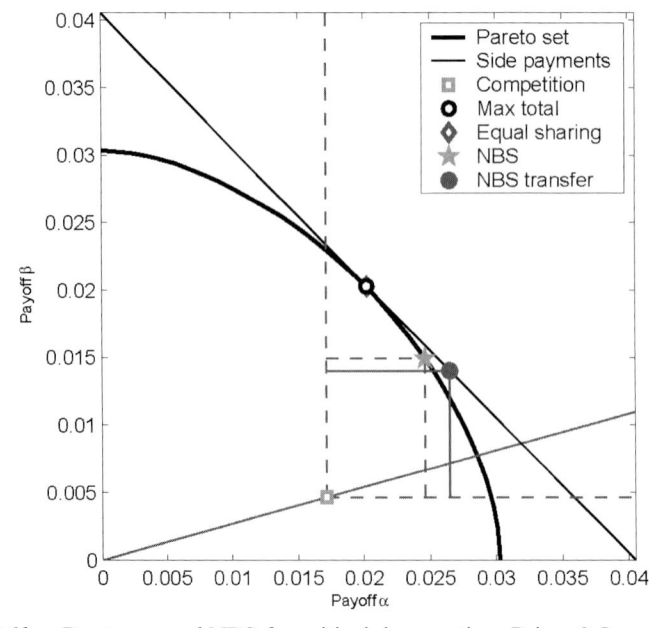

Figure 10.3b   Pareto set and NBS for critical depensation: Price=0.5.

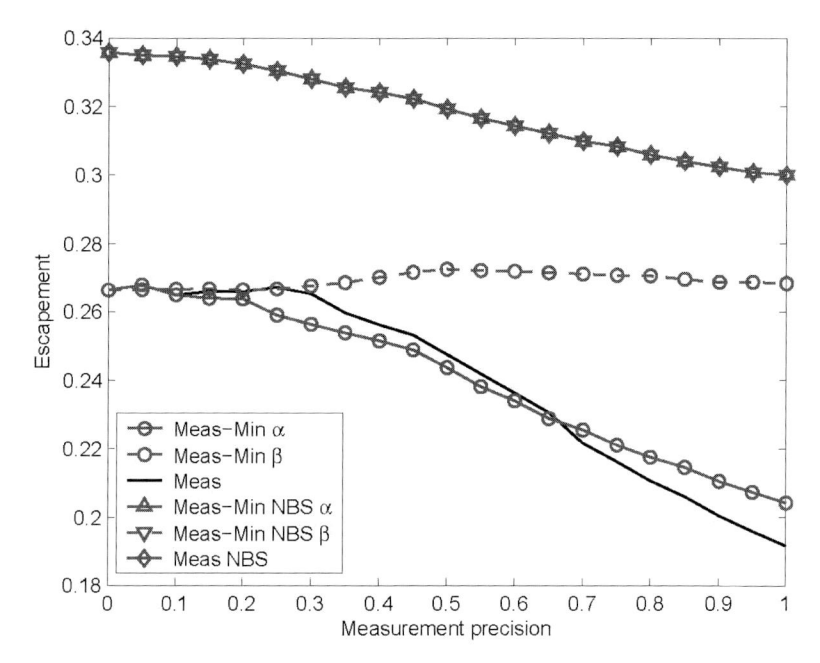

Figure 10.4a   Harvesting with information of varying quality for compensation growth: Payoff.

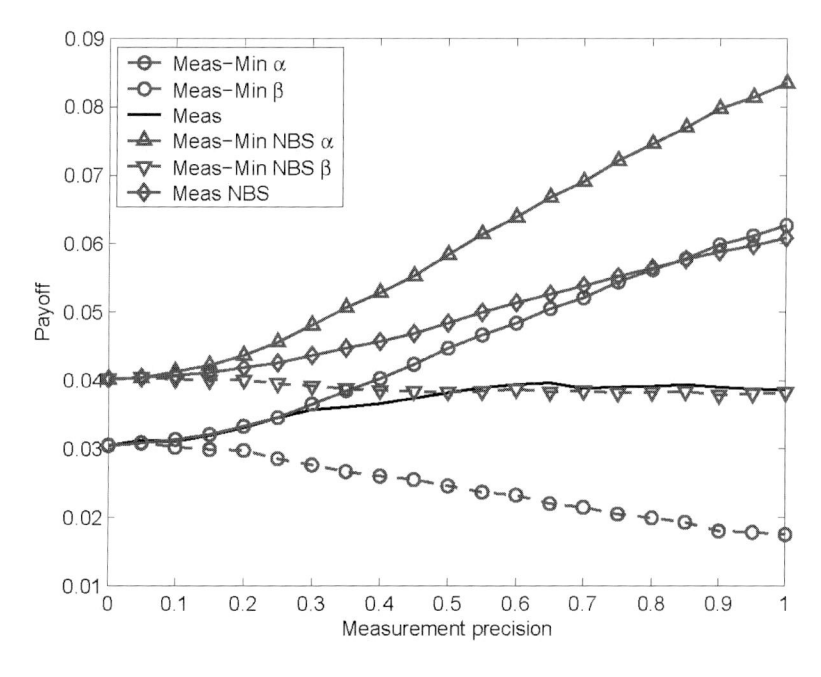

Figure 10.4b   Harvesting with information of varying quality for compensation growth: Escapement.

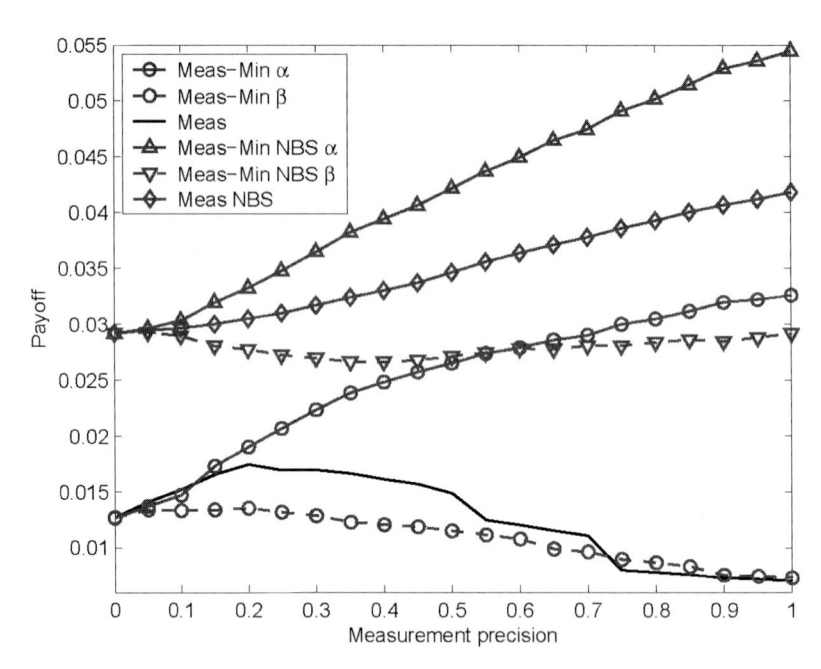

Figure 10.4c Harvesting with information of varying quality for critical depensation growth: Payoff.

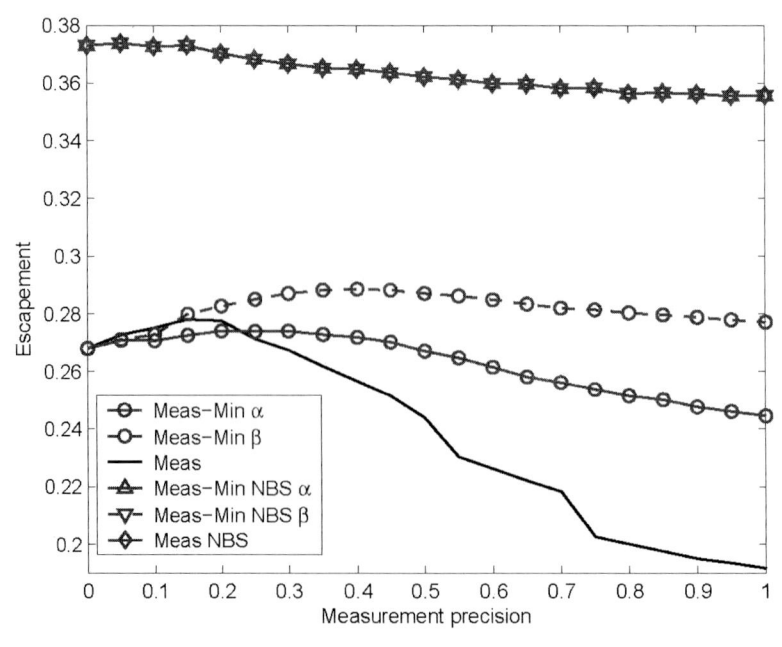

Figure 10.4d Harvesting with information of varying quality for critical depensation growth: Escapement.

10.3.3 Asymmetry in the oceanic environment

In Figure 10.5a-b we see an asymmetry in the substream split, which favours the α-fleet on the right-half of the diagram (where mean$[\theta_\alpha] > 0.5$) and favours the β-fleet on the left. Indeed, if the fleets enjoy *information parity* (both with minimal or full-current knowledge), then the set of graphs will be symmetric about its midpoint, i.e. around mean$[\theta_\alpha] = 0.5$.

Consider first the curves for which the fleets enjoy identical knowledge levels. Under competition(Figure 10.5a), one finds that the *v*-fleet will do better in the region where it has the split-stream advantage in mean$[\theta_v]$, but worse in the region where its opponent has the split-stream advantage. Sometimes the fleets do better when they have minimal knowledge rather than when they have full, or even partial, current knowledge.

This situation does not carry over to the NBS co-operative (Figure 10.5b) game (at least in these examples). Indeed, when knowledge is symmetric, payoffs grow as knowledge levels increase.

The geometrical link between co-operation and competition is revealed in Figure 10.6a-b. When there are fleet asymmetries other than in information state, the Pareto boundary itself will be asymmetric. This is true when mean$[\theta_v]$ is asymmetric between $v = \alpha$ and β, as above. Examples of the determination of the NBS in this situation are shown in Figure 10.6a-b.

10.4 RISK-AVERSE LEVHARI-MIRMAN ASSUMPTIONS

In this section we adopt more specific assumptions concerning fishery objectives and the stochastic stock-recruitment process, so as to carry the mathematical analysis further than in the more general formulation described above. Our assumptions generalise, to a split-stream incomplete-information version of the classical 1981 Levhari-Mirman Fish War.

The stock-recruitment expression is:

$$R = A(\phi)S^\phi, \quad (10.18)$$

where $A(\phi)$ is a specified deterministic function of the random growth parameter ϕ. Furthermore we adopt the risk-averse fleet-utility function:

$$W_v(S, \Psi_v) = \exp E\left\{\ln \Pi_{t=1}^\infty \left[H_v(t)\right]^{\kappa_v(t)}\right\}, \quad (10.19)$$

where

$$\kappa_v(t) = (1-\delta_v)\delta_v^{t-1} \quad (10.20)$$

so that

$$\sum_{t=1}^\infty \kappa_v(t) = 1. \quad (10.21)$$

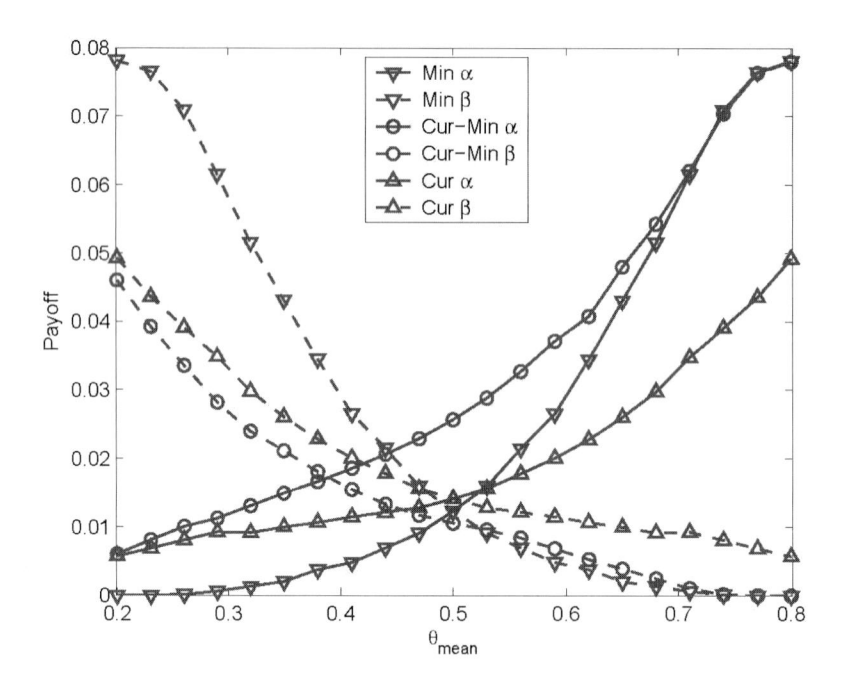

Figure 10.5a   Variable environmental asymmetry (mean of θ): Competition cases.

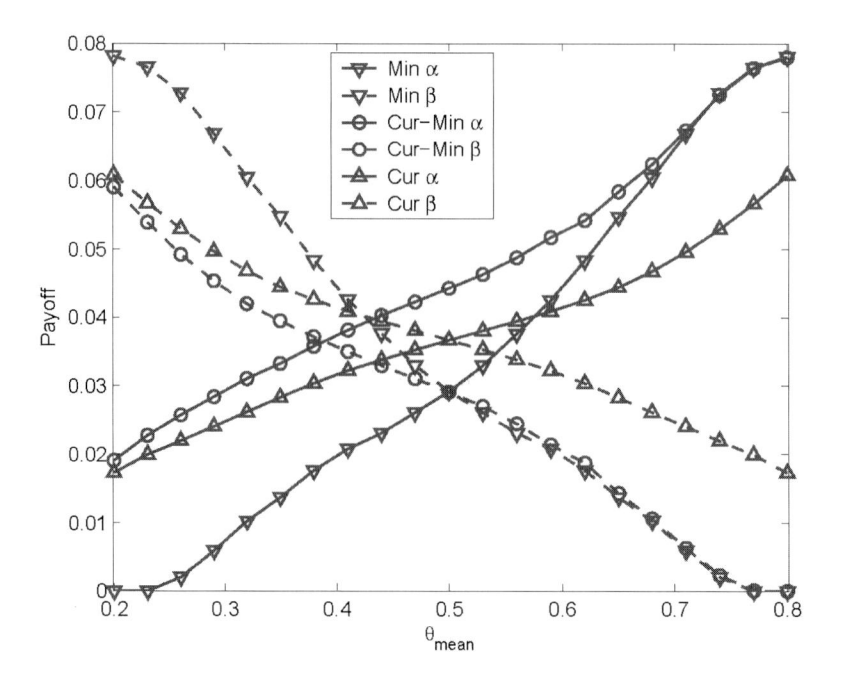

Figure 10.5b   Variable environmental asymmetry (mean of θ): Cooperation cases.

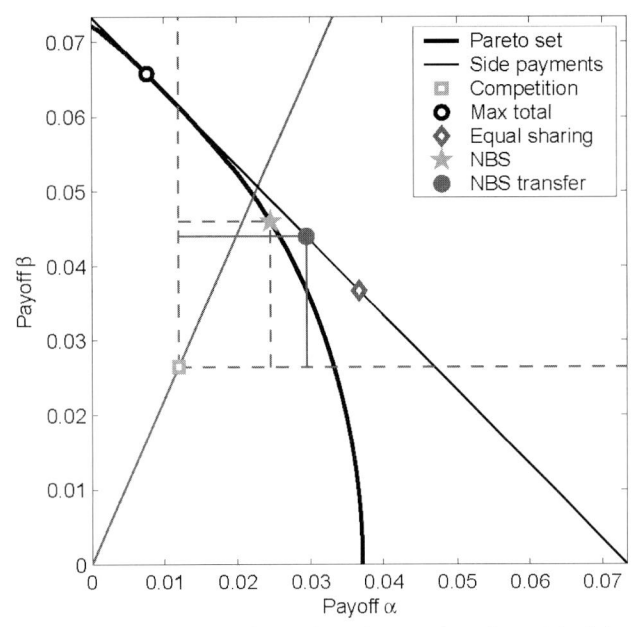

Figure 10.6a Pareto set and NBS for various θ_{mean} values for critical depensation: $\theta_{mean} = 0.3$.

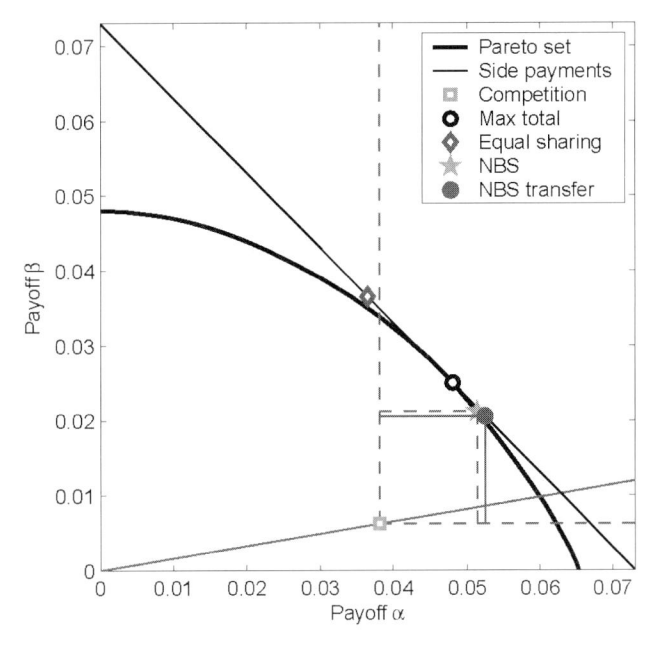

Figure 10.6b Pareto set and NBS for various θ_{mean} values for critical depensation: $\theta_{mean} = 0.6$.

That is, W_v is a (generalised) geometric mean of the annual payoffs. This objective is cardinally equivalent to a generalised Levhari-Mirman objective function:

$$U(S,\Psi_v) = E\sum_{t=1}^{\infty} \delta^{t-1} \ln H_v[t]. \qquad (10.22)$$

in which both unit harvest cost $C(x)$ and landing price p are constant, and $p-C(x)$ is normalised to 1.

Dual to the fleet's objective function is the marginal stock asset elasticity:

$$\Lambda_v(\Psi_v) = \frac{dW_v/W_v}{dS/S} = S\frac{dU_v}{dS}. \qquad (10.23)$$

We now describe the structure of the equilibrium solution to the competitive game with private information. Parallel characterisations apply when one assumes transparent information, or co-ordinated objectives.

10.4.1 Theorem

Let

$$(S,\phi,\theta_\alpha,\theta_\beta) \qquad (10.24)$$

be the state of the natural system at the beginning of time period t, and suppose that, at every time period the measurement realisation-pair:

$$[\varphi_v(t), \vartheta_v(t)] \qquad (10.25)$$

is private information, held by the v-fleet manager. Then the harvest control rule for each fleet is of the form:

$$H_v(t) = h_v(t)R_v(t) = h_v(t)\theta_v(t)R(t) \qquad (10.26)$$

where

$$h_v(t) = h_v[\varphi_v(t), \vartheta_v(t)] \qquad (10.27)$$

independent of $S(t)$, $\varphi_{-v}(t)$, and $\vartheta_{-v}(t)$. In particular, the chosen harvest fraction $h_v(t)$ is independent of $R_v(t)$.

For horizon $T=1$, each v-fleet's equilibrium policy is to maximise harvest within the given constraint ($h_v \leq h_v^{max}$) by setting:

$$h_v = h_v^{max}, \qquad (10.28)$$

so that the v-fleet's marginal asset value is:

$$\Lambda_v[\varphi_v] = E[\phi_v \mid \varphi_v], \qquad (10.29)$$

independent of S and ϑ_v.

For a horizon $T>1$, the two fleet's optimal harvests $h_\alpha(\varphi_\alpha, \vartheta_\alpha)$ and $h_\beta(\varphi_\beta, \vartheta_\beta)$ are independent of S and are determined simultaneously by iteration on T of the pair of equations, for $v = \alpha$ and β:

$$E\left[\frac{\theta_v h_v}{1-\theta_v h_v - \theta_{-v} h_{-v}} \delta_v \Lambda_v^+(\varphi_v^+) \mid h_v, h_{-v}\right] = 1, \quad (10.30)$$

valid unless the constraint $h_v \leq h_{max}$ bind. Here the dual asset value is also determined iteratively:

$$\Lambda_v(\vartheta_v) = E\left[\phi\left(1 + \delta_v \Lambda_v^+[\vartheta_v^+]\right)\right] \quad (10.31)$$

for $v = \alpha$ and β.

Note the economic interpretation:

$$\frac{\theta_v h_v}{1-\theta_v h_v - \theta_{-v} h_v} \Lambda_v^+(\varphi_v^+) = \frac{\text{value of the marginal unit of escapement}}{\text{value of the marginal unit of harvest}}. \quad (10.32)$$

10.4.2 An illustrative simulation of the LM model

Here we consider a fully-symmetric example, where in particular:

$$\theta_1^v + \theta_2^v = 1, \quad (10.33)$$

and $\theta_1^\alpha > \theta_2^\alpha$, so that the first realization $\theta_\alpha = \theta_1^\alpha$ favours the α-fleet.

Both players observe the current realization of θ_v with the same (imperfect) accuracy:

$$\text{prob}(\vartheta_v = \theta_v) = \xi. \quad (10.34)$$

However, we shall compare cases where ξ takes on a range of values from $\xi = 0.5$ (minimal knowledge) to $\xi = 1$ (fully-accurate knowledge).

The growth parameter ϕ (called b on the figures) also takes on two realisations, with ϕ_1 corresponding to high growth and ϕ_2 to low growth.

Both players observe the current realisation of ϕ with the same fixed accuracy:

$$\text{prob}[\varphi_v = \phi] = 0.8, \quad (10.35)$$

for both $v = \alpha$ and β.

In the figures we contrast circumstances where the fleets compete non-co-operatively (optimising their individual logarithmic utility) versus circumstances where they arrive at an asymmetric bargaining solution (through maximising the sum of their logarithmic utilities). We also compare cases where information is privately held versus cases where information is transparent.

Figures 10.7a-d illustrate the game-specific harvest rules for player α (symmetric rules hold for player β). We show here results for only half of the observational possibilities; the others are similar. On each individual figure are shown 5 curves: a case of private α-information, plus the four cases of transparent information which are consistent with that specific α-fleet data.

The heading at the top of the frame specifies which values of ϑ_α and of φ_α (here denoted b_α) that the α-fleet's instruments record. When information is private, this is the only information that the α-fleet possesses concerning θ_α and ϕ. However when information is transparent the α-fleet manager also knows the measurements, ϑ_β and φ_β, recorded by the β-fleet's instruments.

In the box, the distinct information available to the α-fleet in each of these five cases is specified. For each of the four transparent-information cases the distinct β-fleet data are given following the data from α-fleet's instruments, which the same in all five cases. The frame's heading also specifies whether the fleets' utility functions in the five game variants being presented are competitive, i.e. each reflects only the direct harvest payoff to that individual fleet, or are cooperative, i.e. reflect the total payoff to the two-fleet industry.

Figure 10.7a-b graphs correspond to competitive objectives and Figure 10.7c-d graphs to co-operative objectives. The graph-sets are very similar, except that harvest rates are consistently lower with co-operative objectives than with competition.

Note that, in each frame, if preponderant information indicates that $\text{prob}\,\theta_1^\alpha > 0.5$, i.e. the sub-stream split favours the α-fleet), then harvest fraction h_α drops as θ_v information accuracy grows. The curve *rises* if preponderant information indicates that $\text{prob}\,\theta_2^\alpha > 0.5$, (which is bad for the α-fleet), and it is *level* when the available knowledge is ambiguous, i.e. when $\text{prob}\,\theta_1^\alpha = \text{prob}\,\theta_2^\alpha = 0.5$.

Note too that, two parallel curves result from identical realisations of ϑ_v and ϑ_{-v} but differing levels of knowledge of the growth parameter ϕ. In that case the harvest rate is higher for the curve for which the available knowledge indicates the stronger growth.

Finally, note that the harvest-rate curve for private information is not simply a weighted average of the four transparent-information curves which are consistent with that private knowledge. It may lie above (or below) all four of them.

Figure 10.8 shows how these various harvest rules interact stochastically to determine the average time-discounted payoffs to the α-fleet. (The results are symmetric for the β-fleet.) The results displayed are for 20,000 simulations. Each curve shows that superior payoffs (in the geometrically-discounted geometric mean of harvests W_α) result from increased knowledge, by both fleets, of θ_v. However, comparing the various information-specific curves, we see that inferior payoffs result from increased information of the growth parameter ϕ, i.e. result from possessing transparent knowledge rather than only private knowledge.

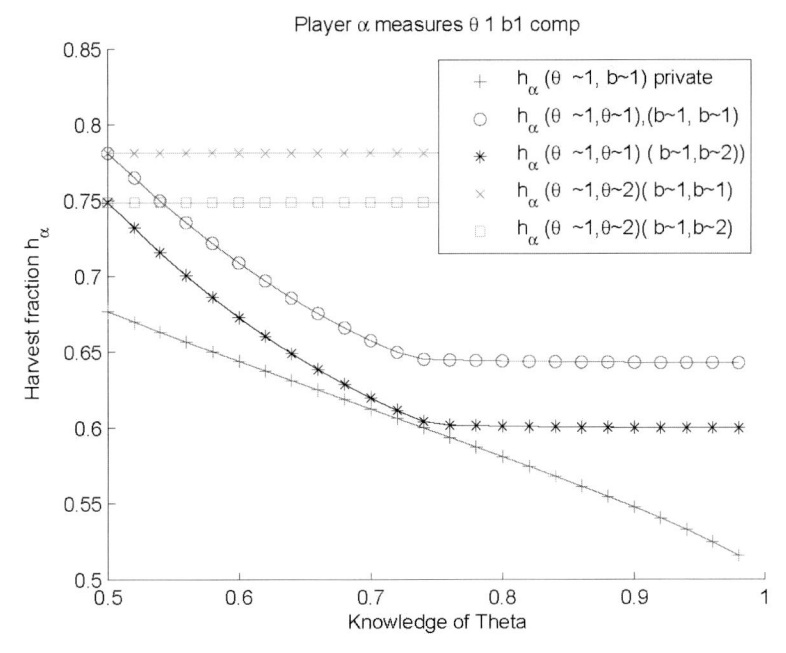

Figure 10.7a Game-specific harvest rules for player α (Symmetric rules hold for player β): Competitive objectives (1,1).

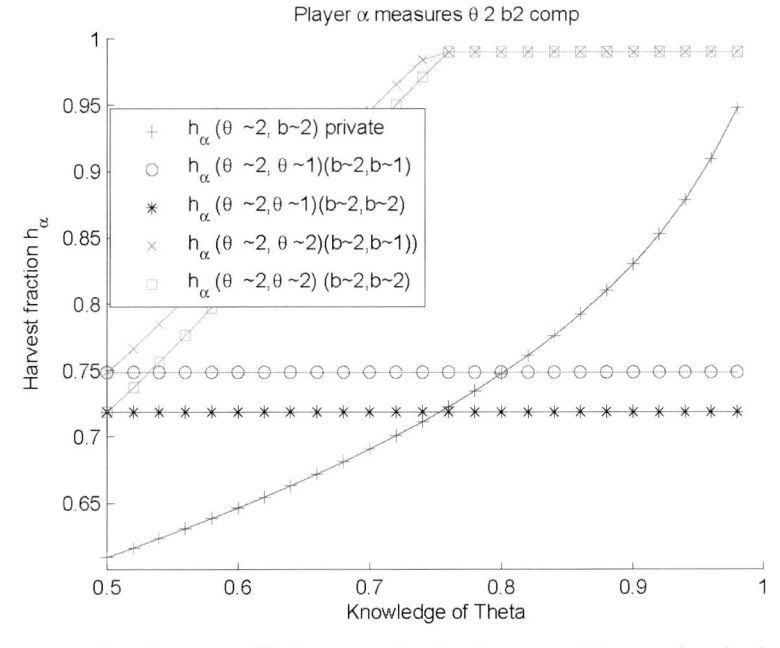

Figure 10.7b Game-specific harvest rules for player α (Symmetric rules hold for player β): Competitive objectives (2,2).

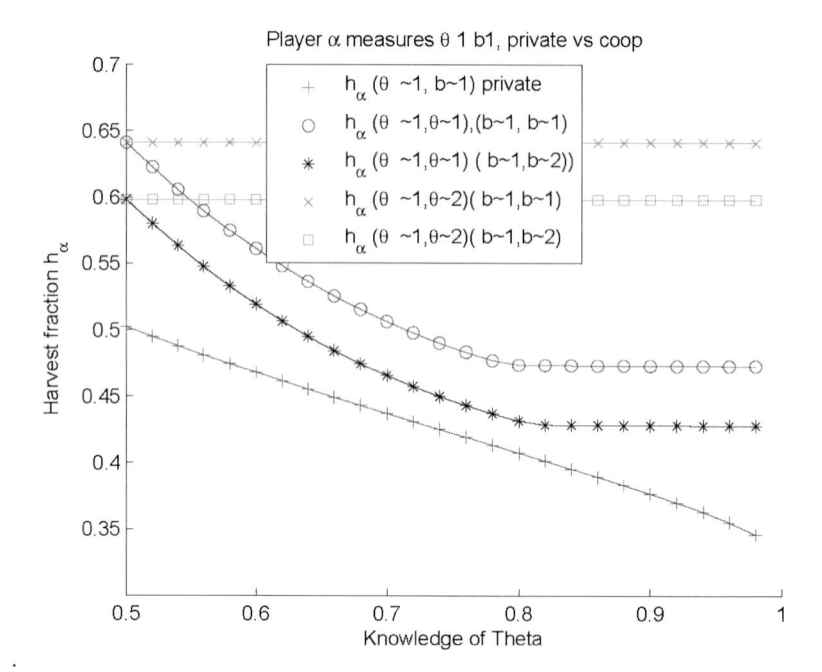

Figure 10.7c Game-specific harvest rules for player α (Symmetric rules hold for player β): Cooperative objectives (1,1).

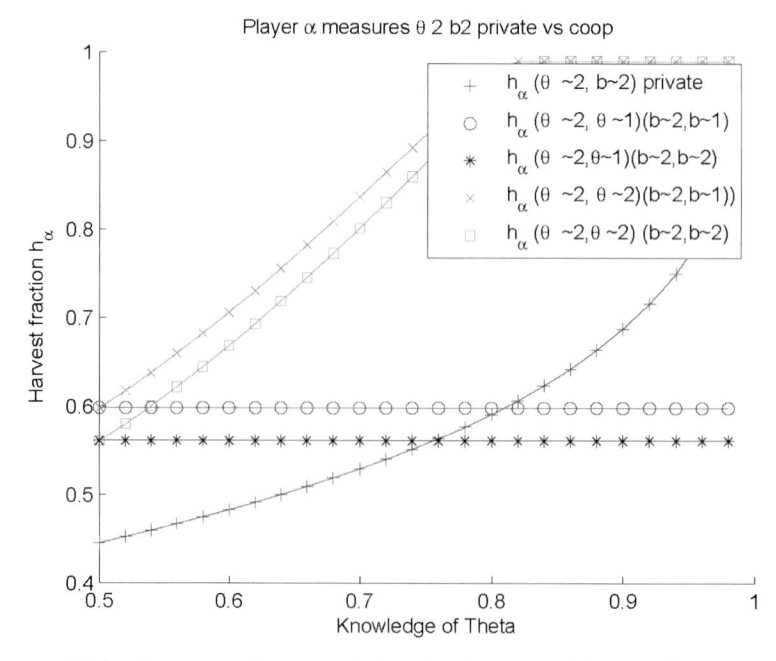

Figure 10.7d Game-specific harvest rules for player α (Symmetric rules hold for player β): Cooperative objectives (2,2).

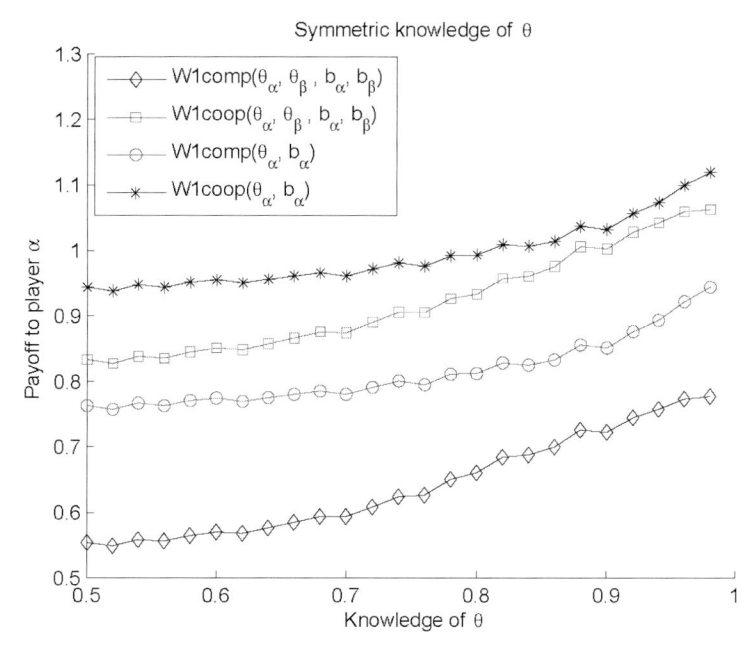

Figure 10.8 Average time-discounted payoffs to the α-fleet. (The results are symmetric for the β-fleet.)

Thus, the pattern that potentially corrosive results may result from increasing knowledge, a phenomenon that we observed in the risk-neutral case, is seen to carry over to this informationally more complex situation, with two distinct stochastic parameters and risk averse fleet managers.

10.5 CONCLUSIONS AND FURTHER WORK

Our work shows, in a conceptually natural context, that, unlike for a single player optimisation process, information enrichment may in principle have destructive implications for payoffs. In this example, the result is a consequence of an unbalance in valuation of present versus future returns, hence an elaboration of the well-known tragedy of the commons. Our models are idealised, and do not attempt to mimic in any quantitative way the conditions in a real-world marine fishery. But they do show what may be possible. However, an elaboration of the mechanisms in the model, as described in the previous section, does demonstrate certain persistence in the general pattern we have displayed. In its details this elaboration may have raised as many questions as it settles.

We have been engaged in developed the split-stream model further, in several significant directions. One extension is to understand the implications, in determining the outcome of the split-stream harvesting game, of the fleet managers' attitudes toward risk. Until now we have assumed identical risk-neutral or logarithmically-risk-averse perspectives by fleet managers. With more general risk-sensitive objectives, our initial simulated results remain counter-intuitive, especially when the competing fleets have different attitudes toward risk. We shall continue to examine these questions in our ongoing studies.

Another channel for future research involves elaborating the network structure of the model, from the simple split-stream structure. In particular we intend to construct an incomplete-information stochastic version of our 'hit-and-run' game model, which can be used to simulate the 'new-member problem', concerning negotiations between a multilateral regional fisheries management organisation (RFMO) and a distant-water fleet (DWF) wishing to enter the regional fishery (Bjørndal and Munro, 2003). This bargaining process becomes particularly interesting when the RFMO controls access to a major portion of the fishing grounds, so that, without co-operative agreements, the DWF is confined to international waters of the high seas. Deterministic versions of the Hit-and-Run model have been published by McKelvey *et al.* (2002, 2003).

Finally, we wish to apply appropriate versions of our incomplete-information stochastic harvest games to a variety of marine fisheries operating across international boundaries.[1] (Note again our initial application to the North American Pacific salmon management controversy; see, McKelvey *et al.* 2003.)

It seems clear that the outcomes of such harvesting games will depend heavily on particular circumstances in the fisheries involved. These differences will sometimes relate to the cyclic patterns and intensities of oceanic environmental conditions, sometimes to the biological characteristics of the harvested fish stock or stocks, and usually upon the economic interests of the nations involved in the fishery, either as harvesters or as countries exercising control over their coastal waters.

Plainly, a 'one size fits all approach' will not be adequate here – the models must be adapted to particular circumstances. On the other hand, the models we are building will remain, at the most, as highly stylised abstractions from reality: beyond the usual abstractions met in bio-economic models, non-co-operative game models must make even more heroic assumptions about human aspirations and behaviour. Their role, then, is not prescriptive in the physical science mode. Rather, they must remain merely suggestible: as a window into an artificial world – one which, we hope may, in some ways resemble our own.

Acknowledgements

This research has received grant funding during 2003-2005 from the Decision, Risk and Management Science Program of the US National Science Foundation.

References

Barnett, T.P., D.W. Pierce and R. Schnur. 2001. Detection of anthropogenic climate change in the world's oceans. *Science* 292 (5515): 270-74.

---

[1] Concrete examples of the destructive effects of incomplete information of climatic stochasticity in transboundary fisheries can be found in articles cited in section one, above, and in the forthcoming volume *Climate Change and the Economics of the World's Fisheries*, M. Barange, R. Hannesson and S. Herrick (eds), to be published by Edgar Elgar Press.

Bjørndal, T. and G. Munro. 2003. The management of high seas fisheries resources and the implementation of the UN Fish Stocks Agreement of 1995. In H. Folmer and T. Tietenberg (eds) *The International Yearbook of Environmental and Resource Economics 2003/2004.* Edward Elgar, UK.

Clark, C.W. 1980. Restricted access to common-property fishery resources: A game-theoretic analysis. In P.T. Liu (ed.) *Dynamic Optimization and Mathematical Economics.* Plenum, New York: 117-32.

Gordon, S. 1954. The economic theory of a common property resource: The fishery. *Journal of Political Economy* 62**:** 124-42.

Hare, S.R. and N.J. Mantua. 2000. Empirical evidence for north Pacific regime shifts in 1977 and 1989. *Progress in Oceanography* 47: 103-45.

Levhari, D. and L.J. Mirman. 1980. The great fish wars: An example using a dynamic Cournot-Nash solution. *Bell Journal of Economics* 11: 322-44.

McFarlane, G.A., J.R. King and R.J. Beamish. 2000. Have there been recent changes in climate? Ask the fish. *Progress In Oceanography* 47: 147-69.

McKelvey, R.W., L.K. Sandal and S.I. Steinshamn. 2002. Fish wars on the high seas: A straddling stock competition model. *International Game Theory Review* 4: 53-69.

McKelvey, R. and P.V. Golubtsov. 2002. The effects of incomplete information in stochastic common stock harvesting games. To appear in *International Game Theory Review.* A preliminary version appeared in the *Proceedings of the International Society for Dynamic Games, St. Petersburg, Russia,* 8-11 July, 2002.

McKelvey, R., K.A. Miller and P. Golubtsov. 2003. Fish-wars revisited: Stochastic incomplete-information harvesting game. In J. Wesseler, H.P. Weikard and R.D. Weaver (eds) *Risk and Uncertainty in Environmental and Natural Resource Economics,* Edward Elgar, UK.

McKelvey, R.W., L.K. Sandal and S.I. Steinshamn. 2003. Regional fisheries management on the high seas: The hit-and-run interloper model. *International Game Theory Review* 5: 327-45.

Miller, K.A., G. Munro, T. McDorman, R. McKelvey and P. Tydemers. 2001. The 1999 Pacific salmon agreement: A sustainable solution? *Occasional Papers: Canadian-American Public Policy*, No. 47, Canadian-American Center. University of Maine, Orono.

Miller, K.A. and G.R. Munro. 2004. Climate and cooperation: A new perspective on the management of shared fish stocks. *Marine Resource Economics* 19(3): 367-93.

Scott, Anthony. 1955. The fishery: The objectives of sole ownership. *Journal of Political Economy* 63: 116-24.

Stenseth, N.C., A. Mysterud, G. Ottersen, J.W. Hurrell, K.S. Chan and M. Lima. 2002. Ecological effects of climate fluctuations. *Science* 297: 1292-96.

11

Coalition Games in Fisheries Economics

Marko Lindroos
Veijo Kaitala
Lone Grønbæk Kronbak

11.1 INTRODUCTION

The focus of this chapter is the analysis of coalitions, a topic that has received increasing attention in fisheries economics in the last ten years (Kaitala and Lindroos, 1998; Arnason *et al.*, 2000; Lindroos and Kaitala, 2000; Lindroos, 2004a; Burton, 2003; Brasão *et al.*, 2000). The need for the research of co-operative fisheries management arises from the current practice of international negotiations and implementation of multi-country fisheries agreements. In several locations worldwide, fishing nations aim at organising local multi-nation fisheries such that the outcome would satisfy all nations interested in participating in the fishery. In this process, it becomes pertinent to not only consider strategic importance of the international fisheries agreements to individual countries, but also to groups of countries. The 1995 United Nations Agreement on Straddling and Highly Migratory Fish Stocks (United Nations, 1995), known as the UN Fish Stocks Agreement, is a practical example of the importance of coalitional game theory. This agreement calls for the establishment of regional fisheries management organisations to manage straddling and highly migratory fish stocks. These organisations are formed by groups of countries – coastal states and relevant distant water fishing nations, i.e. coalitions.

In the recent literature there have been two main directions following the traditional division in game theory. Firstly, non-co-operative coalition games have concentrated on the endogenous formation of coalitions and coalition structures. Secondly, co-operative fisheries coalition games have concentrated on the allocation of co-operative benefits to different countries involved. We shall follow this division in the present review.

The content of the chapter is as follows. In section 11.2 we demonstrate why coalition games are needed and recall the first applications of coalition fisheries games. Section 11.3 discusses the most recent developments in finding equilibrium coalition structures for non-co-operative coalition fisheries games. Section 11.4 then discusses and contrasts the co-operative coalition games applied in the theory and applications of international fisheries management. Finally, section 11.5 concludes and in particular discusses the need for merging these two approaches. Indeed, it is often difficult to make distinction between purely co-operative or non-co-operative modes of games. Therefore, we conclude that the related game theory is not fully mature, and developing a unifying framework would be essential.

11.2 THE NEED FOR COALITION GAMES

Many important contributions in fisheries economics can be traced back to Professor Gordon Munro's exceptional research activity (see Chapter 1). The analysis of coalition games is not an exception in this respect. He was alert when the negotiations processes on the United Nation's New Law of the Sea were active (Kaitala and Munro, 1993). Observing the processes from outside of negotiation process, Professor Munro foresaw that many interesting scientific problems arose during this process. One of these was the emergence of different interest groups in the negotiations. This observation leads to the conclusion that the analysis of coalitions should be brought into fisheries economics (Kaitala and Munro, 1995). The emergence of a new international fisheries agreement on straddling and highly migratory fish stocks (UN, 1995) confirmed that Professor Munro was right in advertising that the new legal environment for the management and utilisation of marine resources left behind a bunch of economic problems that need to be fully understood (Kaitala and Munro, 1997; Kaitala and Lindroos, 1998; Bjørndal et al., 2000; Brasão et al., 2000).

The theory of fisheries games in 1995 had mainly been built on two-player games (Munro, 1979; Clark, 1980; for reviews, see Munro, 1990 and Sumaila, 1999), which has its roots in completing the economic theory related to the property-right problems created by the establishment of the exclusive economic zones and the United Nation's Law of the Sea (1982). The emphasis was in defining fair strategies for exploiting shared fisheries where the number of countries exploiting the stocks was fairly low and in many cases equal to two. However, the agreement on straddling and highly migratory fish stocks made it necessary to study game-theoretic models of more than two players. This was because the potential number of countries involved in high-seas fisheries agreements can be very large.

The interloper problem (Bjørndal and Munro, 2003) is another serious issue threatening efficient and sustainable use of high seas fisheries. The fishing nations having interest in specific fisheries do not necessarily have to be members of regional fisheries management organisations. Coalitional game theory can explain the economic origin of the interloper problem and suggest ways to overcome the problem.

Further, overexploitation is an extremely important issue in international fisheries. Without clear international agreements there is hardly hope for sustainable use of marine fisheries. The negotiations on high-seas fisheries thereby created the foundation for discussing coalition formation and setting harvesting strategies.

11.3 NON-CO-OPERATIVE COALITION GAMES: COALITION FORMATION AND SETTING HARVESTING STRATEGIES

The merits of applying coalition games in analysing fisheries economics is next illustrated by a simplified example. Let us assume that we have four countries exploiting a common fish stock. (This is close to the case of exploitation of Norwegian spring spawning herring (Bjørndal et al., 2000).) The countries aim at maximising their individual net present values from the fishery. Each country participates in international negotiations where countries may or may not be successful in achieving either bilateral or multilateral agreements.

This kind of multi-agent decision-making problem can be formulated as a coalition game (Mesterton-Gibbons, 2000). In a four-player example there are 15 possible coalitions that can be formed. Further, there are also 15 coalition structures that describe all possible outcomes of the negotiations. Coalition structures can be characterised as follows:

Let us assume that countries 1 and 2 find it beneficial to form a coalition (1,2). Thus, countries 1 and 2 act as a co-operative unit. The remaining two countries now have the option to either sign a bilateral agreement with each other and thus form a coalition (3,4) or formulate their own unilateral harvest strategies. Thus, in this case we have two possible coalition structures:

$$(1,2), (3,4)$$
$$(1,2), (3), (4).$$

Clearly, the benefits of coalition (1,2) must depend on the harvest decisions of the outside countries 3 and 4 since externalities are present in fisheries. If countries 3 and 4 sign a bilateral agreement then the fish stock and benefits are typically larger than in the case where the outside countries 3 and 4 choose their fishing strategies attempting to unilaterally maximise their own benefits individually.

Generally, there are three types of coalition structures: non-co-operation, partial co-operation and full co-operation. There is only one non-cooperative coalition structure where all countries maximise their own self interest. Similarly, there is only one full co-operative coalition structure where all countries maximise their joint benefits from the fishery. In between we have the partial co-operative cases. In our four-player case we have nine different coalition structures with two-country coalitions and four coalition structures with three-country coalitions.

The coalition game above can be solved as follows: we can think of the coalition game in two stages. In the first stage, the countries form coalitions with one another. In the second stage countries and coalitions compute their best strategies. The game is solved backwards.

Countries within coalition (1,2) maximise their joint benefits taking into account that they have to play against either coalition (3,4) or against two individual countries (3) and (4). The result will be a Nash equilibrium of the game where it is not profitable for any of the coalitions or countries to unilaterally deviate from the equilibrium strategies.

The procedure for solving the game has to be repeated for each possible coalition structure. Having solved all these games allows us to proceed to stage 1 to find the equilibrium coalition structure. Of course there may exist several equilibrium coalition structures, one equilibrium, or none at all.

The coalition is said to be stable if there is no country that finds it optimal to join the coalition (external stability) and if no country within the coalition finds it optimal to leave the coalition (internal stability). When determining the stability properties of the grand coalition it is sufficient to check for internal stability if there are no potential entrants in the fishery. Other ways of defining stability of coalitions also exist, but these are not addressed in the current review (see e.g. Finus, 2001).

The general framework of coalition fisheries games has been studied in particular by Pintassilgo (2003) who brought the theory a major leap forwards. He introduced the partition function approach to these games and hence formalised and generalised the existing applications in the literature.

Arnason *et al.* (2000), Lindroos and Kaitala (2000) and Lindroos (2004b) analysed Norwegian spring spawning herring fishery as a coalition game. Arnason *et al.* (2000) showed that Norway is a veto-country in the stability sense in the herring fishery. This follows from the fact that all stable coalitions include Norway as a member. Lindroos and Kaitala (2000) were the first to compute Nash equilibria for coalition fisheries games. Finally, Lindroos (2004b) studied the connections between safe minimum biological levels (SMBL or B_{lim}) and stability of full co-operation.

11.3.1 Example 1

We use the analysis of the Norwegian spring spawning herring fishery to illustrate the use of coalition games in fisheries economics (Lindroos and Kaitala, 2000). Table 11.1 summarises the main results, indicating that full co-operation is not stable. It follows that it is not possible for all countries to sign a multilateral agreement since the incentives to free-ride are large.

The Lindroos and Kaitala (2000) model is based on an age-structured model with Beverton-Holt stock recruitment function. The catch in numbers for country i and for a specific cohort is given by

$$Y_y^i = \sum_{a=1}^{a=16} Y_{a,y}^i = \sum_{a=1}^{a=16} CW_a N_{a,y} \frac{S_a F_y^i}{M_a + S_a F_y}(1-e^{-m_a - S_a F_y}) \qquad (11.1)$$

Thus, we have 16 age classes. Subscript a denotes the age classes whereas subscript y the simulation period (year). Parameter CW gives the weight of each age class caught, N is the number of fish, S is selectivity, M natural mortality and F fishing mortality for each country. The net present value each country or coalition of countries is maximising is:

$$J^i(F^i) = \sum_y \frac{pY_y^i - Q_y^i}{\rho_y} \qquad (11.2)$$

where $\rho_y = (1+r)^{y-y_1}$ is the discount factor and r the discount rate. Further, Q is total costs and p is price per kg of herring.

Nash equilibria are calculated for each possible coalition with respect to the fishing mortality. The first three values in Table 11.1 correspond to the equilibrium of a three-player game among all three countries. The next three values correspond to three different games where a two-player coalition plays against the free-rider country. Finally, the last value corresponds to a case where the three countries maximise their joint benefits from the spring spawning herring fishery. In Table 11.1, value corresponds to the net present value for each country or coalition from the fishery (see equation 11.2).

Table 11.1 illustrates the value of the different coalitions. It is clearly seen that by co-ordinating effort (joining a coalition) the payoff to the group increases compared to a non-co-operative solution. This is the main motivation for discussing coalitions. What is of particular interest in Table 11.1 are the values, which the singleton outside a two-player coalition, called free rider, is able to receive. This differs from the values the players can receive, when they all act as singletons. The difference is, as discussed earlier, caused by externalities being present in fisheries. If a player receives a payoff in the grand coalition which is smaller than what the player can receive from free riding on the grand coalition, then the player will not join the grand coalition (assuming rationality). Therefore, if and only if the benefits from the grand coalition exceed the sum of benefits from free riding, then there are enough benefits in the grand coalition to be distributed in a satisfactory way, such that the grand coalition is stable (Pintassilgo, 2003). Thus in the example shown in Table 11.1 the full co-operation is not stable.

Table 11.1 Full co-operation is not stable. Norwegian spring spawning herring fishery.

Coalition	Value	Free rider value
(1)	4,878	
(2)	2,313	
(3)	896	
(1,2)	19,562	14,534 (country 3)
(1,3)	18,141	17,544 (country 2)
(2,3)	17,544	18,141 (country 1)
(1,2,3)	44,494	50,219 (sum of above)

Note: Values are in million NOK (Lindroos and Kaitala, 2000).

Brasão et al. (2000) applied the coalition game approach to Northern Atlantic bluefin tuna fisheries. They constructed a dynamic model where they assumed that non-co-operative behaviour was equal to open access. Further, Pintassilgo and Duarte (2000) studied the problem of new members (Kaitala and Munro, 1993 introduced the new-member problem) into regional fisheries management organisations.

The new-member problem follows from the 1995 UN Fish Stock Agreement (UN, 1995) on straddling and highly migratory fish stocks. According to the agreement any country with real interest in the fishery can participate and thereby receive some share of the benefits. However, Kaitala and Munro (1993) showed that this could lead to a situation close to open access and clearly there must be some mechanisms to protect the fishery from too many countries exploiting it. Among others a membership fee and waiting period have been suggested as this kind of a mechanisms (Bjørndal and Munro, 2003).

To illustrate the new member problem in the context of straddling stocks, assume that a potentially valuable stock has been depleted as a consequence of unrestricted exploitation. Assume further that, following the advice of the UN 1995 Fish Stocks Agreement, a fixed number of fishing nations decide to establish a regional fisheries management organisation (RFMO) to manage the fish stock and among other things, to improve the economic efficiency of the fishery. Thus, the RFMO represents a coalition of countries each of which have a real interest to enter this fishery. Typically the members of the RFMOs are coastal states with a long tradition in exploiting the fish stock in question. The real interest to enter any fish stock is not, however, a static label of any fish stock or the fishing nations as such, but it is the economic interest which may vary with the condition (e.g. size) of the fish stock or the local management policies. Thus, it may be expected that when the RFMO succeeds in increasing the size of the fish stock and increasing the potential economic income from the fishery, real interest to enter the fishery will arise within countries that do not belong to the RFMO. These late-comers are referred to as new entrants (Kaitala and Munro, 1993), which in many cases are distant water fishing nations. Since any increase in the number of the members in the RFMO will decrease the income of the members from the fishery, this process may create conflicts among the fishing nations unless RFMOs are allowed to develop mechanisms to prevent new members entering.

The northern Atlantic bluefin tuna fishery presents a very striking example. Currently the fish stock is in a depleted state and between 25-30 countries participate in the fishery. Historically, nearly 60 countries have been active. If and when the stock increases, the geographical distribution of the stock will widen and it will enter the coastal waters of a number of countries where it has been absent for decades. Obviously this is likely to give rise to new entry to the fishery (Bjørndal and Brasão, 2006).

In the context of coalitions, allowing for new members would mean that the stability properties of various coalitions would change if a new member entered the fishery. Lindroos (2002) has shown that this may even lead to a situation where a new member improves stability of co-operation within a regional fisheries management organisation. However, it may also be that the appearance of a new member would lead to another competing organisation that would play against the original one, if one or several of the original members would join the new member. Hannesson (1997) showed that generally the incentives to deviate from full co-operation increase with the number of players.

Kronbak and Lindroos (2003) combined the enforcement of regulation with the coalition formation by setting up a one-shot game with four stages. This involves a model where two different groups of agents are present, namely the authorities and the fishermen. These two groups act sequentially as in two ordinary coalition games. The model is formulated as a Schäfer-Gordon type of game with logistic growth. Firstly, a two stage model for authorities is solved, where they form coalitions and then compute their best level of enforcement effort to be applied. Secondly, a two stage model for fishermen is solved; fishermen determine their coalition formation and then their best harvest strategies are computed.

11.3.2 Example 2

This example illustrates the importance of also including the management aspects when discussing stability of coalition formation. The example illustrates some of the main results from combining enforcement of regulation and coalition formation (Kronbak and Lindroos, 2003). Figure 11.1 shows how the profits change for respectively a grand coalition (GC), the sum of profits from free riding (F) and the sum of profits from a non-co-operative game (N) between three players as the control effort (Z) varies from 0% to 100%. Control effort means here the actions government undertakes to increase the probability of illegal fishermen getting caught.

Figure 11.1 illustrates how the applied level of control effort affects the stability conditions for a grand coalition. The figure shows that only for low levels of control effort do the benefits from the grand coalition exceed the benefits from playing a Nash game or free riding. Thus, the simulation model shows that for low values of control effort the fishermen will organise adequate control themselves by joining together. For high values of control effort fishermen will act as singletons. The intuition is that the gain from cooperation is much larger if there is no control. This example shows the great importance of also including management issues when discussing the stability of cooperative games.

11.4 CO-OPERATIVE COALITION GAMES: SHARING OF BENEFITS

The previous section described typical non-co-operative two-stage games. However, a third stage could also be incorporated in fisheries coalition games. This is the sharing of co-operative benefits. Typically, in these models co-operation between all countries involved gives the largest overall benefits. However, the tragedy of the commons and prisoner's dilemma types of games predict that it will be extremely difficult or even impossible to escape non-co-operation. Often this can be avoided by allowing for side payments and thereby a redistribution of benefits inside the coalition. Munro (1979) showed for a two-player game the usefulness of side payments.

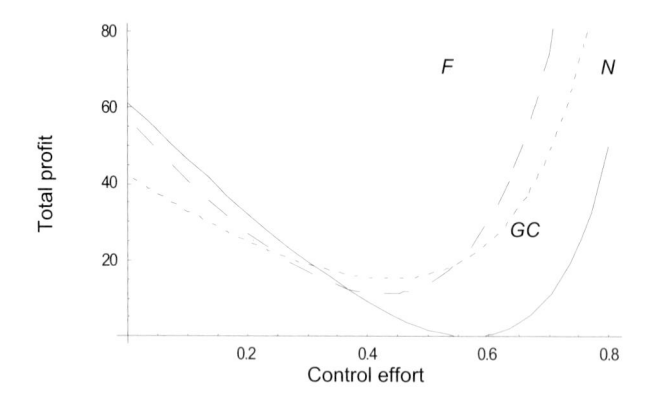

Figure 11.1 Benefits from the grand coalition (*GC*), sum of benefits from free riding (*F*), sum of benefits from three singletons (*N*) (Kronbak and Lindroos, 2003).

Sharing of co-operative benefits can be introduced in coalition games by using co-operative game theory. In particular, co-operative game theory offers several axiomatic solution concepts that can be used to compute the shares of co-operative benefits for each country. These solutions include Shapley value (probably the most used, see Shapley, 1953) and nucleolus (Schmeidler, 1969). The Shapley value for a single player is defined as the potential to change the worth of the coalition by joining or leaving it, that is the expected marginal contribution. The nucleolus minimises the dissatisfaction of the most dissatisfied coalition.

Having decided upon the co-operative solution it is then clear that it will affect the decisions made in earlier stages 1 and 2. Note that stage 3 can not be optimised; there is no optimal way to share benefits. However, it is possible to compare different solution concepts with respect to their stability properties, i.e. which co-operative solutions achieve full co-operation.

Kaitala and Munro (1995) analysed a dynamic co-operative game allowing for coalitions between three countries. They correctly anticipated the results that were later formalised by Kaitala and Lindroos (1998). Kaitala and Lindroos (1998) applied two co-operative solutions to the game proposed by Kaitala and Munro (1995).

In Kaitala and Lindroos (1998) the objective of each of these three countries is given in equation (11.6), namely maximising their net present values from the fishery subject to the resource constraint.

$$\max J_i = \int_0^\infty e^{-rt}\left[p - \frac{c_i}{x}\right]h(t)dt \qquad (11.6)$$

$$\text{s.t. } \dot{x}_t = F(x) - h_C(t) - h_{D1}(t) - h_{D2}(t)$$

Term r is the discount rate, p is the constant price, x is the fish stock, $F(x)$ is biological growth of the stock. Finally, $h(t) = E(t)x(t)$ is the production function, i.e. harvest. C denotes a coastal state, while $D1$ and $D2$ are distant water fishing nations (DWFNs).

Assume further that countries have successfully engaged in negotiations and an agreement is binding to each country. The remaining question is how to share benefits according to some reasonable and fair co-operative solution so that each country will be satisfied.

Before proceeding to sharing rules a characteristic function must be constructed. The characteristic function assigns a value to each possible coalition. The value measures the normalised net present value of each coalition. Value 1 is equal to the maximised net present value of the most efficient fishing nation (coastal state). In the current game the normalised characteristic function is such that only one coalition in addition to grand coalition has a positive value:

$$v(\{C, D_2\}) > 0. \qquad (11.7)$$

In this case a coalition including the coastal state and the more efficient distant water fishing nation is able to maintain the stock level at a level higher than the non-cooperative stock level.

Employing the Shapley value yields a co-operative solution to each country. Shapley value is one alternative and commonly used cooperative solution to the game:

$$z_C^S = z_{D_2}^S = v(\{C, D_2\})/6 + 1/3 \qquad (11.8)$$
$$z_{D_1}^S = [1 - v(\{C, D_2\})]/3.$$

Thus, the two most efficient countries should receive a share higher than one third of co-operative benefits. This is due to their higher contribution to the overall coalitional values. No other two-player coalition is able to obtain a positive value.

This means that the co-operative benefits should be shared differently than in the earlier two-player games. The dominating coastal country of the high-seas fishery game would require more than a third of the benefits since its contribution to each possible coalition is the largest.

Lindroos (2004a) extended the model to four players and assumed that coalition formation is restricted such that coastal states and distant water fishing nations only negotiate as a group. Lindroos (2004a) applied the concept of veto-coalitions, that is, countries or coalitions that are essential to the game either to make a coalition stable or to create a positive value for a coalition. The concept of veto-coalitions is connected to the study of Arnason *et al.* (2000) where Norway was found to be a veto-country in stability sense. However, in Lindroos (2004a) veto-coalition means a coalition with positive bargaining strength.

We concentrate here on a particular case where:

$$c_{C_1} < c_{C_2} < c_{D_1} = c_{D_2}. \tag{11.9}$$

The restricted Shapley values are now:

$$z_{C_1}^{S_R} = z_{C_2}^{S_R} = \frac{1}{4} + \frac{v(\{C_1, C_2\})}{12} \tag{11.10}$$

$$z_{D_1}^{S_R} = z_{D_2}^{S_R} = \frac{1}{4} - \frac{v(\{C_1, C_2\})}{12}. \tag{11.11}$$

The corresponding unrestricted Shapley values are:

$$z_{C_1}^{S} = z_{C_2}^{S} = \frac{1}{4} + \frac{v(\{C_1, C_2\})}{4} \tag{11.12}$$

$$z_{D_1}^{S} = z_{D_2}^{S} = \frac{1}{4} - \frac{v(\{C_1, C_2\})}{4} \tag{11.13}$$

Comparing the unrestricted and restricted Shapley values (equations 11.11 and 11.13) we see that restrictions are beneficial to the DWFNs. However, the restricted Shapley values are in the core only if:

$$\frac{3}{5} \geq v(\{C_1, C_2\}). \tag{11.14}$$

Thus, in the case of superior coastal states coalition restrictions are not stable. This is because core conditions are violated and hence for one or both of the coastal states it is better to leave full co-operation.

Kennedy (2003) applied coalition games to the international mackerel fishery. He applied the coalition-proof Nash equilibrium approach for the first time to fisheries coalition games and in addition various co-operative solutions.

Among other previous empirical work is the determination of the Shapley value to the players participating in the coalition game of the Norwegian spring spawning herring (Arnason *et al.*, 2000) and the evaluation of the Shapley value and the nucleolus to players in the Northern Atlantic bluefin tuna fishery (Duarte *et al.*, 2000). Brasão *et al.* (2000) also apply the coalition game to the Northern Atlantic bluefin tuna. They recognise the instability of the Shapley value due to free-rider incentives and they instead suggest a non-co-operative feedback Nash with side payments. Exactly the instability of the sharing rules due to free rider incentives leads to the future challenges.

11.5 FUTURE CHALLENGES: MERGER BETWEEN NON-CO-OPERATIVE AND CO-OPERATIVE FISHERIES GAME THEORY

What distinguishes fisheries coalition games from many other coalition games is that externalities are present in fisheries. If externalities are not present the decisions made by the coalitions can be assumed to be independent of the actions by the non-members

(Greenberg, 1994). If, however, externalities are present, as in fisheries coalition games, then the most important challenge in general is the merger between non-co-operative and co-operative coalition games. There is a large literature on both sides of coalition theory and it is fairly well understood, but there are several aspects that need a combined analysis. Kronbak and Lindroos (2005) provide an analysis of the stability properties of nucleolus and the Shapley value by showing that the basic idea for the two solution concepts, namely the ability to make a credible threat, can be undermined if externalities are present. Kronbak and Lindroos (2005) develop an alternative sharing rule by merging the non-co-operative and the co-operative game applying the free-rider values as threats. They use the Baltic cod as an illustrative example. The underlying biology is an age-structured model as in e.g. Lindroos and Kaitala (2000) above.

11.5.1 Example 3

This example illustrates, by applying the results from the Baltic Sea cod fishery (Kronbak and Lindroos, 2005), how instability can be imposed by the sharing rules on an otherwise stable grand coalition.

Table 11.2 shows in percentages how large a share of the benefits in the grand coalition each player should receive given different sharing rules or free riding. The grand coalition in itself is stable since the benefits exceed the sum of benefits from free riding.[1] Instability does, however, occur if the benefits in the grand coalition are shared according to the Shapley value or the nucleolus. The reason for this is that they are not taking externalities into account, which changes the belief of what is a credible threat. Applying the free rider value as a credible threat instead of the singleton value allows Kronbak and Lindroos (2005) to develop an alternative sharing rule, the satisfactory nucleolus, which is stable to free riding.

Table 11.2 The sharing functions in the Baltic Sea cod fishery.

Player	Shapley value	Nucleolus	Free rider shares	Satisfactory nucleolus
1	35.9 %	33.3 %	38.1 %	40.3 %
2	32.3 %	33.3 %	28.2 %	30.4 %
3	31.8 %	33.3 %	27.1 %	29.3 %

Note: The results are indicated in percentages of the benefits in the grand coalition. The numbers are subject to rounding (source: Kronbak and Lindroos, 2005).

Comparing Table 11.2 with Table 11.1 we immediately notice an important difference with respect to stability. All previous fisheries coalition games have predicted an unstable grand coalition (full co-operation) like Lindroos and Kaitala (2000) illustrated in Table 11.1. However, Kronbak and Lindroos (2005) predict a stable grand coalition. This raises an important question for future research: what factors influence stability of full co-operation? These may include biological or economic factors or a combination of these.

---

[1] The stability of the grand coalition can be verified by the fact that there exists a sharing rule (satisfactory nucleolus) where the shares to each player exceed the free rider shares.

The problem of new entrants or new members deserves much more attention in the analysis of coalition formation in regional fisheries management organisations. It would be important to gain more knowledge about cases in which new members are a serious problem and cases in which new members do not threaten successful international co-operation. Another important future challenge is time consistency in dynamic coalition games. Dementieva (2004) provides new methods for co-operative multi-stage games. However, there is a strong need to incorporate these methods into the theory of non-co-operative coalition games. Further, allowing for uncertainty and renegotiations are among the important extensions of the coalition fisheries models.

Acknowledgements

We gratefully acknowledge the comments of two anonymous referees.

References

Arnason, R., G. Magnusson and S. Agnarsson. 2000. The Norwegian spring-spawning herring fishery: A stylized game model. *Marine Resource Economics* 15: 293-319.
Bjørndal, T. and A. Brasão. 2006. The northern Atlantic bluefin tuna fisheries: Management and policy implications. *Marine Resource Economics* (forthcoming).
Bjørndal, T., V. Kaitala, M. Lindroos and G. Munro. 2000. The management of high seas fisheries. *Annals of Operations Research* 94: 183-96.
Bjørndal, T. and G.R. Munro. 2003. The Management of high seas fisheries. In H. Folmer and T. Tietenberg. (eds) *The International Yearbook of Environmental and Resource Economics 2003/2004* Elgar, Cheltenham, UK: 1-35.
Brasão, A., C.C. Duarte and M.A. Cunha-e-Sá. 2000. Managing the northern Atlantic bluefin tuna fisheries: The stability of the UN fish stock agreement solution. *Marine Resource Economics* 15(4): 341-60.
Burton, P.S. 2003. Community enforcement of fisheries effort restriction. *Journal of Environmental Economics and Management* 45 (2): 474-91.
Clark, C.W. 1980. Restricted access to common property fisheries resources: A game-theoretic analysis. In P. Liu (ed.) *Dynamic Optimization and Mathematical Economics*. Plenem Press, New York.
Dementieva M. 2004. *Regularization in Multistage Cooperative Games*. PhD thesis. University of Jyväskylä, Finland.
Duarte, C.C., A. Brasão and P. Pintassilgo. 2000. Management of the northern Atlantic bluefin tuna: An application of C-games. *Marine Resource Economics* 15: 21-36.
Finus, M. 2001. *Game Theory and International Environmental Cooperation*. Edward Elgar, Cheltenham, UK.
Greenberg, J. 1994. Coalition Structures. In R.J. Aumann and S. Hart (eds) *Handbook of Game Theory with Economic Applications*, Vol 2, Chapter 37. Elsevier, Amsterdam.
Hannesson, R. 1997. Fishing as a supergame. *Journal of Environmental Economics and Management* 32: 309-22.
Kaitala, V. and M. Lindroos. 1998. Sharing the benefits of cooperation in high sea fisheries: A characteristic function game approach. *Natural Resource Modeling* 11: 275-99.

Kaitala, V. and G.R. Munro. 1993. The management of high seas fisheries. *Marine Resource Economics* 8: 313-29.

Kaitala, V. and G.R. Munro. 1995. The economic management of high seas fishery resources: Some game theory aspects. In C. Carraro and J.A. Filar (ed) *Annals of the International Society of Dynamics Games: Control and Game-Theoretic Models of the Environment.* Birkhauser, Boston: 299-318.

Kaitala, V. and G.R. Munro. 1997. The conservation and management of high seas fishery resources under the new Law of the Sea. *Natural Resource Modeling* 10: 87-108.

Kennedy, J. 2003. Scope for efficient multinational exploitation of north-east Atlantic mackerel. *Marine Resource Economics* 18: 55-80.

Kronbak, L.G. and M. Lindroos. 2003. An enforcement-coalition model: Fishermen and authorities forming coalitions. *IME Working Paper no. 50/03*, Department of Environmental and Business Economics, University of Southern Denmark.

Kronbak, L.G. and M. Lindroos. 2005. Sharing rules and stability in coalition games with externalities: The case of the Baltic Sea cod fishery. *Department of Economics and Management discussion paper no. 7*, University of Helsinki, Finland.

Lindroos. M. 2002. Coalitions in fisheries. *Working paper W-321*, Helsinki School of Economics, Finland.

Lindroos, M. 2004a. Restricted coalitions in the management of regional fisheries organisations. *Natural Resource Modeling* 17: 45-70.

Lindroos, M. 2004b. Sharing the benefits of cooperation in the Norwegian spring-spawning herring fishery. *International Game Theory Review* 6: 35-54.

Lindroos, M. and V. Kaitala. 2000. Nash equilibria in a coalition game of the Norwegian spring-spawning herring fishery. *Marine Resource Economics* 15: 321-40.

Mesterton-Gibbons, M. 2000. *An Introduction to Game-Theoretic Modelling*, 2nd edition. American Mathematical Society, RI.

Munro, G.R. 1979. The optimal management of transboundary renewable resources. *Canadian Journal of Economics* 12: 355-76.

Munro, G.R. 1990. The optimal management of transboundary fisheries: Game theoretic considerations. *Natural Resource Modeling* 4: 403-26.

Pintassilgo, P. 2003. A coalition approach to the management of high seas fisheries in the presence of externalities. *Natural Resource Modeling* 16 (2): 175-97.

Pintassilgo, P. and C.C. Duarte. 2000. The new-member problem in the cooperative management of high seas fisheries. *Marine Resource Economics* 15 (4): 361-78.

Schmeidler, D. 1969. The nucleolus of a characteristic function game. In S. Hart and A. Neymann (eds) *Game and Economic Theory - Selected Contributions in Honor of Robert J. Aumann.* University of Michigan, Ann Arbor: 231-38.

Shapley, L. 1953. A value for n-person games. In A. Roth (ed.) *The Shapley value - Essays in Honor of Lloyd S. Shapley.* Cambridge University Press: 31-40.

Sumaila, U.R. 1999. A review of game-theoretic models of fishing. *Marine Policy* 23: 1-10.

United Nations. 1982. Convention on the Law of the Sea. United Nations, New York.

United Nations. 1995. United Nations Conference on Straddling Fish Stocks and Highly Migratory Fish Stocks. Agreement for the Implementation of the United Nations Convention on the Law of the Sea of 10 December 1982 Relating to the Conservation and Management of Straddling Fish Stocks and Highly Migratory Fish Stocks. UN Doc. A/CONF./164/37. United Nations, New York.

12

Incentive Compatibility of Fish-sharing Agreements

Rögnvaldur Hannesson

12.1 INTRODUCTION

The view that the sharing of fish stocks can be solved easily, fairly, and 'scientifically', by looking at how much of each stock is located in each country's exclusive economic zone (EEZ) is not uncommon among fisheries biologists and others.[1] This is sometimes referred to as the 'zonal attachment principle'. Unfortunately, the issue is more complicated than that. The basic lesson of the game theoretic approach is that each country's share in a co-operative solution must be at least as good as its threat point, that is, what it would get on its own.[2] This chapter is concerned with whether or not the zonal attachment principle does pass this test. This question has not, to my knowledge, been addressed in the literature. Bjørndal and Lindroos (2004) consider a bargaining solution for the North Sea herring fishery and show that the Norwegian share, which is based on the zonal attachment principle (cf. below), is not compatible with the bargaining solution.

Even if, as we shall see, the incentive compatibility of the zonal attachment principle is not a foregone conclusion, it has in fact been successfully applied. In the late 1970s Norway and the European Union agreed to share seven transboundary stocks in the North Sea according to the zonal attachment of each stock.[3] The stocks are managed by setting an overall total allowable catch (TAC), which is then divided between Norway and the European Union as determined by the zonal attachment.

Zonal attachment can be defined and measured in various ways, and precisely how this is done can be controversial. Some fish may be spawned in the economic zone of one country while not becoming fishable until they have moved into the zone of another. Other types of fish may feed in the zone of one country but be fishable mainly in the zone of another. In the agreements between the European Union and Norway, zonal attachment was based on the presence of the fishable part of the stocks in each party's zone in the years 1974-78 (Engesæter, 1993). In other contexts different approaches have been applied. One such uses biomass multiplied by the time

---

[1] See, for example, Engesæter (1993).
[2] Gordon Munro (1979) was one of the first to apply game theory to the problem of sharing fish stocks, see also Vislie (1987).
[3] These stocks are cod, haddock, saithe, plaice, whiting, sprat and herring (Engesæter, 1993). On the concept of zonal attachment, see ICES (1978) and Engesæter (1993).

migrating stocks spend in each country's zone (Hamre, 1993). This was applied in the sharing of the capelin stock that migrates between the zones of Greenland, Iceland, and Jan Mayen, an island under Norwegian sovereignty (Engesæter, 1993). Instead of biomass this approach could be based on the growth of the stock (Hamre, 1993).

With the exception of North Sea herring, the sharing agreement for the North Sea stocks has held up well. Like other herring stocks, the North Sea herring stock fluctuates considerably in size because of environmental factors, and it changes its migratory behaviour as it becomes more abundant. When the stock recovered in the 1980s from the breakdown in the 1970s it started to migrate further north and to a greater extent into the Norwegian EEZ. This made Norway unhappy with the four percent share she was being offered on the basis of the previous zonal attachment of the stock. For some time no agreement was in force, and Norway fished the stock at will within its EEZ after the herring moratorium was lifted in 1984. In 1986 a new agreement was concluded, giving Norway a share of 25, 29 or 32 percent, depending on the size of the spawning stock, the more the larger the stock is (Engesæter, 1993).

There have been other and less successful attempts at applying the zonal attachment principle. No agreement on sharing the total permitted catch has yet been obtained for blue whiting and mackerel in the north-east Atlantic, and an agreement on sharing the Norwegian spring spawning herring fell apart in 2003. A complicating factor is that these stocks migrate into the high seas outside the EEZs of coastal countries where no single country has jurisdiction and international agreements are therefore difficult to enforce.

However, there are other problems with the zonal attachment principle. Here it is shown that countries with a minor share in a stock could be better off by exploiting the fish in their own zone as they best see fit than by co-operating on the basis of the zonal attachment principle. This is particularly likely to happen for stocks where the unit cost of fish is only weakly related to the size of the exploited stock. This apparently is the case for the said stocks for which no agreement is in force.[4] These incentive problems could be the reason, rather than the fact that the stocks involved are accessible on the international high seas. What is perhaps more surprising is that the sharing agreements for the stocks in the North Sea have held up so well despite being based on the principle of zonal attachment.

Here it will be assumed that the zonal attachment of fish stocks is constant and independent of how intensively they are exploited. This is probably the case most favourable for the incentive compatibility of the zonal attachment principle. Controversies over how to share fish catches would seem more likely to happen if the zonal attachment varies randomly, or if it depends on the intensity of exploitation. As to the latter, it could, for example, be the case that a more intensive exploitation by one particular country would prevent the stock from migrating into the EEZs of other countries, as may be the case for the Norwegian spring spawning herring (see Hannesson, 2005 and Bjørndal et al., 2004). The incentive to exploit such a stock co-operatively would in that case obviously be less than otherwise. Another complication could arise from age-dependent migration patterns, such as young fish being recruited from country A's zone into country B's zone, from where they would migrate back into country A's zone as they grow older. These settings will not be further discussed

---

[4] Bjørndal (1987) has estimated a production function for herring that implies a weak dependence of the unit cost of fish on the size of the stock. This is due to fishing on fish aggregations (schools) that are relatively easily detected. This also characterises the blue whiting fishery.

here, as we are concerned with analysing the zonal attachment principle on its most favourable terms.

Below two cases are considered. In the first, the unit cost of fish is treated as a constant and independent of the size of the exploited stock. In the second case the unit cost is inversely proportional to the size of the exploited stock. This latter case serves to illustrate, in a simple and transparent way, the importance of stock-dependence of unit cost. It turns out, not unexpectedly perhaps, that this can be critical for the non-co-operative solution.

12.2 STOCK-INDEPENDENT UNIT COST

12.2.1 The co-operative solution

The model to be used is formulated in discrete time, with fishing and growth being regarded as sequential processes and fishing taking place prior to growth. Let X denote the stock at the beginning of each period and S the stock when the fishing is over, i.e. $S = X - H$ where H is the catch of fish. The initial stock in period t therefore is:

$$X_t = G(S_{t-1}) + S_{t-1} \qquad (12.1)$$

where $G(.)$ is the surplus growth of the fish stock.

At $t = 0$ we start with some stock X_0 inherited from the immediate past. If the initial stock is sufficiently great, the net price[5] (p) of fish is constant, and there is sufficient fishing capacity, it would be optimal to deplete the stock in the first period to the optimal level (S^o), to be maintained ever after.[6] With an infinite time horizon and a constant discount rate (r), the present value of the future rents from the fishery would be $pG(S^o)/r$. Assuming that the objective is to maximise the present value (V) of rent, the co-operative solution to the stock management problem can be formulated as:

$$\text{maximise } V = p(X_0 - S) + pG(S)/r. \qquad (12.2)$$

The optimum solution is given by:

$$G'(S^o) - r = 0 \qquad (12.3)$$

that is, in the optimal steady-state the marginal return on the stock should be equal to the return on alternative assets, given by the discount rate.

---

[5] That is, the market price less the cost per unit of fish.
[6] In the numerical examples below, the initial stock is set equal to the returning stock in equilibrium, i.e. $X_0 = G(S) + S$. In this paper, S is taken as the strategic variable. Other candidates for strategic variables are catch and effort. Using S as a strategic variable implies that there is sufficient fleet capacity to catch whatever is needed to bring the stock down to the desired level. In a deterministic model like the one used here, this would only be a problem if the initial adjustment of the stock requires a greater catch than in equilibrium and would in any case be transient. Catch would be a candidate for strategic variable if the price depends on the catch volume, a possibility not dealt with here.

12.2.2   The non-co-operative solution

Now let us assume that there are two players (countries) sharing the stock, both having the same net price of fish. If the co-operative solution is realised, they must have agreed on a sharing parameter α, defined as the share of the dominant player in whose zone more than half of the stock is located. Here dominant player means the one with the largest share of the stock, according to the zonal attachment, and not first mover advantage. The question now is how α will compare to the zonal attachment parameter β reflecting the share of the stock actually in the economic zone of Player 1, the dominant player.

Since the stock is common for both players, the growth of the stock depends in some way on how much both players leave behind after fishing (growth is assumed to take place after the fishing is over). This is simplified to assuming that the stock growth depends on the sum of the stock components left behind in both players' zones. In the non-co-operative solution, each player maximises the net present value of his rents, for any given strategy of the other player. Player 1's maximisation problem therefore is:

$$\text{maximise } V_1 = p[\beta X_0 - S_1] + p[\beta G(S_1 + S_2) + \beta(S_1 + S_2) - S_1]/r \qquad (12.4)$$

and analogous for Player 2. Analogous to the optimal solution above, each player will drive the stock down to his long term optimal level in the first period, but the stock emerging at the beginning of each period is assumed to be distributed among the two players' zones according to the zonal attachment parameter β.

The optimal escapement (S_1^o) for Player 1 is given by:

$$\beta G'(S_1^o + \bar{S}_2) + \beta - 1 - r = 0 \qquad (12.5a)$$

and for Player 2

$$(1-\beta)G'(\bar{S}_1 + S_2^o) - \beta - r = 0 \qquad (12.5b)$$

with the bar over S indicating that the player takes the stock level in the other player's zone as given. These first order conditions could only be satisfied simultaneously for both players if $\beta = 0.5$. Otherwise, if the condition for the first player is satisfied, the condition for the second player must be negative. This is consistent with an equilibrium where the second player leaves nothing behind. If the condition for the second player is zero, the condition for the first player would be positive, and he would leave more fish behind until his condition becomes zero and that for the second player negative.[7] Hence we conclude that the first order condition can only be satisfied for the dominant player. For the minor player the expression will be negative, implying that he will take all of the stock that he finds in his zone. He will nevertheless be able to free ride on the dominant player and get some fish in every period, as some of the stock growth realised due to the dominant player leaving behind some of the stock in his zone will spill over into the minor player's zone.

---

[7] Provided $G(S)$ is well behaved; i.e. $G''(S) < 0$ for all S.

It may, furthermore, be noted that even the dominant player could have an incentive to wipe out the stock in his own zone in a competitive solution. Provided that $G''(S) < 0$, the critical value of β is (from [12.5a]):

$$\beta > \frac{1+r}{G'(0)+1}. \quad (12.6)$$

This is a variant of Clark's classic result for viability of a stock under time discounting (Clark, 1973). The dominant player gets a share β of the marginal growth of the stock. Therefore, $\beta[G'(0)+1] > 1+r$ is necessary in order to make the stock an interesting investment object for the dominant player. It may, furthermore, be noted that the discount rate is not the only issue. Setting $r = 0$ we note that $\beta > 1/(G'(0)+1)$. Hence, the share of the dominant player must exceed some minimum in order for the stock to be viable, even in the absence of discounting. For example, with a low marginal growth rate of five percent, the share of the dominant player would have to exceed 95% in order for the stock to be viable. This is a dramatic result and underlines that extinction of a fish stock becomes much more likely the greater the number of countries or management units sharing the stock.

12.3 Incentives to achieve the co-operative solution

Now to the question what it would take to persuade the minor player to accept the co-operative solution. Would a share $1 - \alpha = 1 - \beta$ be sufficient? Giving the minor player this share of the cooperative solution, his sustained profit would be:

$$(1-\beta)pG(S^o) \quad (12.7)$$

while his sustained profit in the non-cooperative solution is:

$$(1-\beta)p\left[G(S^*)+S^*\right] \quad (12.8)$$

where S^* is the stock level the dominant player leaves behind in the non-co-operative solution (the other player leaves nothing behind, as already demonstrated). From the first order condition for the dominant player (12.5a) and the condition for the globally optimal solution (12.3) we have:

$$G'(S^*) = \frac{1-\beta+r}{\beta} > r = G'(S^o) \quad (12.9)$$

which implies $S^* < S^o$ and $G(S^*) < G(S^o)$, but not necessarily $G(S^o) > G(S^*) + S^*$. Hence we cannot say anything in general about whether $1 - \alpha = 1 - \beta$ will ensure that the minor player will prefer the co-operative solution.[8]

---

[8] To fully compare the non-co-operative and co-operative solutions, we would need to include also the initial adjustment of the stock, i.e. $X_0 - S^o$ and $X_0 - S^*$. Since the latter is greater than the former, the non-co-operative solution becomes somewhat more attractive, but this does not

12.4 A numerical example

For a further illustration, consider a numerical example based on the well-rehearsed logistic growth equation:

$$G(S) = a\left[S_1 + S_2 - \left(S_1^2 + 2S_1S_2 + S_2^2\right)\right] \quad (12.10)$$

$$G'(S) = a\left[1 - 2(S_1 + S_2)\right] \quad (12.11)$$

where the carrying capacity of the environment has been normalised at unity. Let $a = 1$ and $r = 0.05$. In order to make the co-operative solution attractive, the minor player has to be offered a share in the profit from the co-operative solution which gives him at least as much as what he would get from the non-co-operative solution. Figure 12.1 shows what his share $(1 - \alpha)$ has to be.

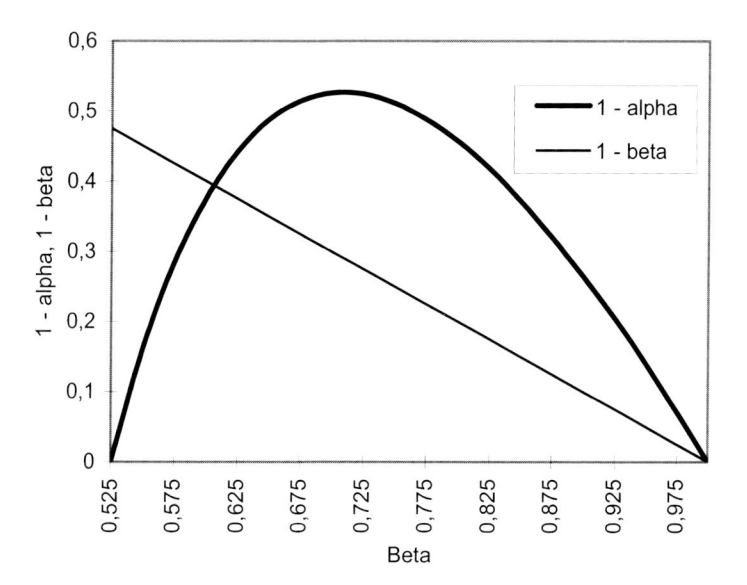

Figure 12.1 Minimum profit share $(1 - \alpha)$ of the minor player in the co-operative solution compared to his share $(1 - \beta)$ of the stock. Stock-independent unit harvesting cost.

For the most part it is higher than the minor player's zonal attachment coefficient $(1 - \beta)$, but for low values of β (note that $\beta \geq \frac{1}{2}$) it is in fact lower. The reason for the latter is that the dominant player would deplete the stock to a very low level if he does not have much more than half of it, making the non-co-operative solution a very unattractive one. A small share of the much more profitable co-operative solution would then be attractive for the minor player. An implication of this is that the zonal

---

change the general conclusion that sharing based on zonal attachment is not necessarily incentive-compatible.

attachment principle is most likely to work if the shares of the two players are not very different. In this example the minor player would be happy to accept a share of the co-operative profits that is equal to or even less than his share of the stock if the dominant player's share of the stock is less than 0.6.

Figure 12.2 shows the optimal escapement for the dominant player. With $\beta = 1$ it is identical to the sole owner solution. Furthermore we see that the critical value of β is 0.525. For the assumed value of a the dominant player must have at least 52.5% of the stock in his zone in order to ensure its viability in the non-co-operative solution, even if the minimum growth rate of the stock passes the minimum rate of return test by a wide margin. Another way to look at this is to say that the viability of shared stocks with unit costs independent of the stock size will be assured in a competitive equilibrium only if the dominant player controls more than 50% of the stock, and the more so the lower is the growth rate of the stock.

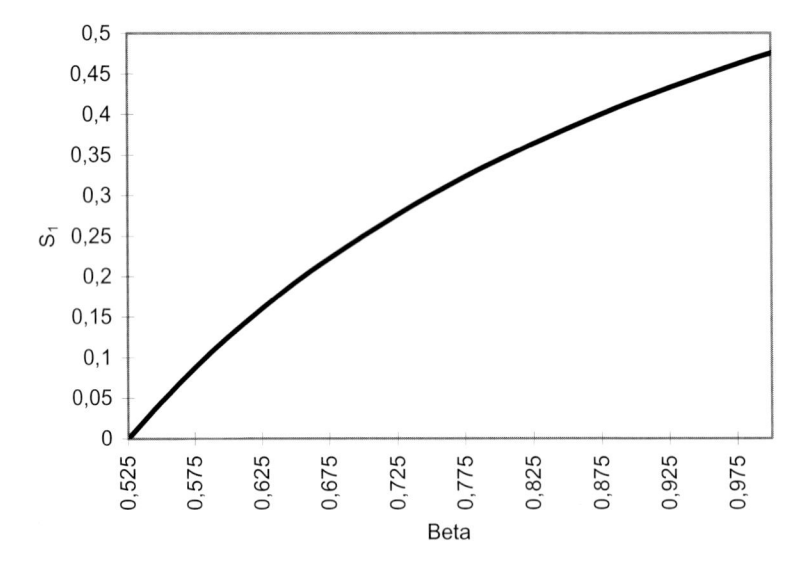

Figure 12.2   How the stock (S_1) optimal for the dominant player changes with his share (β) of the stock; stock-independent unit cost of fish.

12.3   STOCK-DEPENDENT UNIT COST

When the cost per unit of fish depends on the size of the exploited stock the co-operative solution is more likely to be achieved. The general reason is that both parties become more interested in fishing from a large stock in order to keep unit costs down. Furthermore, the non-co-operative solution will not result in a total depletion of the stock in any player's zone.

Here we shall look at a special but nevertheless interesting case where the cost per unit of fish is inversely proportional to the stock. This particular case comes from two assumptions; (i) the cost (c) per unit of fishing effort is constant, and (ii) the instantaneous catch is the product of effort and the stock times eventually a scaling parameter. We normalise effort (E) so that the scaling parameter is equal to one, in which case the instantaneous cost per unit of fish becomes $cE/ES = c/S$. With p now denoting the market price of fish, the net revenue (rent) from reducing the stock from X to S over a fishing season will be:

$$\int_S^X (p - c/s)\,ds = p(X - S) - c(\ln X - \ln S). \tag{12.12}$$

The co-operative solution now involves:

$$\begin{aligned}\text{maximise } V &= p(X_0 - S) - c(\ln X_0 - \ln S) \\ &+ \left[pG(S) - c(\ln(G(S) + S) - \ln S)\right]/r\end{aligned} \tag{12.13}$$

which yields the following first order condition for maximum:

$$p\left[G'(S^o) - r\right] + (r+1)c/S^o - c\left(G'(S^o) + 1\right)/\left(G(S^o) + S\right) = 0. \tag{12.14}$$

The above cost function implies that the stock is evenly distributed over the area it occupies. Fishing down the stock to its break-even level implies $S = c$. Hence, extinction is never compatible with profit maximisation irrespective of what the growth rate, the discount rate, or the dominant player's share of the stock might be.

With the same costs characterising both countries, the break-even stock levels in the two countries' areas will be βc and $(1 - \beta)c$ respectively. Hence the non-co-operative solution involves, for Player 1:

$$\begin{aligned}\text{maximise } V_1 &= p(\beta X_0 - S_1) - \beta c\left[\ln \beta X_0 - \ln S_1\right] \\ &+ \{p[\beta G(S_1 + S_2) + \beta(S_1 + S_2) - S_1]/r\} \\ &- \{\beta c\left[\ln \beta(G(S_1 + S_2) + (S_1 + S_2)) - \ln S_1\right]\}/r\end{aligned} \tag{12.15}$$

and analogous for the other player. From this we get the following first order condition for Player 1:

$$\begin{aligned}&p\left[\beta G'(S_1^o + \bar{S}_2) + \beta - 1 - r\right] \\ &+ \frac{\beta c(r+1)}{S_1^o} - \frac{\beta c\left[G'(S_1^o + \bar{S}_2) + 1\right]}{G(S_1^o + \bar{S}_2) + S_1^o + \bar{S}_2} = 0\end{aligned} \tag{12.16a}$$

and for Player 2:

$$\begin{aligned}&p\left[(1-\beta)G'(S_2^o + \bar{S}_1) - \beta - r\right] + \frac{(1-\beta)c(r+1)}{S_2^o} \\ &- \frac{(1-\beta)c\left[G'(S_2^o + \bar{S}_1) + 1\right]}{G(S_2^o + \bar{S}_1) + S_2^o + \bar{S}_1} = 0\end{aligned} \tag{12.16b}$$

where a bar over the other player's S means that it is taken as given. In contrast to the previous case of stock-independent unit cost, the first order conditions are now satisfied simultaneously for both players, and it is not optimal for either of them to wipe out the stock in his zone.

To analyse the incentive compatibility we resort to the same example as in the previous section, with $p = 1$ and $c = 0.2$, which implies a break-even stock level at 20% of the pristine state biomass. Figure 12.3 summarises the results. Also in this case it could be necessary to give the minor player a larger share in the co-operative profits than corresponds to his zonal attachment parameter. The difference between the zonal attachment parameter $(1 - \beta)$ and the required share in the co-operative profits $(1 - \alpha)$ is less here, however, than with stock-independent unit cost (cf. Figure 12.1). As in the previous case, the zonal attachment principle is most likely to work if the shares of the dominant and the minor player are not very different.

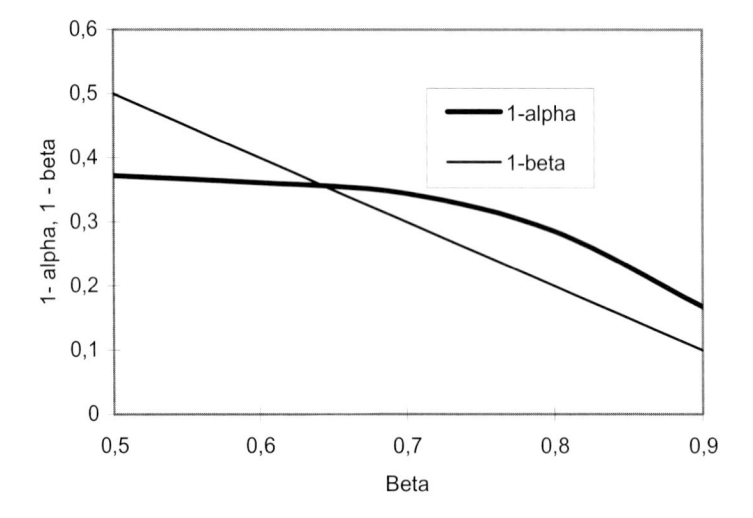

Figure 12.3   Minimum profit share $(1 - \alpha)$ of the minor player in the cooperative solution compared to his share $(1 - \beta)$ of the stock, with stock-dependent unit cost of fish.

12.4   CONCLUSION

It has been shown that fish stock sharing agreements based on zonal attachment need not be incentive compatible. It may be necessary to give a minor player a larger share in the co-operative profits, or the total permitted catch, than corresponds to his share of the stock, in order to make him better off than he would be in the absence of co-operation. This is all the more likely to happen the smaller is the minor player's share of the stock. The minor player will be able to free ride on the conservation efforts of the dominant player, and the conservation incentives for the dominant player will be stronger the larger is his share of the stock. Stock-dependent unit costs of fish make it less likely that the minor player will need to be enticed with a larger share in the co-operative profits than his share of the stock, or will at any rate diminish the necessary 'overcompensation.'

It is perhaps surprising, given these findings, that the stock sharing agreements for the North Sea stocks have been unchallenged for 20 years or more. For these stocks the unit costs of fish are, however, probably more sensitive to the stock size than for the stocks for which no agreements have been reached. This could explain why the agreements on the North Sea stocks have been so resilient. The North Sea herring is an interesting exception in this regard, for two reasons. First, the unit cost is insensitive to the stock size, which tends to widen the difference between the zonal attachment of the stock and the share a country has to be offered in the co-operative solution. Second, the zonal attachment depends on the size of the stock, which makes the principle more difficult to apply.

The results of this chapter could possibly explain why it has been difficult to reach or to sustain agreements on some stocks in the north-east Atlantic where the zonal attachment principle apparently is strong. The sharing agreement on the Norwegian spring spawning herring broke down recently, and no agreements have been reached for mackerel and blue whiting, despite repeated attempts. In these fisheries the unit costs are probably insensitive to the size of the stock. In addition, these stocks straddle into the high seas where the zonal attachment principle does not apply and the number of players is indeterminate. The problems associated with the high seas could be a sufficient reason why it has been difficult to reach or to maintain agreements on these stocks;[9] the fact that there is an agreement in place on North Sea herring would support that hypothesis. Attempts to base agreement on the said three stocks on the zonal attachment principle and the incompatibility of that principle with the incentives of the individual countries could, however, also be a reason.

This chapter has not analysed the problems arising from increasing the number of players. From other settings (see Hannesson, 1997) we know that this makes it more difficult to sustain co-operative solutions to the stock management problem. As to the number of players and the share of the co-operative profits they have to be offered compared to their zonal attachment parameters, there are two opposite effects. With more minor players and a strong conservation interest by the major player, each minor player would have to be offered a share in excess of his zonal attachment parameter. On the other hand, increasing the number of players would dilute the conservation incentives for the dominant player, and he might elect to fish down the stock to a low and unproductive level, as his share of the stock tends to be smaller the more players there are. In a situation like that the minor players might be happy to accept a smaller share in the co-operative solution than their zonal attachment parameters would suggest, because the non-co-operative solution would be so unattractive.

References

Bjørndal, T. 1987. Production economics and optimal stock size in a north Atlantic fishery. *Scandinavian Journal of Economics* 89: 145-64.
Bjørndal, T. and M. Lindroos. 2004. International management of North Sea herring. *Environmental and Resource Economics* 29: 83-96.

---

[9] On the management of high seas fisheries, see Munro (1990, 2000).

Bjørndal, T., D.V. Gordon, V. Kaitala and M. Lindroos. 2004. International management strategies for a straddling fish stock: A bio-economic simulation model of the Norwegian spring-spawning herring fishery. *Environmental and Resource Economics* 29: 435-57.

Clark, C.W. 1973. Profit maximization and the extinction of animal species. *Journal of Political Economy* 81: 950-61.

Engesæter, S. 1993. Scientific input to international fish agreements. *International Challenges. The Fridtjof Nansen Institute Journal* 13 (2): 85-106.

Hamre, J. 1993. A model of estimating biological attachment of fish stocks to exclusive economic zones. *International Council for the Exploration of the Sea*, C.M. 1993/D:43, Copenhagen.

Hannesson, R. 1997. Fishing as a supergame. *Journal of Environmental Economics and Management* 32: 309-22.

Hannesson, R. 2005. Sharing the herring: Fish migrations, strategic advantage, and climate change. In R. Hannesson, M. Barange and S. Herrick (eds) *Climate Change and the Economics of the World's Fisheries: Examples of Small Pelagic Stocks*. Edward Elgar, Cheltenham, UK.

ICES. 1978. The biology, distribution and state of exploitation of shared stocks in the North Sea area. *Cooperative Research Report* No. 74, International Council for the Exploration of the Sea, Copenhagen.

Munro, G.R. 1979. The optimal management of transboundary renewable resources. *The Canadian Journal of Economics* 12: 355–76.

Munro, G.R 1990. The optimal management of transboundary fisheries: Game theoretic considerations. *Natural Resource Modeling* 4: 403-26.

Munro, G.R. 2000. The United Nations Fish Stocks Agreement of 1995; History and problems of implementation. *Marine Resource Economics* 15. 265-80.

Vislie, J. 1987. On the optimal management of transboundary renewable resources: A comment on Munro's paper. *The Canadian Journal of Economics* 20: 870–75.

13

Fish Stew: Uncertainty, Conflicting Interests and Climate Regime Shifts

Kathleen A. Miller

13.1   INTRODUCTION

Fishery managers, harvesters and scholars have long recognised that competitive harvesting of shared fish stocks tends to dissipate potential resource rents, and may lead to such unsatisfactory outcomes as severely depleted stocks and impoverished fishing communities. The desire to avoid those outcomes drove the establishment of national-level fishery management programmes. It has also contributed to the development of international law governing marine resources, including transboundary fishery resources. In the past half century, technological progress and growing demand for fishery products fuelled both increased harvesting pressure on most marine fish stocks, and major international efforts to control the unwelcome effects of competitive harvesting. Efforts to more clearly delineate national rights and responsibilities with respect to these resources include the Law of the Sea Convention (United Nations, 1982), which provided international legal recognition of 200 mile coastal-state exclusive economic zones (EEZs), as well as a variety of bi-lateral and multi-lateral treaties detailing cooperative agreements regarding harvest levels and joint management arrangements.

Some of these agreements have provided a stable base for mutually satisfactory harvest management, while others have seen a troubled history of recurring disputes, ineffective control of harvesting activities, and degradation of the shared resource stocks. In many cases, one can trace this inability to maintain stable cooperation to the fact that the agreements lack flexibility to adapt to changing circumstances. Misunderstandings and frustrated expectations have sometimes developed into serious conflicts when there have been unanticipated, often climate-driven, changes in stock levels or their physical distribution across various EEZs and high-seas areas. In this regard, the report of the Norway-FAO Expert Consultation on the Management of Shared Fish Stocks emphasised: '... the need for cooperative management arrangements to be resilient through time, in the sense that they be able to absorb the impact of unpredictable shocks stemming from natural variability, climate change or other unpredictable ecological or economic disruptions' (FAO, 2002, p. *iv*).

Here, I argue that while such natural shocks are likely to remain unpredictable, a better understanding of their nature and their role in previous fishery disputes can provide valuable input for current efforts to design co-operative fishery management regimes. To that end, this chapter begins with a brief description of the current state of scientific understanding regarding the impacts of some important modes of climate variability on the marine environment and on commercially important fish stocks. Next, I describe cases in which limited understanding and poor predictability of these biological effects contributed to the dysfunction or breakdown of existing cooperative management arrangements (Krovnin and Rodionov, 1992; Miller *et al.,* 2001; Sissener and Bjørndal, 2005). These cases highlight the significance of frustrated expectations and shifting incentives to cooperate, in the context of overly-narrow conceptions of bargaining options. They also demonstrate that conflicting interpretations of available scientific evidence can lead to dangerous delays in taking appropriate management actions. This evidence points to a few practical suggestions to improve the resilience and effectiveness of international fishery management regimes. For example, if harvesting nations agree in advance on a formula to adjust allocations on the basis of an objectively determined indicator of stock status, they may be able to forestall costly bickering over shares when there is a sudden change in abundance, even if that change cannot be precisely forecast. In addition, various sorts of side payments and mutual access agreements can promote both more efficient harvesting and perceptions of fairness when the spatial distribution of a stock fluctuates as a result of natural variability.

13.2 CLIMATE AND FISH

Climatic variability affects marine fish populations in a variety of ways, and fisheries biologists have devoted considerable effort to understanding the linkages between physical changes in the ocean and changes in biological processes that ultimately affect the recruitment of harvestable fish. These processes can be quite complex, sometimes involving subtle physical changes with large impacts on survival and growth at critical life stages. Furthermore, variability in marine populations rarely takes the form of simple 'white noise'. Rather, occasional dramatically good or bad recruitment years may drive the dynamics of commercially important fish populations, and abrupt and persistent changes in patterns of recruitment or migratory behaviour may occur at unpredictable intervals. There is mounting evidence that such biological 'regime shifts' may be linked to large-scale, persistent changes in atmospheric circulation that alter ocean temperature and circulation patterns (Stenseth *et al.,* 2002). While improved understanding of such linkages could potentially lead to improved predictions of stock abundance (Bakun, 1998), many observers caution that predictability may be inherently limited by chaotic dynamic behaviour and by limits on our ability to observe important variables in the system (Ludwig *et al.,* 1993).

The waxing and waning of El Niño (warm) and La Niña (cool) events in the tropical Pacific as a result of the El Niño – Southern Oscillation (ENSO) phenomenon has well-documented effects on the abundance and spatial distribution of several commercially important fish stocks (Lehodey *et al.*, 1997; Lluch-Cota *et al.,* 1997; Yáñez *et al.,* 2001). ENSO warm and cool events do not follow a regular periodicity, but they occur frequently enough – roughly every three to seven years – that their biological effects are becoming reasonably well-known.

Other biologically significant climatic phenomena undergo abrupt, but infrequent changes that are best described as climate regime shifts. In particular, two recently recognised climatic phenomena in the north Pacific and the north Atlantic appear to flip abruptly between distinct modes of atmospheric circulation that have substantial, long-term impacts on fish populations. These are the Pacific Decadal Oscillation (PDO) and the North Atlantic Oscillation (NAO). Regarding the PDO, many observers have noted an abrupt climatic and biological shift in the North Pacific, commencing in 1977 and lasting, with minor breaks, for more than two decades. This shift to a positive phase of the PDO (Figure 13.1a,b) entailed a pronounced warming of coastal sea surface temperatures (SSTs) along the west coast of North America, suppressed plankton productivity in the California current system, intensified onshore winds, increased winter storminess in the Gulf of Alaska and sharp changes in the recruitment of a number of important fish stocks (Mantua *et al.*, 1997; Hare and Mantua, 2000; and Hollowed *et al.*, 2001).

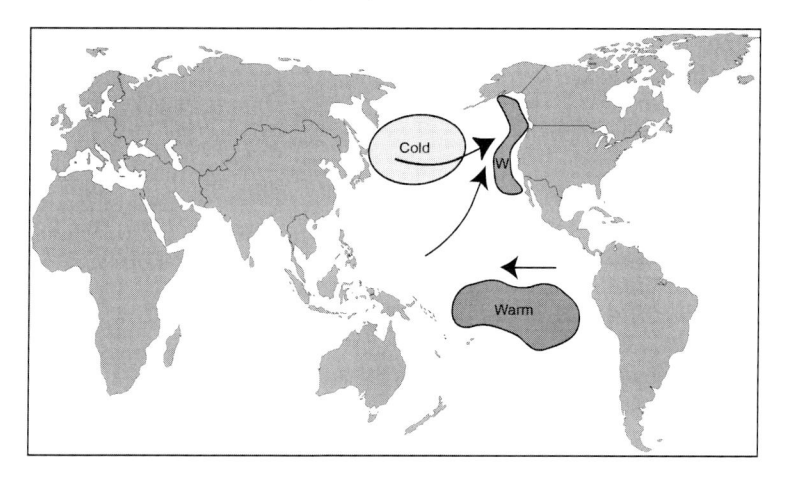

Figure 13.1a   Pacific Decadal Oscillation: Sea surface temperature and wind stress anomalies – coastal warm phase (positive values).

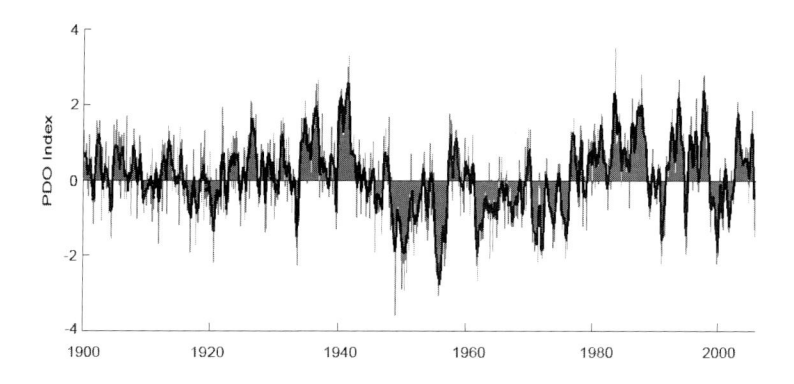

Figure 13.1b   Pacific Decadal Oscillation: Monthly values – PDO Index. The solid line is the time series smoothed with a low pass filter. Figure reprinted, with permission, from Miller and Munro, 2004. Data source: Dr. Nathan Mantua, JISAO, University of Washington. Accessible online http://www.jisao.washington.edu/pdo/

Major changes in Pacific salmon stocks have been linked to changes in the PDO, with Alaskan salmon stocks generally thriving during the coastal warm period, while many stocks spawning in rivers along the US west coast experience poor survival and growth. During coastal cool periods, southern stocks tend to fare well, while Alaskan stocks are less productive (Hare *et al.*, 1999). The PDO captures large-scale changes in the physical condition of the north Pacific, but it is an imperfect predictor of biological changes that also depend heavily on lagged effects and local-scale processes (McFarlane *et al.*, 2000; Hare and Mantua, 2000).

The North Atlantic Oscillation is an analogous phenomenon that drives climatic variability over much of the northern hemisphere. When the NAO Index is positive, the winter pressure difference between the low-pressure cell centred over Iceland and the high-pressure cell centred over the Azores is larger than normal. This pattern drives strong, westerly winds over northern Europe, bringing warm stormy winter weather, while cold winter temperatures prevail over Greenland, the Labrador Sea and north-eastern Canada. In the negative phase, the pressure differential is smaller than average and winter conditions are unusually cold over northern Europe and milder than normal over Greenland, north-eastern Canada and the north-west Atlantic (Hurrell and Dickson, 2004). The NAO was generally low throughout the 1950s and 1960s, and then abruptly switched to an extreme positive state for most of the period from 1970 to the present, except for a sharp drop in the winter of 1996 (Figure 13.2a,b). There is substantial evidence that the NAO affects recruitment success and migratory behaviour for several important fish stocks in the north Atlantic and adjacent seas (Parsons and Lear, 2001; Alheit and Hagen, 1997, 2001; Ottersen and Stenseth, 2001). For example, Parsons and Lear (2001) argue that the unusually cold winter conditions in the north-west Atlantic during recent decades may have contributed to the collapse of Canadian Atlantic groundfish, west Greenland cod, and northern cod stocks, although overfishing also clearly played a significant role. In the North Sea, they link a period of high abundance for gadoid stocks (cod, haddock, whiting, and saithe) to cold conditions associated with extreme negative values of the NAO Index in the 1960s, and subsequent poor recruitment of those stocks to the shift to strong positive NAO conditions.

Other analysts have noted that warm conditions in the Norwegian Sea (associated with positive NAO conditions) increase the likelihood of good recruitment for Norwegian spring spawning herring (Krovnin and Rodionov, 1992; Alheit and Hagen, 1997, 2001). In addition, this stock has displayed major long-term shifts in migratory behaviour that appear to be linked to changes in stock size and environmental conditions.

Although there has been significant recent scientific progress toward understanding the effects of large-scale climatic processes on marine fish populations, predictability will likely remain elusive. Nevertheless, a clear understanding of the nature and *possibility* of climate-related biological regime shifts may prove useful in designing more robust co-operative management agreements. In the case studies to follow, failure to recognise or understand the extent of climate-driven changes in biological processes led representatives of the competing harvesting nations to make decisions that resulted in severe loss of goodwill, foregone economic returns, and biological damage to the resources.

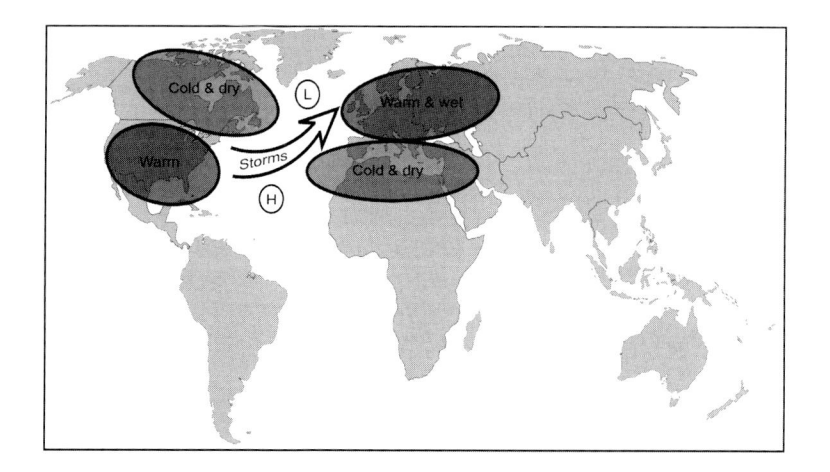

Figure 13.2a North Atlantic Oscillation temperature and precipitation anomalies – positive phase.

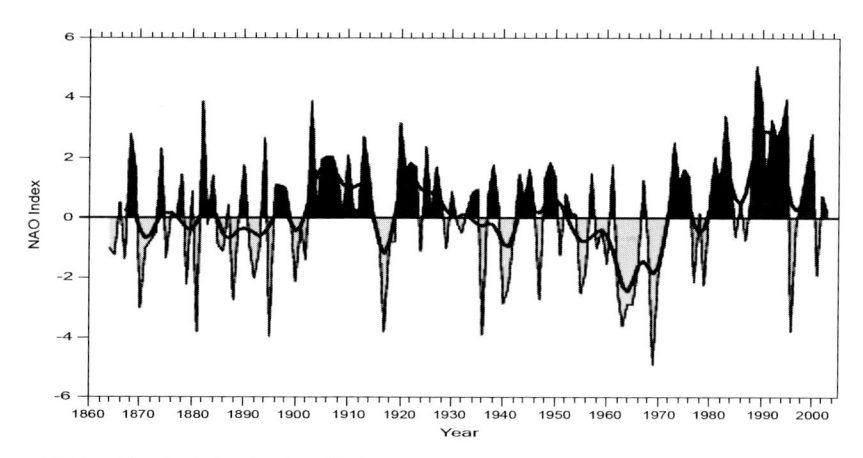

Figure 13.2b North Atlantic Oscillation Dec-Mar values – NAO Index. The solid line is the time series smoothed with a low pass filter. Figure reprinted, with permission, from Miller and Munro 2004. Data source: Dr James Hurrell, National Center for Atmospheric Research. Accessible online http://www.cgd.ucar.edu/~jhurrell/nao.html

13.3 CLIMATE AND THE PACIFIC SALMON DISPUTE

The history of conflict between Canada and the United States over their Pacific salmon harvests illustrates how unanticipated and poorly understood climate-related changes in stock abundance and migratory behaviour can contribute to the breakdown of a cooperative harvesting agreement (Miller *et al.*, 2001). Despite their commitment to science-based management, the fishery managers of both nations failed to recognise, or anticipate, the impacts of the mid-1970s climate regime shift on their shared salmon resources. In fact, they generally did not accept the fact that the shift had occurred until long after it had contributed to the collapse of existing co-operative management arrangements (Hare and Mantua, 2000; Miller and Munro, 2004).

Pacific salmon are anadromous fish that feed and mature in the ocean before returning to their natal streams to spawn and die. Most of the commercial harvest of salmon occurs in coastal waters where the stocks from both nations intermingle. Thus, harvesters from each jurisdiction inevitably 'intercept' some of the salmon heading to spawn in the rivers of other jurisdictions. The 1985 Pacific Salmon Treaty addressed the interceptions issue by articulating the following equity objective: '... each Party shall conduct its fisheries and its salmon enhancement programmes so as to ... provide for each Party to receive benefits equivalent to the production of salmon originating in its waters' (Pacific Salmon Treaty, Article III).

When the Treaty went into effect, the two nations failed to agree on a specific formula to measure the equity balance (Memorandum of Understanding to the Pacific Salmon Treaty, 1985), but both sides assumed that they would fulfill that objective by maintaining a rough balance between US interceptions of Canadian Fraser River salmon and Canadian interceptions of Washington and Oregon coho and chinook salmon, which had historically accounted for the vast majority of the interceptions (DFO, 1985). Little attention was given to salmon originating in south-eastern Alaska and northern British Columbia, which also were covered by the Treaty.

The treaty negotiators never imagined that large *natural* changes in stock abundance could interfere with efforts to maintain the equity balance, but that is exactly what happened. In their view, the Commission's primary task was to encourage enhancement and conservation efforts by guaranteeing that the *expected* increase in production would benefit the party making the investment. The regimes established by the Commission relied heavily on the use of 'ceilings', based on the notion that capping harvests in the intercepting fishery would allow any increase in run strength to primarily benefit the nation of origin – whose hatchery or habitat restoration investments had presumably caused the increase (Huppert, 1995). However, negotiators on both sides underestimated the potential impacts of natural environmental fluctuations, and relied on optimistic assumptions that proved to be grossly incorrect.

The bargaining framework established in 1985 called for frequent renegotiation of the fishing regimes and gave effective veto power to Canada as well as to each of three voting US Commissioners representing Alaska, Washington/Oregon and the Treaty Indian Nations (US Senate, 1985; Schmidt, 1996). That arrangement proved to be destructive when incentives to continue co-operating changed over time. Another source of difficulty was the fact that some of the Commissioners and other senior policy makers adopted a narrow definition of what was being shared – focusing on balancing 'fish' as opposed to 'benefits from the fishery'.

The 1977 climatic regime shift had striking impacts on the relative productivity of the various salmon stocks shared by Canada and the United States. Significant warming of coastal waters and associated changes in patterns of upwelling, nutrient transport and related physical and biological processes led to favourable survival and growth conditions for salmon in the Gulf of Alaska, while survival rates plummeted for stocks that enter the marine environment along the US west coast. These climate-related changes contributed to a nearly ten-fold increase in Alaskan salmon harvests, with harvests rising from fewer than 22 million salmon (of all species) in 1974 to three successive record highs in 1993, 1994, and 1995. At the 1995 peak, Alaska harvested close to 218 million salmon. Another high was attained in 1999 when Alaska harvested almost 217 million salmon. In particular, pink salmon harvests increased dramatically in south-eastern Alaska, where those stocks are intermingled with Canadian salmon. In the southern border region, the effects of the climatic regime shift were profoundly different. There, chinook and coho salmon abundance declined to the point that some stocks faced a significant risk of extinction, prompting the US National Marine Fisheries Service to list a number of these stocks as 'threatened' under the Endangered Species Act.

The dramatic increase in pink salmon abundance in south-eastern Alaska led Alaskan harvesters to fish harder in that area, so that Alaskan interceptions of Canadian salmon increased. The Canadians proved unable to redress the growing interceptions imbalance because declining southern coho and chinook stocks prevented Canadian harvesters from reaching the agreed-upon ceilings for harvests of those stocks along the west coast of Vancouver Island. In frustration, Canada turned to aggressive competitive tactics with respect to its harvests of Fraser sockeye and its interceptions of southward-bound chinook and coho salmon. The southern US jurisdictions offered to make further concessions on their harvests of Fraser River salmon, but the offers were insufficient in Canadian eyes. Rather, Alaska's harvests had become the major source of contention. Alaska, in turn, was unwilling to make the concessions requested by Canada because they appeared to entail only uncompensated costs. By 1993, co-operation collapsed, and the parties proved unable to agree on a full set of fishing regimes. While binding in a legal sense, the treaty-based cooperative resource management regime had nonetheless foundered, because it had not met the test of resiliency through time.

One of the necessary conditions for a stable and efficient co-operative solution to an international fishery game is that it must not be possible for any player to do better by refusing to co-operate (Munro *et al.*, 1998). In the case at hand, this individual rationality constraint was no longer satisfied for at least one player, namely Alaska, which now found that it had little, or nothing, to gain from the treaty.

The dispute festered for six years, during which time the two federal governments made several efforts to resolve the impasse. They achieved a solution only after there was a significant shift in bargaining objectives coupled with a new-found willingness to try more flexible tools to achieve equity. Deterioration in the condition of Canada's autumn chinook and coho stocks during the 1990s (Pacific Fisheries Resource Conservation Council, 1999; McFarlane *et al.*, 2000) caused the Canadian focus to shift radically from insistence on an equitable interceptions balance to the need to tailor harvesting efforts to protect the stocks that had become severely depleted. This shift facilitated the negotiation of amendments to the Treaty, concluded in 1999.

The 1999 agreement replaces the expired short-term harvest management regimes with new longer-term arrangements in which harvest shares are to be defined as a function of indices of the abundance of each salmon species in the areas covered by the Treaty. For example, the new arrangements for chinook, which will be in effect for ten years, take account of the fact that the various fisheries along the coast differ considerably in the extent to which they rely on healthy or depressed stocks. Accordingly, the agreement designates two types of fisheries: 1) aggregate abundance-based management (AABM) fisheries will be managed based on indices of the aggregate abundance of chinook present in the fishery, without specific reference to any individual stock; 2) individual stock-based management (ISBM) fisheries, which are primarily located in fishing areas near the spawning rivers, will be managed based on the status of individual stocks or groups of stocks (e.g. on the basis of the evolving status of currently endangered or threatened stocks). This new approach will better protect weak stocks by limiting the parties' ability to aggressively fish 'up to the ceiling' when the resource is in a fragile state. As such, it serves to enhance the resiliency of the cooperative resource management arrangement. The parties are still working to define reliable measures of abundance for all of the major stocks, but they have agreed to delegate that task to joint technical committees.

The agreement accommodates the Canadian position on the equity imbalance by further decreasing the US share of the Fraser sockeye harvest. It also accommodates Alaska's position, in that Alaskan harvests will remain relatively unchanged under the new abundance-based rules. Another major feature of the agreement is its provision for two endowment funds, financed almost entirely by the United States. These funds provide an implicit side-payment to Canada in the form of financing for research and enhancement activities. In short, the 1999 agreement represents a significant effort to come to grips with some of the major sources of instability in previous efforts to co-operate.

13.4 THE NAO AND NORWEGIAN SPRING-SPAWNING HERRING

In the north Atlantic, long-term climatic regime shifts have complicated the management of other internationally shared fisheries. As in the Pacific salmon case, the role of climate was only recently recognised. Notably, changes in oceanographic conditions associated with variations in the NAO have affected both recruitment success and the migratory behaviour of Norwegian spring spawning herring (Bjørndal et al., 1998; Sissener and Bjørndal, 2005).

Norwegian spring-spawning herring is, historically, the largest fish stock in the north Atlantic, and it has been an important source of food for centuries. Historical records show dramatic fluctuations in harvests of this stock over the past 500 years (Alheit and Hagen, 1997). Recent research suggests that the NAO may have played a major role in driving these fluctuations, while modern levels of fishing pressure clearly can exacerbate natural downturns in abundance.

The NAO was mostly positive during the first half of the twentieth century, followed by a decrease and a negative index after 1952 (Figure 13.2b). From the 1930s through the 1950s the stock appeared robust despite rapidly increasing harvests, facilitated by new technology. Then during the 1960s, poor recruitment, coupled with continued intense harvesting pressure caused the stock to collapse. Although evidence of the declining spawning stock biomass was available from the International Council for the Exploration of the Sea (ICES) as early as the mid 1950s (Kronvin and Rodionov, 1992), it was initially disputed and/or ignored. In the short run, the competitive harvesting race remained profitable due to the effectiveness with which these tightly-schooling fish could be captured using the new technology (including sonar and powerblocks). In 1966, the total catch of adult herring reached a maximum of nearly two million tonnes, and exploitation rates increased to the point of scooping up a large fraction of the spawning stock biomass – a pattern that continued as the stock crashed over the next few years (Bjørndal et al., 2004; ICES, 2004; Marine Institute, 2005).

The drastic declines in catches in the late 1960s finally led all participants in the fishery to conclude that exploitation had to be reduced, and they introduced a sequence of strict control measures, with Norway playing the leading role. The stock only began to recover significantly during the late 1980s, after several years of severely restricted fisheries, and a series of strong recruitment years, particularly after 1990.

In addition to the changes in abundance, there were significant changes in the migratory behaviour of this straddling stock. Historically, Norwegian spring-spawning herring have been harvested by Norway, Iceland, Faeroe Islands, countries of the EU, Russia and distant water vessels fishing in the 'Loop Ocean' (Churchill, 2001), which is a high seas enclave surrounded by the EEZs of Norway, Iceland, the EU and Faeroe Islands. The harvesting nations have strong incentives to harvest the population before it moves elsewhere (Bjørndal *et al.*, 1998). If they cannot agree on an equitable division of the catch, they may resort to 'strategic over-fishing' that could jeopardise the health of the resource. Changes in migratory behaviour have made it difficult for these countries to maintain agreement on what constitutes an equitable distribution of harvest. In particular, the notion that harvest shares should be determined by 'zonal attachment' becomes difficult to interpret when the physical location of the stock changes dramatically in concert with changes in abundance.

Prior to the period of cold temperatures and poor recruitment commencing in the mid-1950s, the herring migrated over a wide area that included areas both west and north of Norway's current EEZ. After the stock collapsed at the end of the 1960s, the herring stayed in Norwegian coastal waters all year. That pattern prevailed until 1994, when the herring migrated outside the Norwegian EEZ for the first time in 26 years.

Beginning in 1987, the Norwegian government set the total allowable catch (TAC) for Norwegian spring-spawning herring, giving Russia a share of the TAC after the yearly fishery negotiations between the two countries. However, international disagreements arose as soon as the herring resumed a pattern of migration to feeding grounds outside of Norwegian waters. In 1994, for example, Iceland demanded a share of the TAC. The EU, Russia and the Faeroe Islands also began to pressure Norway to engage in negotiations for a multi-national management arrangement, based on the argument that the new migratory pattern created new conditions for the fishery and for the management of the stock. Norwegians feared that their strict regime during the last decade would be wasted and that the involvement of more nations in the fishery would cause another collapse of the stock.

Starting in 1995, Norway and the other harvesting nations negotiated a series of agreements regarding the size of the TAC and its distribution among the parties. These agreements have been imperfect – for example, harvests in 1995 were almost twice the quantity recommended by the Advisory Committee on Fishery Management (ACFM) of ICES (Bjørndal *et al.*, 1998). Nevertheless, in spite of these high catch levels, the herring spawning stock continued to increase due to good growth and recruitment. In effect, good luck forestalled significant adverse impacts from renewed competitive fishing.

In 1996, with a four-party-agreement in force, Norway, Russia, Iceland and the Faeroe Islands shared a quota of 1,107,000 tonnes, but the EU was not yet part of the agreement and fished at full capacity in international waters. From 1997 through 2002, Norway, Russia, Iceland, Faeroe Islands and EU jointly set a yearly quota based on the recommendations of ICES, and negotiated the allocation of shares to their respective countries. The North East Atlantic Fisheries Commission (NEAFC) exercises formal responsibility for the distribution and the fixation of the TAC in international waters.

The yearly negotiations on the TAC continue to cause conflict. There also has been discontent over the distribution of shares of the TAC. A recent report of the ICES Advisory Committee on Fisheries Management notes, for example, that the parties failed to agree on either the TAC or share allocations in 2003 and 2004 (ICES, 2004), but nevertheless appeared to be maintaining harvest levels within the recommended range.

13.5 LESSONS FOR MANAGING NATURALLY-VARYING SHARED FISH STOCKS

The actions of the harvesting nations in these two case studies conform to the predictions of simple game theoretic models in which the parties have fallen into a competitive harvesting race as a result of an unanticipated change in circumstances. In the Pacific salmon case, the climate regime shift caused asymmetrical changes in the threat-point payoffs of the three primary players in the game. The original terms of the bargain embodied in the 1985 Treaty simply no longer met the individual rationality constraint for a stable cooperative solution to the game. Alaska, in particular, found that it had little to gain – and potentially much to lose – from the Treaty, while Canada's initial insistence on a fish-for-fish balancing rule artificially constrained the scope for bargaining and left fishery managers in Washington and Oregon with few options. In addition, uncertainty and conflicting interpretations of the causes of the biological changes tended to prolong the conflict. In an international fishery game, one can easily demonstrate that negotiations for a co-operative solution may fail if the parties have disparate perceptions of their threat point payoffs and potential gains from co-operation. Scientific consensus (which is distinct from predictability) thus plays an important role by fostering a common understanding of the nature of the game and the payoffs that are likely to result from alternative strategies.

In the case of Norwegian spring spawning herring, intense competitive fishing was initially touched-off by technical progress and rapidly declining harvesting costs in the absence of effective co-operative management arrangements. Predictably, the race quickly dissipated the rents that could have been generated by the herring resource. The coincidental decline in recruitment rates associated with adverse climatic conditions greatly accelerated the deleterious effects of the harvesting race, while failure to recognise the natural changes delayed efforts to negotiate an effective management regime until the stock had been nearly destroyed. Good luck in the form of a dramatic shift to favourable climatic conditions played a large role in the subsequent restoration of the resource. While a co-operative agreement is now in place, it appears to be somewhat fragile in that the various harvesting nations are continuing to squabble over the size of the TAC and the allocation of national shares.

Given the role of uncertainty and misperception in these two case studies, one might be tempted to conclude that better scientific information is the essential key to efficient and congenial multi-national management of shared fishery resources. Certainly, better information can play an important role, but it also can be a two-edged sword. In a 'split-stream' harvesting game, McKelvey *et al.* (2003) demonstrated that improvements in the ability to forecast changes in the migratory path of a shared fish stock would always be valuable if co-operation prevails, but the same information could do more harm than good in a competitive fishery.

This suggests that the optimism displayed by many natural scientists regarding the potential value of better predictions of climate impacts on fish populations must be tempered by an understanding of how such information may be used by the parties exploiting the resource. Here, it is important to note that improved predictions and improved understanding are two different types of information. When it comes to designing a workable and resilient international fishery management regime, the latter type of information may be the most valuable.

A recent paper by Miller and Munro (2004) draws a distinction between the terms 'prediction' and 'anticipation'. Prediction implies foretelling an event with some level of detail and precision. Anticipation, on the other hand, suggests expecting the possibility of an event at some unspecified time in the future. There are likely to be inherent differences in predictability for the various climatic influences on different fish populations. At one extreme, there may be cases in which modest investments in data collection and modelling can yield reliable predictions. At the other extreme, complex dynamics or unobservable components of the system may make accurate predictions virtually impossible. Even in the latter case, however, one can anticipate the possibility of significant change. In this light, recent advances in scientific understanding of climate regime shifts and their impacts upon fisheries have created new opportunities for managers of co-operative resource arrangements to anticipate the possibility of such shifts and take appropriate precautionary measures.

Economists can contribute to the analysis of effective management arrangements. The existing literature on shared fishery games demonstrates that co-operation is, in most cases, essential to maintain a sustainable stream of benefits from fishery resources that are exploited by two or more nations (Munro, 1990, 2003). The literature further demonstrates that if co-operation is to be maintained, all parties must perceive some benefit from continued adherence to the agreement. In the face of changeable circumstances, workable agreements must incorporate both flexibility and mechanisms to reduce the transaction costs needed to make adjustments (Barzel, 1989; North, 1990; Miller and Munro, 2004).

The eventual solution to the Pacific salmon impasse affirms the value of both *flexibility* and *anticipation* of the possibility of natural changes in the condition of the shared stocks. The case further affirms that side-payments, broadly construed, are an indispensable tool for achieving flexibility (FAO, 2002). Specifically, contributions by the US to the endowment funds under the 1999 Pacific Salmon Agreement effectively function as side payments to Canada, and have thus provided some of the needed flexibility. The second requirement for a more resilient agreement – anticipation of the potential impacts of natural regime changes – is now being served by the new abundance-based allocation approach in the 1999 Pacific Salmon Agreement. The intent of the approach is to begin articulating clear rules for adjusting quotas and allocations as a function of mutually accepted indicators of the status of the shared stocks.

The possibility of natural regime shifts also affects the usefulness of concepts such as 'zonal attachment' that have been widely applied to negotiate national shares in transboundary fisheries. If negotiators falsely construe zonal attachment as a stable long-term natural condition, the concept could prove to be a source of more confusion than enlightenment. Geographical distributions can change abruptly, and when that happens, threat point payoffs will also shift. To maintain co-operation, participants in a fishery agreement would need to recognise and accommodate the reality of the shift in threat point payoffs.

Again, flexibility in the form of various types of side payments is likely to be useful. While monetary side payments are certainly among the possible options, a variety of more subtle and indirect transfers among the parties are possible as well. Mutual access agreements and provisions allowing quota trading also can be tailored to provide implicit side payments (Stokke *et al.*, 1999). The literature suggests other strategies that can be used to maintain co-operation in the face of significant changes in the abundance or availability of fish stocks, including pre-agreements that outline actions to be taken under a variety of contingencies (Hilborn *et al.*, 2001). For example, the control rule approach adopted by the International Pacific Halibut Commission (IPHC) has allowed that organisation to easily reduce or raise catch in response to fluctuations in the condition of the stock. Clearly, *anticipation* of possible changes in the condition or distribution of the stock is a necessary condition for workable pre-agreements.

In many cases, neighbouring nations may share in the harvest of several fish species, each of which may be affected in different ways by a climatic change. In cases involving exploitation of multiple species, Hilborn *et al.*, (2001) suggest adopting a portfolio management approach to managing the risks associated with uncertain fluctuations in abundance by promoting fleet diversification and flexible licensing that would allow vessels to move more easily between fisheries. While that paper focuses on the issue of efficient reallocation of effort across a set of domestic fisheries, a similar approach could be applied to international fisheries as well. To some extent, the practice of multi-species quota swapping that has developed in the Baltic and Barents Seas (Stokke *et al.*, 1999; Ranke, 2003) can be seen as a tool for achieving the flexibility needed for this sort of risk management. Perhaps greater attention to the need to manage risks associated with significant regime shifts could allow such an approach to be refined and extended to other international fishery management agreements, where the focus is now largely confined to management of a single species or a closely related group of species.

In order to properly address the effects of climatic regime shifts, further development of the theory is needed. At present most attempts to incorporate uncertainty in fishery game models, with a few exceptions, treat uncertainty only as add-on disturbances. The possibility of large and persistent regime shifts suggests the inadequacy of that approach. In particular, attention should be given to developing models of co-operative games in which the players anticipate the possibility of a regime shift that radically alters the strength of their relative bargaining positions.

Finally, sustainable management of climate-sensitive transboundary fishery resources will require adaptive and resilient management institutions. These can best be designed and operated by interdisciplinary teams of managers, analysts, and skilled negotiators, whose efforts are informed by ongoing contributions from the biological and physical science communities as well as from economists and other policy scientists.

Acknowledgements

This chapter is based upon work supported by the National Science Foundation under Grant No. 0323134. Any opinions, findings, and conclusions or recommendations expressed in this material are those of the author(s) and do not necessarily reflect the views of the National Science Foundation.

References

Alheit, J. and E. Hagen. 1997. Long-term climatic forcing of European herring and sardine populations. *Fisheries Oceanography* 6(2): 130-39.

Alheit, J. 2001. The effect of climatic variation on pelagic fish and fisheries. In P.D. Jones, A.E.J. Ogilvie, T.D. Davies and K.R. Briffa (eds) *History and Climate: Memories of the Future?* Kluwer, New York: 247-65.

Bakun, A. 1998. Ocean triads and radical interdecadal variation: bane and boon to scientific fisheries management. In T.J. Pitcher, P.J.B. Hart and D. Pauly (eds) *Reinventing Fisheries Management,* Kluwer Academic Publishers, Dordrecht: 331-58.

Barzel, Y. 1989. *Economic Analysis of Property Rights.* Cambridge University Press.

Bjørndal, T., A. Drange Hole, W.M. Slinde, F. Asche and S. Reite. 1998. *Norwegian Spring Spawning Herring – Some Biological and Economic Issues: An Update*, Working Paper 46, Centre for Fisheries Economics, SNF, Bergen.

Bjørndal, T., A.A. Ussif and U.R. Sumaila. 2004. A bioeconomic analysis of the Norwegian spring spawning herring (NSSH) stock. *Marine Resource Economics* 19: 353-65.

Churchill, R.R. 2001. Managing straddling fish stocks in the north-east Atlantic: A multiplicity of instruments and regime linkages – but how effective a management? In O.S. Stokke (ed.) *Governing High Seas Fisheries: The Interplay of Global and Regional Regimes.* Oxford University Press: 235-72.

DFO. 1985. *Information Bulletin,* No. 1-HQ-85-1E: 2. Department of Fisheries and Oceans, Canada.

FAO. 2002. *Report of the Norway-FAO Expert Consultation on the Management of Shared Fish Stocks, Bergen, Norway, 7-10 October 2002*, FAO Fisheries Report No. 695. Food and Agricultural Organization, Rome.

Hare, S.R., N.J. Mantua and R.C. Francis. 1999. Inverse production regimes: Alaska and west coast Pacific salmon. *Fisheries* 24(1): 6-14.

Hare, S.R. and N.J. Mantua. 2000. Empirical evidence for north Pacific regime shifts in 1977 and 1989. *Progress in Oceanography* 47: 103-45.

Hilborn, R., J. Maguire and A. Parma. 2001. The precautionary approach and risk management: Can they increase the probability of successes in fishery management? *Canadian Journal of Fisheries and Aquatic Sciences* 58: 99-107.

Hollowed, A.B., S.R. Hare and W.S. Wooster. 2001. Pacific Basin climate variability and patterns of northeast Pacific marine fish production. *Progress in Oceanography* 49: 257-82.

Hurrell, J.W. and R.R. Dickson. 2004. Climate variability over the north Atlantic. In Stenseth, N.C., G. Ottersen, J.W. Hurrell and A. Belgrano (eds) *Ecological Effects of Climate Variations in the North Atlantic.* Oxford University Press: 15-31.

Huppert, D.D. 1995. *Why the Pacific Salmon Treaty Failed to End the Salmon Wars.* School of Marine Affairs, University of Washington, SMA 95-1, Seattle.

ICES. 2004. Report of the ICES Advisory Committee on Fishery Management and Advisory Committee on Ecosystems, 2004. ICES Advice 1(2). International Council for the Exploration of the Sea. Accessible online http://www.ices.dk/products/icesadvice/Book2Part3.pdf

Krovnin, A.S. and S.N. Rodionov. 1992. Atlanto-Scandian herring: A case study. In M.H. Glantz (ed.) *Climate Variability, Climate Change and Fisheries.* Cambridge University Press: 231-60.

Lehodey, P., M. Bertignac, J. Hampton, A. Lewis and J. Picaut. 1997. El Niño Southern Oscillation and tuna in the western Pacific. *Nature* 389: 715-18.

Lluch-Cota, D.B., S. Hernández and S. Lluch-Cota. 1997. *Empirical investigation on the relationship between climate and small pelagic global regimes and El Niño-Southern Oscillation (ENSO).* FAO Fisheries Circular No. 934, FIRM/C9334. FAO, Rome.

Ludwig, D., R. Hilborn and C. Walters. 1993. Uncertainty, resource exploitation and conservation: Lessons from history. *Science* 260: 17, 36.

Mantua, N.J., S.R. Hare, Y. Zhang, J.M. Wallace, and R.C. Francis. 1997. A Pacific interdecadal climate oscillation with impacts on salmon production. *Bulletin of the American Meteorological Society* 78(6): 1069-79.

Marine Institute. 2005. *Norwegian Spring Spawning Herring.* Accessible online http://www.marine.ie/NR/rdonlyres/0FECC71D-77B7-431B-8965-D519FD5107B6/0/HerringNorwegianSpringSpawners05.pdf

McFarlane, G.A., J.R. King and R.J. Beamish. 2000. Have there been recent changes in climate? Ask the fish. *Progress In Oceanography* 47: 147-69.

McKelvey, R., K. Miller and P. Golubtsov. 2003. Fish-wars revisited: Stochastic incomplete-information harvesting game. In J. Wesseler, H.P. Weikard, and R.D. Weaver (eds) *Risk and Uncertainty in Environmental and Natural Resource Economic.* Edward Elgar, Cheltenham: 93-112.

Memorandum of Understanding to the Pacific Salmon Treaty. 1985. Pacific Salmon Treaty, March 18, 1985, US-Can., 99 Stat. 7 [codified at 16 U.S.C. 3631-44 (1997)].

Miller, K., G. Munro, T. McDorman, R. McKelvey and P. Tydemers. 2001. The 1999 Pacific Salmon Agreement: A sustainable solution? *Occasional Papers: Canadian-American Public Policy*, No. 47, Canadian-American Center. University of Maine, Orono.

Miller, K.A. and G.R. Munro. 2004. Climate and cooperation: A new perspective on the management of shared fish stocks. *Marine Resource Economics* 19: 367-93.

Munro, G.R. 1990. The optimal management of transboundary fisheries: Game theoretic considerations. *Natural Resource Modeling* 4(4): 403-26.

Munro, G.R. 2003. *Discussion Paper on the Management of Shared Fish Stocks.* FAO Fisheries Report No. 695 Supplement: Papers Presented at the Norway-FAO Expert Consultation on the Management of Shared Fish Stocks, Bergen, Norway, 7-10 October 2002. FAO, Rome: 2-29.

Munro, G.R., T. McDorman and R. McKelvey. 1998. Transboundary fishery resources and the Canada-United States Pacific Salmon Treaty. *Occasional Papers: Canadian - American Public Policy,* No. 33, Canadian - American Center, University of Maine, Orono.

North, D.C. 1990. *Institutions, Instutional Change and Economic Performance.* Cambridge University Press.

Ottersen, G., and N.C. Stenseth. 2001. Atlantic climate governs oceanographic and ecological variability in the Barents Sea. *Limnology and Oceanography* 46(7): 1774-80.

Pacific Fisheries Resource Conservation Council. 1999. *1998-1999 Annual Report.* PFRCC, Vancouver.

Parsons, L.S., and W.H. Lear. 2001. Climate variability and marine ecosystem impacts: A north Atlantic perspective. *Progress In Oceanography* 49: 167-88.

Ranke, W. 2003. *Co-operative Fisheries Management Issues in the Baltic Sea*. FAO Fisheries Report No. 695 Supplement: Papers presented at the Norway-FAO Expert Consultation on the Management of Shared Fish Stocks, Bergen, Norway, 7-10 October 2002. FAO, Rome: 123-32.

Schmidt, R.J., Jr. 1996. International negotiations paralyzed by domestic politics: Two-level game theory and the problem of the Pacific Salmon Commission. *Environmental Law* 26: 95-139.

Sissener, E.H. and T. Bjørndal. 2005. Climate change and the migratory pattern for Norwegian spring-spawning herring – implications for management. *Marine Policy* 29(4): 299-309.

Stenseth, N.C., A. Mysterud, G. Ottersen, J.W. Hurrell, K.S. Chan and M. Lima. 2002. Ecological effects of climate fluctuations. *Science* 297: 1292-96.

Stokke, O.S., L.G. Anderson and N. Mirovitskaya. 1999. The Barents Sea fisheries. In O. Young (ed.) *The Effectiveness of International Environmental Regimes: Causal Connections and Behavioral Mechanisms.* MIT Press, Cambridge, MA: 91-154.

United Nations. 1982. United Nations Convention on the Law of the Sea. UN Doc. A/Conf. 62/122, United Nations, New York.

US Senate. 1985. Pacific Salmon Treaty Act of 1985, Pub. Law no. 99-5, 99 Stat. 7.

Yáñez, E., M.A. Barbieri, C. Silva, K. Nieto and F. Espíndola. 2001. Climate variability and pelagic fisheries in northern Chile. *Progress In Oceanography* 49: 581-96.

14

A Dynamic Game on Renewable Natural Resource Exploitation and Markov Perfect Equilibrium

Shinji Kobayashi

14.1 INTRODUCTION

This chapter studies oligopolistic firms' exploitation of renewable natural resources in a dynamic setting. Specifically, we present a differential game to examine oligopolistic firms' harvesting behaviour regarding a renewable natural resource in a continuous time infinite horizon model. Although differential games have been widely used in the economics literature, it is well known that finding feedback equilibrium is notoriously difficult except in linear-quadratic games.[1] Nevertheless, for models of exploitation of renewable natural resources, we need to analyse games that are not linear-quadratic since reasonable growth functions regarding the stock of natural resources are not linear[2] In this chapter, we analyse a differential game model that is not linear-quadratic and derive a Markov perfect equilibrium for the game. One prominent feature of the model is that demand for the harvest of a natural resource is assumed to depend upon the stock of the natural resource.

In the context of fishery economics, Levhari and Mirman (1980) examine competition of harvesting in a dynamic setting. They study utility maximisation, but do not consider profit maximising firms. In this chapter, we will consider oligopolistic firms and analyse the firms' exploitation of a renewable natural resource in a differential game. We will examine feedback strategies and under certain conditions, derive a Markov perfect equilibrium. We also examine open-loop equilibrium in our model. For this purpose, we consider two settings regarding firms' harvesting decisions. One setting is the case where the firms make their decision non-co-operatively, and the other is the one where the firms can co-operate.[3] The firms are assumed to undertake Cournot competition in the product market.

---

[1] See Basar and Olsder (1995) for more on differential games and for differential game models in economics, see Fudenberg and Tirole (1991) and Kamien and Schwartz (1991).
[2] For natural resource economics, see for instance, Clark (1990), Conrad and Clark (1987), and Munro (1982).
[3] For analyses of cooperative exploitation of natural resources, see Munro (1979) and Kaitala and Pohjola (1988) among others.

The chapter is organised as follows. In section 14.2, we will describe the basic model. In section 14.3, we will examine Markov feedback strategies and derive Markov perfect equilibrium. In section 14.4, we will examine open-loop equilibrium for games both under the case of non-co-operation and under the case of co-operation. In section 14.5, we will discuss the effects of taxation on equilibrium exploitation of the renewable natural resource. Section 14.6 concludes.

14.2 THE MODEL

There are n firms in the industry, indexed by $i \in I = \{1, \cdots, n\}$. The firms harvest a renewable natural resource and sell it in the product market. We assume that the firms engage in Cournot competition in the product market.

Let $X \in \Re_+$ be the stock of the renewable natural resource. Let $g(X)$ be the growth function of the natural resource. We assume that $g(X)$ is a strictly concave function such that $g(0) = 0$, $g(\tilde{X}) = 0$ for some $\tilde{X} > 0$.

Let x_i be the harvest of firm i. Then the change in the stock at time t, $\dot{X}$, is given by:

$$\dot{X} \equiv \frac{dX}{dt} = g(X) - \sum_{i=1}^{n} x_i. \tag{14.1}$$

The initial stock of the natural resource is denoted:

$$X(0) = X_0. \tag{14.2}$$

For each firm, let $C(X)$ be the unit cost of harvesting. We assume that the unit cost function $C(X)$ satisfies:

$$\frac{\partial C}{\partial X} < 0 \text{ and } \frac{\partial^2 C}{\partial X^2} > 0.$$

Note that the unit cost of harvesting is decreasing in the stock of the natural resource.

Let the inverse demand function be given by:

$$p = f(\sum_{i=1}^{n} x_i, X),$$

where p is the price of the product, x_i the harvest, that is, the product of firm i. Note that demand for the product is assumed to depend on the stock of the natural resource. To put it differently, we assume that consumers are concerned with the environment, that is, the stock of the natural resource, when they determine their demand.

We also assume that for each X, $f''x_i + f' < 0$, where $f' \equiv \partial f / \partial x_i$ and $f'' \equiv \partial^2 f / \partial x_i^2$. This decreasing marginal revenue assumption is a standard assumption for the existence of Cournot equilibrium in the product market.

The objective of each firm is to maximise the discounted sum of its profits over an infinite time horizon. Let $r > 0$ be a common discount rate. Then the objective function of firm i, J_i, is given by:

$$J_i = \int_0^\infty \left[px_i - C(X)x_i \right] e^{-rt} dt . \tag{14.3}$$

Then the game is described as follows: for $\forall i \in I$, given other firms' strategies x_{-i}, firm i chooses a strategy x_i in such a way that it maximises (14.3) subject to (14.1) and (14.2).

In this chapter, we will consider feedback strategies in section 14.3 and open-loop strategies in section 14.4.

14.3 MARKOV PERFECT EQUILIBRIUM

In this section, we will analyse the differential game described in section 14.2. First we examine feedback strategies. Feedback strategies and Markov perfect equilibrium are defined as follows:

Definition 1 *The feedback strategy space for firm i is the set*
$$S_i^F = \{x_i(X,t) \text{ is continuous in } (X,t), x_i(X,t) \geq 0, \text{ and } X \geq 0\} .$$

Definition 2 *A Markov perfect equilibrium is an n-tuple of feedback strategies such that for each $i \in I$,*
$$J_i(x_i^*, x_{-i}^*) \geq J_i(x_i, x_{-i}^*) \text{ for every } x_i \in S_i^F .$$

In general, deriving a Markov perfect equilibrium is extremely difficult except for linear-quadratic games. The game considered in this chapter, however, is not a linear-quadratic game. In order to derive a closed form solution, we assume that the inverse demand function, the unit cost function, and the growth function take specific forms described below. First, we assume that the inverse demand function is given by:

$$p = f(\sum_{i=1}^n x_i, X) = \frac{a}{X} - \frac{b\sum_{i=1}^n x_i}{X^2}, \ a > 0, \ b > 0 . \tag{14.4}$$

Note again that the inverse demand function depends on the stock of the natural resource. Notice also that given X, this inverse demand function is a standard linear inverse demand function. This functional form is set out to capture the idea that consumers are concerned with the natural resource as environment.

We next assume that the unit cost function takes the following form:

$$C(X) = \frac{\overline{c} - \gamma \ln X}{X}, \ \overline{c} > 0, \ \gamma > 0 . \tag{14.5}$$

Note that the unit cost function depends upon the stock of the resource and satisfies:

$$C'(X) = \frac{-\gamma - (\overline{c} - \gamma \ln X)}{X^2} < 0 \text{ and } C''(X) = \frac{\gamma + 2[\gamma + (\overline{c} - \gamma \ln X)]}{X^3} > 0 \,^4$$

Thus the larger the stock of the natural resource, the lower the unit cost.

Finally we assume that the growth function of the stock of the natural resource takes the following form:

$$g(X) = X(\alpha - \beta \ln X), \ \alpha > 0, \ \beta > 0. \tag{14.6}$$

The growth function (14.6) is shown in Figure 14.1.

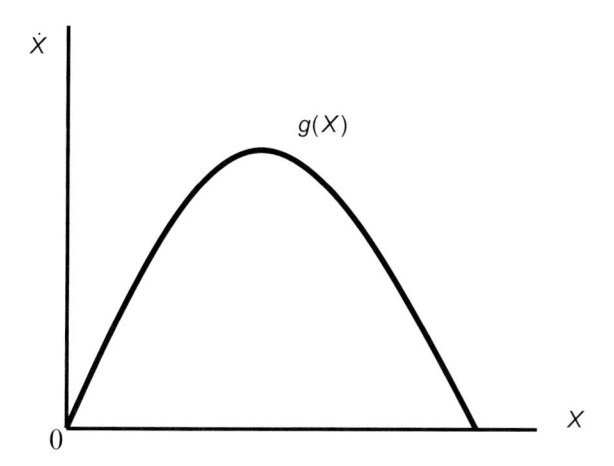

Figure 14.1 Growth function, natural resource stock.

There are many possible functional forms for the growth function. One of the simplest yet often adopted growth functions is a logistic function. The above growth function could be thought of as a slightly modified functional form of a logistic curve.[5] Under these specific functional forms, we may explicitly derive a Markov perfect equilibrium. In what follows, for simplicity, we consider the case of two firms, i.e. we assume $n = 2$. Let us define $Y \equiv \ln X$ and $y_i \equiv x_i/X$, $i = 1, 2$. Then we have the following theorem for the existence of a Markov perfect equilibrium.

---

[4] Since $\overline{c} - \gamma \ln X > 0$ by $C(X) = (\overline{c} - \gamma \ln X)/X$.
[5] For a logistic model of the growth function, see, for instance, Conrad and Clark (1987).

Theorem 1 There exists a Markov perfect equilibrium given by:

$$y_i^* = \frac{a - \bar{c} - F + (\gamma - G)Y}{3b}, \quad i = 1, 2,$$

where the constants F and G are given in (14.44) and (14.40) below and the stability condition is given by $3b\beta + 2\gamma - 2G > 0$.

The proof of Theorem 1 is in Appendix 14.1. In the following example the set of solutions stated in Theorem 1 is not empty. That is, for the following set of the parameter values, there exists a stable Markov perfect equilibrium.

Example 1 Let $a = 5, b = 1, \bar{c} = 2, r = \frac{1}{10}, \alpha = 5, \beta = 1$, and $\gamma = \frac{1}{20}$. Then the coefficients G and F are $G = \frac{1}{16}\left[\frac{97}{5} - 3\sqrt{\frac{209}{5}}\right]$ and $F = \frac{3 + 300G}{104 - 80G}$. The stability condition $3b\beta + 2\gamma - 2G > 0$ is satisfied.

We have derived a Markov perfect equilibrium for the case of two firms. For $n \geq 3$, we may derive a Markov perfect equilibrium by conducting a similar analysis above for the symmetric case, $V^i = V = \hat{E} + \hat{F}Y + 0.5\hat{G}Y^2$ for every i. Thus we have:

$$y_i^* = \frac{a - \bar{c} - \hat{F} + (\gamma - \hat{G})Y}{(n+1)b}, \quad \forall i.$$

14.4 OPEN-LOOP EQUILIBRIUM

In the previous section, we have derived a Markov perfect equilibrium. In this section, we examine open-loop strategies. Open-loop strategies and open-loop Nash equilibrium are defined as follows:

Definition 3 The open-loop strategy space for firm i is the set
$$S_i = \{x_i(t) : x_i(t) \text{ is piecewise continuous and } x_i(t) \geq 0 \text{ for every } t\}$$

Definition 4 An open-loop Nash equilibrium is an n-tuple of open-loop strategies $x^* = (x_i^*, x_{-i}^*)$ such that for each $i \in I$, $J_i(x_i^*, x_{-i}^*) \geq J_i(x_i, x_{-i}^*)$, for $\forall x_i \in S_i$.

We consider two settings regarding the firms harvesting decisions. First, we consider the case where the firms choose harvesting strategies non-co-operatively. Then we analyse the case where the firms co-operate in their harvesting decisions.

For firm i, the sum of its discounted profits is given by:

$$\int_0^\infty [a - b\sum_{i=1}^n y_i - (\bar{c} - \gamma Y)]y_i e^{-rt} dt. \tag{14.7}$$

Recall that the change in the stock of the natural resource is given by:

$$\dot{Y} = \alpha - \beta Y - \sum_{i=1}^{n} y_i . \tag{14.8}$$

The current value Hamiltonian for firm i is then given by:

$$H_i = \left[a - b\sum_{i=1}^{n} y_i - (\overline{c} - \gamma Y) \right] y_i + \lambda_i \left(\alpha - \beta Y - \sum_{i=1}^{n} y_i \right), \tag{14.9}$$

where λ_i is a costate variable.

The necessary conditions for an open-loop Nash equilibrium are:

$$\frac{\partial H_i}{\partial y_i} = -by_i + \left[a - b\sum_{i=1}^{n} y_i - (\overline{c} - \gamma Y) \right] - \lambda_i = 0, \tag{14.10}$$

$$\dot{\lambda}_i = r\lambda_i - \frac{\partial H_i}{\partial Y}$$
$$= r\lambda_i - (\gamma y_i - \beta \lambda_i) \tag{14.11}$$
$$= (r + \beta)\lambda_i - \gamma y_i,$$

and

$$\lim_{t \to \infty} e^{-rt} \lambda_i = 0 . \tag{14.12}$$

Summing equation (14.10) over i, we get:

$$-b\sum_{i=1}^{n} y_i + n\left[a - b\sum_{i=1}^{n} y_i - (\overline{c} - \gamma Y) \right] - \sum_{i=1}^{n} \lambda_i = 0 . \tag{14.13}$$

Let $Q \equiv \sum_{i=1}^{n} y_i$. Then equation (14.13) may be rewritten as:

$$-bQ + n[a - bQ - (\overline{c} - \gamma Y)] - \sum_{i=1}^{n} \lambda_i = 0 . \tag{14.14}$$

Differentiating (14.14) with respect to time t, we have:

$$-b\dot{Q} + n\left[-b\dot{Q} + \gamma \dot{Y} \right] - \sum_{i=1}^{n} \dot{\lambda}_i = 0 . \tag{14.15}$$

Using (14.8) and (14.11), (14.15) may be rewritten as:

$$-b\dot{Q} + n\left[-b\dot{Q} + \gamma(\alpha - \beta Y - Q) \right] - [(r + \beta)\sum_{i=1}^{n} \lambda_i - \gamma Q] = 0 . \tag{14.16}$$

It follows from (14.16) that:

$$\dot{Q} = \frac{[-(n-1)\gamma + (n+1)(r+\beta)b]Q + n\gamma(\alpha - \beta Y) - n(r+\beta)[a - (\bar{c} - \gamma Y)]}{(n+1)b}. \quad (14.17)$$

Thus we have the system of the differential equations (14.8) and (14.17). Then for the locus of $\dot{Q} = 0$, we have:

$$[(n+1)(r+\beta)b - (n-1)\gamma]Q = n[(r+\beta) + \gamma\beta]Y + n(a-\bar{c})(r+\beta) - n\alpha\gamma. \quad (14.18)$$

Thus we have:

$$Q = \frac{(r+2\beta)\gamma Y + (a-\bar{c})(r+\beta) - \alpha\gamma}{(n+1)(r+\beta)b - (n-1)\gamma} \cdot n \equiv \Phi(Y). \quad (14.19)$$

Let $A \equiv (a-\bar{c})(r+\beta) - \alpha\gamma$ and $B \equiv (n+1)(r+\beta)b - (n-1)\gamma$. Then $B > 0$ and $\alpha B > A$ are conditions necessary for the existence of an open loop Nash equilibrium. We then note that $\Phi(Y)$ is upward sloping in the (Y,Q) space, that is, $d\Phi/dY > 0$.

At the steady state, we have $\dot{Q} = 0$ and $\dot{Y} = 0$. Then two loci $\dot{Q} = 0$ and $\dot{Y} = 0$ are $Q = \Phi(Y)$ and $Q = \alpha - \beta Y$ (see Figure 14.2).

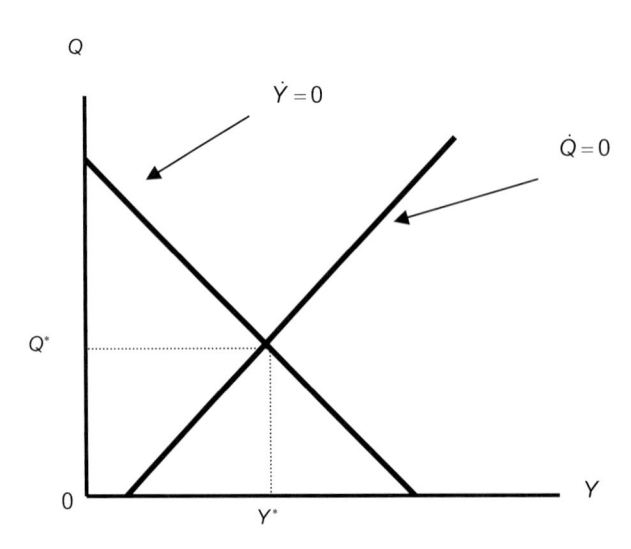

Figure 14.2 Open-loop Nash Equilibrium.

Therefore we may conclude that there exists a unique open-loop Nash equilibrium, characterised by the harvest Q^{N*} and the stock level Y^{N*} such that Q^{N*} and Y^{N*} satisfy $Q = \Phi(Y)$ and $Q = \alpha - \beta Y$. We may also demonstrate that by (14.19), if n increases, then the vertical intercept and the slope of the curve $Q = \Phi(Y)$ in the (Y,Q) space increase. Thus we may conclude that if n increases, then the stock level at the equilibrium decreases.

We may also examine the effects of change in the discount rate on the open-loop equilibrium. Consider the same parameter values as in Example 1 except the discount rate r. Then we may easily show that the larger the discount rate, the smaller the stock level at the open-loop equilibrium.

Using the previous numerical example, we may now look at the stock levels of the resource at the open-loop Nash equilibrium and the Markov perfect equilibrium. Recall that we set $n = 2, a = 5, b = 1, \bar{c} = 2, r = 1/10, \alpha = 5, \beta = 1$, and $\gamma = 1/20$. Then at the open-loop Nash equilibrium, $Y^{N*} \doteq 2.933$. At the Markov perfect equilibrium, $Y^* \doteq 2.922$. Therefore the stock level of the resource is larger at the open-loop Nash equilibrium than the Markov perfect equilibrium for this set of parameter values.

For a general comparison of the stock levels between the open-loop Nash equilibrium and the Markov perfect equilibrium, we need to examine $Q = \Phi(Y)$ obtained by open-loop strategies and $Q = 2[a - \bar{c} - F + (\gamma - G)Y]/3b$ by Markov feedback strategies. Thus we must compare Y^{N*} which solves:

$$Q = \Phi(Y^{N*}) \text{ and } Q = \alpha - \beta Y^{N*} \text{ and}$$

Y^* which solves:

$$Q = \frac{2[a - \bar{c} - F]}{3b} + \frac{2(\gamma - G)}{3b} \cdot Y^* \text{ and } Q = \alpha - \beta Y^*.$$

Therefore we may conclude that a comparison between the open-loop equilibrium and the Markov perfect equilibrium depends upon the slopes and intercepts of the two lines of $Q = \Phi(Y)$ and $Q = 2[a - \bar{c} - F + (\gamma - G)Y]/3b$.

For the parameter values in the previous numerical example, we have $Q \doteq 1.8769 + 0.0646Y$ for open-loop equilibrium and $Q \doteq 1.9803 + 0.0331Y$ for Markov perfect equilibrium. Thus we obtain $Y^* \doteq 2.922 < Y^{N*} \doteq 2.933$.

We have so far analysed the case where the firms choose their strategies non-co-operatively. Now we consider the case where the firms co-operate when they harvest the natural resource.

The objective function in this case is given by:

$$\sum_{i=1}^{n} \int_0^\infty \left[f(\sum_{i=1}^{n} x_i, X) - C(X) \right] x_i e^{-rt} dt. \qquad (14.20)$$

The current value Hamiltonian is then given by:

$$K = \sum_{i=1}^{n}\left[a - b\sum_{i=1}^{n}y_i - (\overline{c} - \gamma Y)\right]y_i + \xi\left(\alpha - \beta Y - \sum_{i=1}^{n}y_i\right), \quad (14.21)$$

where ξ is a costate variable.

The necessary conditions for an open-loop equilibrium are

$$\frac{\partial K}{\partial y_i} = -\sum_{i=1}^{n}by_i + \left[a - b\sum_{i=1}^{n}y_i - (\overline{c} - \gamma Y)\right] - \xi = 0, \quad (14.22)$$

$$\dot{\xi} = r\xi - \frac{\partial K}{\partial Y}$$

$$= r\xi - (\gamma \sum_{i=1}^{n}y_i - \beta\xi)$$

$$= (r + \beta)\xi - \gamma \sum_{i=1}^{n}y_i \quad (14.23)$$

$$= (r + \beta)\xi - \gamma Q,$$

and

$$\lim_{t \to \infty} e^{-rt}\xi = 0. \quad (14.24)$$

Differentiating (14.22) with respect to time t, we obtain:

$$\dot{\xi} = -2b\dot{Q} + \gamma\dot{Y}. \quad (14.25)$$

Substituting (14.8), (14.22) and (14.23) into (14.25), we have:

$$(r + \beta)\left[a - 2bQ - (\overline{c} - \gamma Y)\right] - \gamma Q = -2b\dot{Q} + \gamma(\alpha - \beta Y - Q). \quad (14.26)$$

It follows from (14.26) that we have:

$$\dot{Q} = \frac{\gamma(\alpha - \beta Y) - (r + \beta)a + 2(r + \beta)bQ - (r + \beta)(\overline{c} - \gamma Y)}{2b}. \quad (14.27)$$

Thus we have the system of the differential equations (8) and (27). Then for the locus of $\dot{Q} = 0$:

$$Q = \frac{(r + 2\beta)\gamma Y + (a - \overline{c})(r + \beta) - \alpha\gamma}{2(r + \beta)b} \equiv \Psi(Y). \quad (14.28)$$

Note also that $\Psi(Y)$ is upward sloping in the (Y, Q) space, that is, $d\Psi/dY > 0$.

At the steady state, we have $\dot{Q} = 0$ and $\dot{Y} = 0$. Then two loci $\dot{Q} = 0$ and $\dot{Y} = 0$ are $Q = \Psi(Y)$ and $Q = \alpha - \beta Y$.

Thus we see that there exists a unique open-loop equilibrium, characterised by the harvest Q^{C*} and the stock level Y^{C*} such that Q^{C*} and Y^{C*} satisfy $Q = \Psi(Y)$ and $Q = \alpha - \beta Y$. Note that this cooperative outcome corresponds to the case of $n = 1$ in (14.19).

Then for each Y, we have $\Phi(Y) > \Psi(Y)$. That is, the locus $\Phi(Y)$ lies above the locus $\Psi(Y)$ in the (Y, Q) space. Then we have the following result:

Proposition 1 At the open-loop equilibrium, we have

$$\sum_{i=1}^{n} y_i^{N*} \text{ is increasing in } n \text{ and } Y^{N*} \text{ is decreasing in } n.$$

This proposition then implies that the stock level is smaller under the non-co-operative case than under the co-operative outcome. We note that the proportional harvest relative to the stock level is larger in the non-co-operative equilibrium.

For the parameter values that we used in Example 1, we obtain the stock level of the co-operative solution $Y^{C*} \doteq 3.449$ for open-loop equilibrium, which is larger than that of the non-co-operative equilibrium $Y^{N*} \doteq 2.933$.

14.5 TAXATION

In the above analysis, we have considered a renewable natural resource as common property. It is very often documented that common property resources will be over exploited. In normative analyses of common property resources, the effects of taxes are often discussed. In this section, we will examine the effects of taxation on equilibrium harvest. We consider a specific tax whose rate depends upon the stock of the natural resource. Let θ/X be a tax rate, and $\theta \geq 0$. Thus, given θ, the smaller the stock of the natural resource, the higher the effective tax rate. The objective function of firm i becomes:

$$\int_0^\infty \left[p - C(X) - \frac{\theta}{X} \right] x_i e^{-rt} dt.$$

From the result in section 14.2, for a Markov perfect equilibrium, we have:

$$y^* = \frac{a - c(Y) - \theta - \{F + GY\}}{3b}.$$

Thus we get $\partial y^* / \partial \theta < 0$.

Therefore, if the tax rate parameter θ increases, then the equilibrium harvest rate y^* will decrease.

For deriving an open-loop Nash equilibrium for the game under taxation, conducting a similar analysis as in section 14.4 yields the following loci:

$$\Phi^T(Y) = \frac{(r+2\beta)\gamma Y + (a-\bar{c}-\theta)(r+\beta) - \alpha\gamma}{(n+1)(r+\beta)b - (n-1)\gamma} \cdot n$$

and

$$\Psi^T(Y) = \frac{(r+2\beta)\gamma Y + (a-\bar{c}-\theta)(r+\beta) - \alpha\gamma}{2(r+\beta)b}.$$

When θ increases, the curves $\Phi^T(Y)$ and $\Psi^T(Y)$ shift downward in the (Y,Q) space. Thus we may conclude that if θ increases, then the stock level will be larger for both non-co-operative and co-operative cases.

14.6   CONCLUSION

In this chapter, we have studied firms' exploitation of a renewable natural resource in a dynamic setting. We have constructed a differential game to examine oligopolistic firms' harvesting in a continuous time infinite horizon model. The differential game that we have analysed is not linear-quadratic. As is well known, it is notoriously difficult to derive a Markov perfect equilibrium for differential games that are not linear-quadratic. In this chapter, we have specified particular functional forms and have been able to derive a Markov perfect equilibrium. One salient feature of the model is that an inverse demand function depends on the stock of the renewable natural resource. This implies that consumers are concerned with their environment when they make their purchasing decisions on the product. We have also derived an open-loop equilibrium. We have shown that the stock level is smaller under non-co-operation than under co-operation.

We have conducted a positive analysis of renewable natural resource exploitation, but have not examined a normative analysis. We have shown that with a particular form of taxation that depends on the stock of the resource, there exist a Markov perfect equilibrium and an open loop Nash equilibrium. In particular, we have shown that if the tax rate parameter becomes larger, then the stock level increases. Based upon this positive analysis, we believe that our model could be used in the normative analysis of the problems of oligopolistic exploitation of renewable natural resources.

Since one of the main objectives in this chapter is to derive a Markov perfect equilibrium for a game that is not a linear-quadratic, we have adopted the specific functional forms regarding the demand function, the cost function, and the growth function. Whether these specific functions are relevant to some of the actual renewable natural resources would be an interesting empirical question. Examining more general models than ours is also left for future theoretical research.

APPENDIX 14.1

Proof of Theorem1: The objective function of firm i is:

$$J_i = \int_0^\infty [px_i - C(X)x_i]e^{-rt}dt.$$

Using $Y \equiv \ln X$ and $y_i \equiv x_i/X$, $i = 1, 2$, the objective function of firm i may be rewritten as:

$$\int_0^\infty [a - b(y_i + y_j) - (\bar{c} - \gamma Y)]y_i e^{-rt}dt, \ i, j = 1, 2, \ i \neq j. \tag{14.29}$$

Recall that the change in the stock is given by:

$$\dot{X} = X(\alpha - \beta \ln X) - \sum_{i=1}^n x_i.$$

Then we have:

$$\frac{\dot{X}}{X} = \alpha - \beta \ln X - \frac{\sum_{i=1}^n x_i}{X},$$

which may be rewritten as:

$$\dot{Y} = \alpha - \beta Y - (y_i + y_j). \tag{14.30}$$

Let $V^i(X)$ be the value function for firm i. Then the system of Hamilton-Jacobi-Bellman equation becomes:

$$rV^i(Y) = \max_{y_i} \left\{ \begin{array}{l} [a - b(y_i + y_j) - (\bar{c} - \gamma Y)]y_i + \dfrac{dV^i(Y)}{dY} \\ *[\alpha - \beta Y - (y_i + y_j)] \end{array} \right\}. \tag{14.31}$$

Let $c(Y) \equiv \bar{c} - \gamma Y$. Solving the maximisation problem of the right hand side of (14.31) yields:

$$y_i^* = \frac{a - c(Y) - \frac{dV^i(Y)}{dY}}{3b}.$$

Now we assume that the value function is symmetric. That is, $V^i = V$ for $i = 1, 2$. Next suppose that the value function takes the following form:

$$V(Y) = E + FY + \frac{1}{2}GY^2. \tag{14.32}$$

We will determine coefficients E, F, and G. From (14.32), we have:

$$\frac{dV(Y)}{dY} = F + GY. \tag{14.33}$$

Then we get:

$$y_i^* = y^* = \frac{a - c(Y) - \{F + GY\}}{3b}. \tag{14.34}$$

Substituting (14.32), (14.33) and (14.34) into (14.31), we get:

$$r\left[E + FY + \frac{1}{2}GY^2\right]$$
$$= \left[a - 2b\left\{\frac{a - c(Y) - F - GY}{3b}\right\} - c(Y)\right]\left\{\frac{a - c(Y) - F - GY}{3b}\right\} \tag{14.35}$$
$$+ (F + GY)\left[g(Y) - 2\left\{\frac{a - c(Y) - F - GY}{3b}\right\}\right].$$

The equation (14.35) must hold for any Y, and hence we must have:

$$8G^2 - \{9br + 18b\beta + 10\gamma\}G + 2\gamma^2 = 0, \tag{14.36}$$

$$\{9br + 5\gamma - 8G + 9b\beta\}F - (a - \bar{c})(2\gamma - 5G) - 9b\alpha G = 0, \tag{14.37}$$

and

$$9brE - (a - \bar{c} + 2F)(a - \bar{c} - F) - 3F\{3b\alpha - 2(a - \bar{c} - F)\} = 0. \tag{14.38}$$

It follows from (14.36), (14.37) and (14.38) that we obtain:

$$G = \frac{(9br + 18b\beta + 10\gamma) \pm \sqrt{(9br + 18b\beta + 10\gamma)^2 - 64\gamma^2}}{16}, \tag{14.39}$$

$$F = \frac{(a - \bar{c})(2\gamma - 5G) + 9b\alpha G}{9b(r + \beta) + 5r - 8G}, \tag{14.40}$$

and

$$E = \frac{(a - \bar{c} + 2F)(a - \bar{c} - F) + 3F\{3b\alpha - 2(a - \bar{c} - F)\}}{9br}. \tag{14.41}$$

Next we will derive a stable Markov perfect equilibrium trajectory. Substituting (14.34) into (14.30) yields the following differential equation for Y.

$$\dot{Y} + \{\beta + \frac{2\gamma}{3b} - \frac{2G}{3b}\}Y + \frac{2}{3b}\{a - \bar{c} - F\} - \alpha = 0. \tag{14.42}$$

A particular solution to the differential equation (14.42) is:

$$Y^* = \frac{3b\alpha - 2(a - \bar{c} - F)}{3b\beta + 2\gamma - 2G}.$$

Then the solution of (14.42) is:

$$Y(t) = Y^* + (Y_0 - Y^*)e^{-\left(\frac{3b\beta + 2\gamma - 2G}{3b}\right)t}. \tag{14.43}$$

We must have $3b\beta + 2\gamma - 2G > 0$ in order that this state trajectory is asymptotically stable.

Now we have:

$$\frac{(9br + 18b\beta + 10\gamma) + \sqrt{(9br + 18b\beta + 10\gamma)^2 - 64\gamma^2}}{16} > \frac{36b\beta + 16\gamma}{16} > \frac{3b\beta + 2\gamma}{2}.$$

Thus $G = \frac{(9br + 18b\beta + 10\gamma) + \sqrt{(9br + 18b\beta + 10\gamma)^2 - 64\gamma^2}}{16}$ in (14.39) does not satisfy the above stability condition. Hence we may conclude that the coefficient G must be:

$$G = \frac{(9br + 18b\beta + 10\gamma) - \sqrt{(9br + 18b\beta + 10\gamma)^2 - 64\gamma^2}}{16}. \tag{14.44}$$

We can also show that $\frac{dy^*}{dY} = \frac{\gamma - G}{3b} > 0$. Thus a Markov perfect equilibrium strategy y^* is increasing in Y.

Acknowledgements

I would like to thank Gordon Munro and participants at the Conference on Fisheries Economics and Management in Honour of Professor Gordon R. Munro for their helpful comments. I also gratefully acknowledge the helpful comments and the detailed and insightful suggestions of two anonymous referees.

References

Basar, T. and G.J. Olsder. 1995. *Dynamic Noncooperative Game Theory*, 2nd Ed. Academic Press, San Diego.

Clark, C.W. 1990. *Mathematical Bioeconomics: The Optimal Management of Renewable Natural Resources*, 2nd Ed. Wiley, New York.

Conrad, J.M. and C.W. Clark. 1987. *Natural Resource Economics*. Cambridge University Press.

Fudenberg, D. and J. Tirole. 1991. *Game Theory.* MIT Press, Cambridge, MA.

Kamien, M.I. and N.L. Schwartz. 1991. *Dynamic Optimization: The Calculus of Variations and Optimal Control in Economics and Management*, 2nd Ed. North Holland, New York.

Kaitala, V. and M. Pohjola. 1988. Optimal recovery of a shared resource stock: A differential game model with efficient memory equilibria. *Natural Resource Modeling* 3: 91-119.

Levhari, D. and L.J. Mirman. 1980. The great fish war: An example using a dynamic Cournot-Nash solution. *Bell Journal of Economics* 11: 322-34.

Munro, G.R. 1979. The optimal management of transboundary renewable Resources. *Canadian Journal of Economics* 12: 355-76.

Munro, G.R. 1982. Fisheries, extended jurisdiction and the economics of common property resources. *Canadian Journal of Economics* 15: 405-25.

Section 4

Applied fisheries economics and management

15

The Role of the Fishing Industry in the Icelandic Economy

Sveinn Agnarson
Ragnar Arnason

15.1 INTRODUCTION

According to the conventional wisdom, fishing has been Iceland's most important industry during the twentieth century (Nordal and Kristinsson, 1987; Snævarr, 1993; Arnason, 1994). Indeed, Iceland's rapid economic growth during the twentieth century is generally attributed to the expanding fishing industry (Nordal and Kristinsson, 1987; Jónsson, 1999). Certain macro-economic statistics lend support to this belief. During most of the twentieth century, fish products constituted the bulk of Iceland's exports, reaching as high as 95% of the merchandise exports in the 1940s. At the end of the century fish products still accounted for over 60% of merchandise exports. It may be noted in this context that during the twentieth century exports as a fraction of GDP averaged about 35%. The use of labour also tends to support the conventional wisdom. During the first half of the twentieth century, the fishing industry often employed about a quarter of the working population with an average during the whole period of around 20%. Since then, however, the proportion of the population working in the fishing industry has declined to about 10% in recent years. These developments in merchandise exports and employment are illustrated in Figure 15.1. Finally, more qualitative evidence (Jónsson, 1984; Jónsson, 1999; Snævarr, 1993) strongly suggests that virtually all towns and villages along the Icelandic coastline, where most of the population currently resides, were initially established to take advantage of favourable fishing and fish exports conditions and their fortunes subsequently waxed and waned in step with those of the fisheries.

When it comes to measurements of the direct contribution of the fishing industry to the country's gross domestic product (GDP) as carried out by conventional national accounting, the picture is far less conclusive. These statistics, available for the period 1900-1945 and since 1973, indicate that the fishing industry has in the past three decades only accounted for about 15% of the GDP on average. Moreover, this direct contribution has been declining in recent years and was in 2000 only about 11%. By comparison, the share the fishing industry in GDP averaged about 25% during 1900-1945. This development is further illustrated in Figure 15.2.

So, according to the national accounts statistics, the fishing industry, while certainly significant, seems far from being of overriding economic importance as claimed by the conventional wisdom. Indeed, by the national accounts measure, the banking and insurance, the general service sector, commerce and tourism are all larger than the fisheries sector.

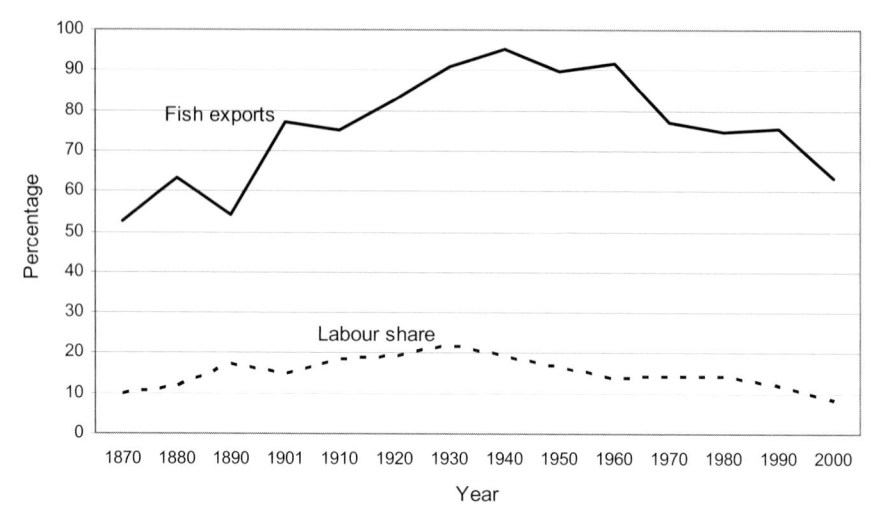

Figure 15.1 Labour employed by the fishing sectors and fish exports as percentage of total labour and merchandise exports 1870-2000. (Jónsson and Magnússon (1997), Table 3.8 and National Economic Institute, Table 5.3.)

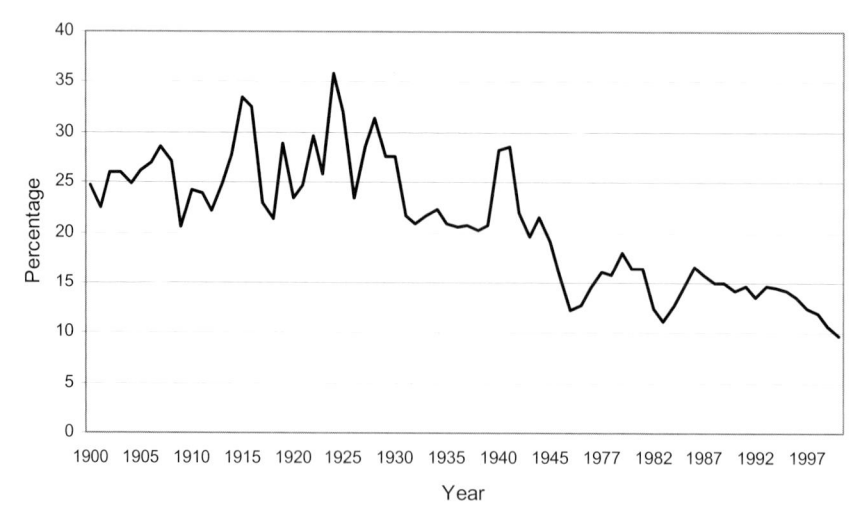

Figure 15.2 Direct contribution of the fishing industry to GDP 1900-2000. Jónsson and Magnússon (1997), Table V.14.2, and National Economic Institute, Table 1.7.)

The contradiction between the conventional wisdom and the national accounts measure of contribution to GDP is, however, perhaps more apparent than real. The national accounts measure economic contribution of industries to GDP by simply aggregating value-added recorded in each industry. This mechanical procedure ignores all links and relationships between industries. In particular, it makes no distinction between industries that may be regarded as primary (such as mining for export) and industries that are derived or secondary (such as hairdressing, manicure and, of course, public services). However, it seems intuitively obvious that expansion or contraction in the former set of industries (primary ones) will have much greater impact on GDP than variations in the latter. If this is true, national accounts measures of contribution to GDP do not necessarily reflect the 'real' economic importance of the respective industries. Indeed they are not designed to do so and should not be interpreted as such.

The above raises the question of the 'real' role of the fishing industry in the Icelandic economy. Is its impact limited to the value added generated within the industry as suggested by the way the national accounts are constructed or is it much greater than this direct contribution in accordance with the conventional wisdom? More generally, is it possible that some industries are more fundamental than others in the sense that if the former are removed the latter will greatly decline while the converse does not apply? In this chapter we try to answer these questions in terms of the role of the fishing industry in the Icelandic economy. Clearly, however, the basic question is of general interest. Do certain industries have GDP impacts far in excess of the value added generated within the industry?

In the next section we analyse further the role the fisheries sector have played in Iceland's economic development during the twentieth century. We then give a brief account of the theoretical basis for our approach, namely the so-called economic base theory. In section 15.4, we introduce the econometric methodology employed in the chapter and describe the data on which this study is based. In section 15.5, we employ the econometric techniques to investigate the importance of the fishing industry for the Icelandic economy. In the final section, we summarise our results and how this approach can be used to determine the role of various industries in other economies.

15.2 HISTORICAL BACKGROUND

15.2.1 Icelandic economic development and the fisheries

The above mentioned contradiction between the conventional wisdom and the national accounts measure of contribution to GDP is particularly well exemplified by the case of the Icelandic fishing industry. In spite of its relatively modest direct contribution to GDP – according to the national accounts – it appears that all five major economic depressions experienced during the twentieth century may be directly related to changes in the fortunes of the fishing sector. Let us now briefly review this evidence.

The first major economic depression in Iceland in the twentieth century begins in the latter stages of World War I, which had catastrophic effects on Iceland as on so many other European countries. The first two war years were, however, favourable for the fishing sector, as increased demand pushed up foreign prices, but in 1916 the international trade structure broke down and Iceland had to accept harsh terms of trade with the Allies. In 1917, Iceland sold half of her trawler fleet to France and demersal fish and herring catches were consequently substantially reduced in 1917 and 1918. The result was a sharp dip in the GDP and a generally depressed economy until 1920 (Figure 15.3).

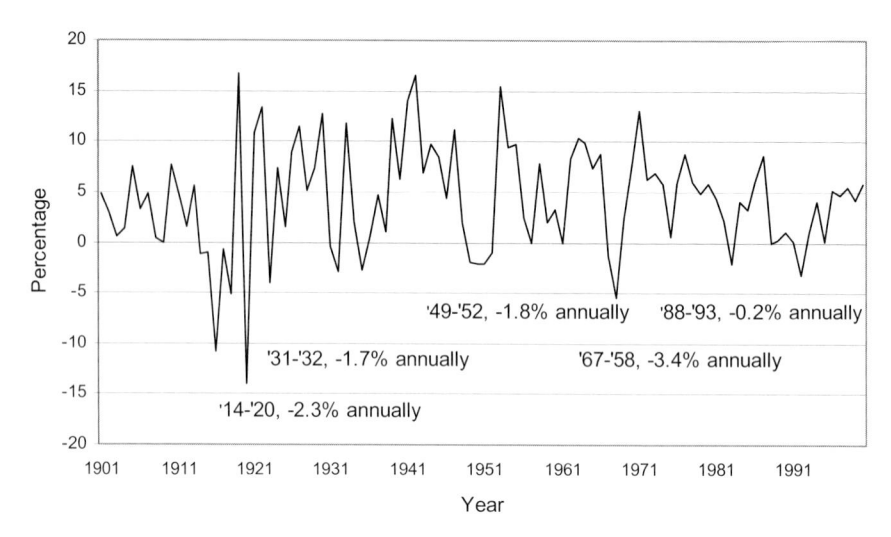

Figure 15.3 GDP growth in Iceland 1901-2000: Major depressions. (National Economic Institute, Table 1.1.)

The effects of the 'Great Depression' were first felt in Iceland in the autumn of 1930, and in the following two years GDP fell by 0.4% and 5% respectively as demand for maritime exports declined sharply. Following a brief recovery, the economy was hit again when the Spanish Civil War broke out in 1936 and closed Iceland's most important market for fish products. Despite these shocks, economic growth still averaged 3% in the 1930s, mostly because of strong rebound in the fisheries, especially the herring fisheries during 1933 to 1939. From this it appears that it was primarily because of the strong performance of the fisheries in the 1930s that the 'Great Depression' was felt less in Iceland than most other western countries.

World War II was a boom period for Iceland led by good catches and very favourable export prices. But in 1947 and subsequent years, herring catches fell considerably and real export prices subsided from the high wartime levels. The result was a prolonged economic contraction from 1949-52.

During 1961-67, the economy exhibited a very respectable growth rate of over 5% on average. This was to a large extent based on extremely good herring fisheries during most of the decade. When the herring stocks collapsed toward the end of the decade the result was a severe economic depression in 1968-69 with GDP declining by 1.3% and 5.5% respectively. Unemployment reached over 2% – a great shock for an economy used to excess demand for labour since the 1930s – and many households moved abroad in search of jobs. Net emigration amounted to 0.6% of the total population in 1969, and 0.8% in 1970.

High economic growth resumed in 1971-80, averaging 6.4%. However, just as in the 1960s, this growth was to a significant extent based on over-exploitation of the most important fish stocks. Reduced fishing quotas and weak export prices reduced fishing industry profitability in the late 1980s. Partly as a consequence of this, the Icelandic economy remained stagnant through the years 1988-1993, with an average annual decline in the GDP of 0.2%.

Since 1993, however, the Icelandic economy has registered steady and quite impressive annual growth rates. One reason for this is a recovery of some of the fish stocks. More importantly, however, are generally more favourable fish export prices and the impact of the individual transferable quota (ITQ) system, which was introduced in the demersal fisheries in 1984 and in 1990 in all Icelandic fisheries. The ITQ system has enabled the fishing industry to increase and stabilise profits and much more easily adjust to changing quotas and fish availability.

15.3 THEORETICAL BACKGROUND

15.3.1 *The concept of the economic base*

Economies may be seen as a collection of industries. Obviously, the contribution of the various industries to GDP may differ. It is equally obvious that taking indirect as well as direct impacts into account, the overall contribution of individual industries to GDP may deviate substantially from their direct contribution as measured by the national accounts. It is even possible that economies depend wholly on certain industries in the sense that they came into being as a result of these industries and would collapse if those industries were removed.

It is easy to think of examples of this. Imagine for instance the case of an oil rig in the middle of frozen tundra. The oil pumping activity requires labour *in situ*. The labour demands a range of local services. This gives rise to local economic activity that with it own linkages and interrelationships may easily amount to a significant multiple of the initial value-added in the oil industry. To be a bit more concrete, let the local services include a barber/hairdresser. Imagine now, that the barber/hairdresser is taken ill and not replaced. As a result the local population will have to spend their income on something else (or save it). Depending on what these alternative expenditures are, the local GDP will normally be reduced by some fraction of the net income of the barber/hairdresser. Now imagine that instead of the hairdresser being taken ill, the oil well runs dry. In this case, the GDP is not simply reduced, the whole economy folds. In this sense the oil pumping activity is more important than hair-dressing and similar services. Its functioning is necessary for the continued existence of the local economy. It constitutes the base or foundation of the local economy. Note that this result is in no way dependent on how large a fraction of the local economy the oil pumping activity constitutes. The point is that the oil-pumping activity is the reason people are living and an economy exists in the area.

Observations of this kind have given rise to the concept of the economic base (Tiebout, 1956a; Schaffer, 1999). The economic base is simply an industry or a collection of industries that are disproportionately important in a region's (or, for that matter, country's) economy in the sense that other economic industries depend on the operation of the economic base but not vice versa, at least not to the same extent. By implication, removing the base industries would reduce the GDP more than their direct contribution to the GDP as measured by the national economic accounts and vice versa. Note that the economic base industries can be regarded as being autonomous (or basic) while the other industries are dependent (non-basic).

The idea of the economic base has a long history. Schaffer (1999) traces the origins of this theory back to the Mercantilists, who regarded any activity conducive to a favourable balance of trade as the nation's economic base, and later the Physiocrats who regarded agriculture as the national economic base. The modern concept of the economic base was initially formulated by the German economic historian, Werner Sombart[1] (Krumme, 1968) but has subsequently been refined by several researchers in the fields of economic history and regional economics including North (1955) and Tiebout (1956a, 1955b, 1962). Interestingly, the economic base theory is closely related to the well-known theory of economic staples developed by Innis (1930) to help explain the process of economic development in newly settled and resource rich regions, such as Canada. The exports of the staple in question – fur or cod in the Canadian case – drive the economy on and set the pace for economic growth. The central concept of the staple theory is the multiplier effects of the export sector and application of the theory involves identifying staples and explaining their economic roles (Watkins, 1963).

It may be helpful to illustrate the idea of a base industry a little more formally. For this purpose, consider a geographical region. Let us, for convenience of exposition, imagine that initially there is no economic activity in the region.[2] Now, assume that a natural resource is discovered in the region and that this resource is brought under exploitation.[3] For simplicity, assume moreover that the output from the resource is exported. Let us refer to the value-added generated by the exploitation by the symbol ϕ. The GDP in the region is now ϕ plus any derived economic activity. More formally:

$$y = yd + \phi, \qquad (15.1)$$

where yd represents the net demand for local goods (i.e. value–added) and y represents the total regional GDP. It should be noted that (15.1) is but a simple rewrite of the usual national income identity. At the same time it is an equilibrium relationship expressing the idea that supply equals demand.

Obviously, demand for local goods can only be generated by local income.[4] Now, let a represent the share of capital in the value-added, ϕ. Thus, the income of capital in the natural resource industry, which we may as well call resource rents, is $a \cdot \phi$. All other income, non-resource rent income, is obviously $y - a \cdot \phi$. On this basis let the demand for local goods be represented by the function:

$$yd = C1(y - a \cdot \phi) + C2(a \cdot \phi), \qquad (15.2)$$

where $C1(y - a \cdot \phi)$ is the demand for local goods out of non-resource rents income and $C2(a \cdot \phi)$ is the propensity to consume out of resource rent income. For analytical convenience, although this is by no means essential, we assume that both of these functions are differentiable and that the first function is increasing in its arguments.

Combining (15.1) and (15.2) yields:

$$y = C1(y - a \cdot \phi) + C2(a \cdot \phi) + \phi, \qquad (15.3)$$

---

[1] Sombart was an interesting fellow. He started his career as a Marxist sympathiser and ended it as a committed Nazi. In between he became the leading member of the last generation of the German Historical School of economics.
[2] It is easy but messier to extend the analysis to assume an initial positive level of GDP.
[3] Note that we do not need to assume the resource can be profitably exploited.
[4] To keep within the spirit of the story, other exports are not possible, although this could easily be included as well.

which, under the appropriate regularity assumptions implicitly defines the reduced form function:

$$y = Y(\phi), \qquad (15.4)$$

where, in accordance with our point of departure, $Y(0)=0$.

Now, differentiating (15.4) or, for that matter, (15.3) yields:

$$\frac{dy}{d\phi} = \frac{[C2'-C1']\cdot a + 1}{1-C1'}, \qquad (15.5)$$

where $C1'$ and $C2'$ represent the first derivatives of the two consumption functions, respectively. On basic economic grounds it seems safe to take both of these derivatives to be non-negative. Also, it seems likely that the marginal propensity to consume locally out of labour income should be higher than the marginal propensity to consume out of out of resource rents, i.e. $C1' > C2'$.

Expressions of type (15.5) are typically referred to as economic base multipliers (Frey, 1989). They represent the response of GDP to a change in the economic base industry, in this case the natural resource industry. Thus, in a sense they provide a measure of the economic importance or rather economic potency of the base industry.

The higher this multiplier, the more economically potent is the economic base industry.

As can be seen from (15.5), the size of economic base multiplier depends critically on the shape of the two consumption functions and, in the case where the two marginal propensities for local consumption are different, the share of capital in the resource industry value-added. The economic base multiplier is a somewhat complicated function of these items. However, given certain apparently reasonable numerical restrictions on their values, we may derive some bounds on the numerical value of the economic base multiplier. Thus, assuming that (i) $1 \geq a \geq 0$, (ii) $C2' \geq 0$ and (iii) $1 > C1' \geq C2'$, it is easy to show that $dy/d\phi \in [1,\infty]$. Note that the minimum of unity means that the GDP is increased by the natural resource value-added only and there are no multiplier effect. This occurs when the share of capital is unity and the capitalists' marginal propensity to consume locally is zero.

More generally, if the marginal propensity to consume locally out of resource rents (capitalists) is less than that of other income (labour), then the economic base multiplier is higher the higher the share of labour in the natural resource industry and vice versa. If the two marginal propensities to consume out of income are equal, i.e. $C1' = C2'$, then the multiplier reduces to the simple Keynesian multiplier $1/(1-C1')$. The same applies if the share of capital in natural resource value-added is zero.

Note that if natural resource exports are proportional to value-added, (15.5) is also proportional to the export base multiplier. More precisely, the export base multiplier would be:

$$\frac{dy}{de} = b \cdot \left(\frac{dy}{d\phi}\right) \qquad (15.6)$$

where e denotes export from the resource industry and b is the factor of proportionality between exports and value added, namely $\phi = b \cdot e$.

Finally, it is convenient to use this opportunity to clarify the relationship between the elasticity of GDP with respect to ϕ, $E(y, \phi)$, say, the economic base multiplier, $dy/d\phi$ and the share of the base industry in GDP, ϕ/y. By the definition of elasticities, this is simply:

$$E(y,\phi) \equiv \frac{dy}{d\phi} \cdot \frac{\phi}{y}. \tag{15.7}$$

According to the theory outlined here, an industry may constitute an economic base in spite of having limited backward and forward linkages, provided the consumption linkages are substantial. Reverting to the example of the oil-in-the-tundra discussed earlier, we can imagine a situation where all the oil is exported. Thus, there would be virtually no forward linkages. We can also assume that all the inputs, save labour, used in the oil industry are imported, so that the backward linkages would be small. However, the income accruing to oil workers and possibly capitalists would to some extent be spent in the area, and it is through this channel and subsequent multiplier effects that the oil industry serves as an economic base driving the local economy.

The base industry approach is different from the 'key-sector' analysis that has been used to identify so-called key industries in economies (Hazari, 1970; Schulz, 1977; Lenzen, 2003). The key sector approach is based on the input-output methodology introduced by Leontief in the 1930s (Leontief, 1936). It defines key sectors as those with the largest backward and forward linkages as specified by the corresponding rows and columns in the relevant input-output table (Harzai, 1970; McGilvray, 1977). Thus, the key sector analysis is production oriented. It ignores the consumption linkages that form an integral part of the base industry theory. In fact, as the oil-in-the-tundra example should make clear, the key sector analysis is liable to miss identifying base industries, i.e. industries without which the regional economy would collapse.

Having outlined the theoretical background underlying the concept of the economic base, we now turn to statistical modelling and estimation.

15.4 STATISTICAL THEORY

Since our primary aim is to analyse the relationship between GDP and the fishing sectors and establish a measure of the size of the economic base, an obvious way to proceed is to use regression techniques. As the model has to be flexible enough to capture both the short- and long-run impacts of the fisheries sector on economic growth we specify the following error correction model (Engle and Granger, 1987):

$$\Delta y_t = \beta_0 + \sum_{i=0}^{k} \beta_i \Delta f_{t-i} + \sum_{i=0}^{k} \delta_i \Delta k_{t-i} + \sum_{i=0}^{k} \gamma_i \Delta l_{t-i} + \lambda \mu_{t-i} + \varepsilon_t \tag{15.8}$$

where y, f, k, and l denote logarithmic transformations of GDP, production of marine products, capital stock and labour respectively, Δ represents the difference operator, i.e. $\Delta y_t = y_t - y_{t-1}$, the βs, δs, and γs are parameters to be estimated, λ is an adjustment parameter, μ_{t-i} is the error correction term, and ε represents a white noise error term. The error correction term is defined as:

$$\mu_t = y_t - (\alpha_0 + \alpha_1 f_t + \alpha_2 k_t + \alpha_3 l_t), \qquad (15.9)$$

which may be rewritten as

$$y_t = \alpha_0 + \alpha_1 f_t + \alpha_2 k_t + \alpha_3 l_t + \mu_t. \qquad (15.10)$$

The function depicted in equation (15.10) can be regarded as a long-run production function, and – provided all variables are in logs – the parameter α_3 thus represents the long-run output elasticity of the fisheries.

For the relationship described by equation (15.10) to hold, it is necessary that all the four variables – y, f, k, and l – are integrated of the same order, and that the error correction term is stationary. If that is the case the four variables are said to be co-integrated. Statistical tests can be used to check for co-integration, and also whether there exist more than one co-integrating relationships.

The short-run effects in equation (15.8) can be gauged by studying the parameters associated with the differenced variables, β_i, δ_i, and γ_i. The long-run effects, on the other hand, are revealed by the λ parameter and the error correction term, μ_{t-1}. Consider first the error term. As defined here, long-run disequilibrium is characterised by a situation where $\mu_{t-1} \neq 0$. The error term will take on a positive value if:

$$y_t > (\alpha_0 + \alpha_1 f_{t-1} + \alpha_2 k_{t-1} + \alpha_3 l_{t-1}), \qquad (15.11)$$

i.e. during booms, and a negative value during slacks. The adjustment parameter, λ, measures how quickly long-run equilibrium is re-established. A positive value for λ indicates that if the system is thrown off balance it will move further and further away from long-run equilibrium in subsequent periods. Positive values are therefore inconsistent with the notion of a stable equilibrium. The same applies to $\lambda < -2$, except the divergence is cyclical. A value of λ in the interval $]0,-2[$ indicates, on the other hand, that the system will revert to long-run equilibrium following a positive or negative shock. A $\lambda > -1$ indicates a non-cyclical adjustment. A $\lambda < -1$ indicates a cyclical adjustment. λ close to minus one indicates a relatively fast adjustment to equilibrium while λ close to zero or -2 indicates that once out of equilibrium, the adjustment to equilibrium will be a sluggish.[5]

Engle and Granger (1987) have developed a three-stage testing methodology for error correction models (ECMs). In the first step, the variables are tested for their order of integration. In the second stage, the long-run relationship outlined in equation (15.10) is estimated and the residuals from that equation, $\bar{\mu}_t$, checked for stationarity. The third step consists of estimating the ECM, in our case as specified in (15.8),

---

[5] The half time (the time it takes to bridge half of the distance to equilibrium) is given by the equation. $t = \ln(0.5)/\ln(1+\lambda)$.

where $\hat{\mu}_{t-1}$ serves as an instrument for the co-integrated variables in the long term relationship:

$$y_t - (\alpha_0 + \alpha_1 f_{t-1} + \alpha_2 k_{t-1} + \alpha_3 l_{t-1}). \tag{15.12}$$

Although straightforward to implement, the Engle-Granger procedure suffers from a rather serious defect concerning the nature and number of the co-integrating relationships. In equation (15.10), GDP is modelled as a function of the variables capital, labour and output of the marine sectors. However, we could just as easily have modelled, say, labour as a function of GDP, capital and marine output. Thus, the results obtained can depend on rather arbitrary assumptions, i.e. which variable we chose to normalise on. Further, when three or more variables are present, there may exist more than one co-integrating relationships, which may escape our attention if we believe that there should only exist a single long-run relationship.

The most popular way to sidestep these problems is to apply the Johansen (1988; 1995; Johansen and Juselius, 1990) maximum likelihood approach. This method uses full-information maximum likelihood (FIML) estimation to estimate the linear space spanned by the co-integrating vectors, and incorporates all prior knowledge about the existence of unit roots in the time series data. Further, as pointed out by Gonzalo (1994), FIML results in coefficient estimates being symmetrically distributed and asymptotically efficient. The Johansen's method also performs better than other procedures even when the error terms are non-normally distributed, and when the dynamics of the model under investigation are unknown. If the Johansen's procedure reveals the existence of one and only one co-integrating relationship, then we can proceed as before, and use the residuals from that relationship as our error correction term. If, on the other hand, more co-integration vectors are discovered, the Engle-Granger single-equation ECM approach cannot be applied.

15.5   ESTIMATION

Our model (see (15.8)) includes four variables; GDP, maritime production, capital and labour, denoted by y, f, k and l, respectively. The first three variables are all measured in constant (1990) prices, while the fourth variable, labour, is measured in man-years.[6]

Descriptive statistics for the four variables in levels are given in Table 15.1. Henceforth we will use the series in logarithmic form, which we will refer to as levels. Figures 15.4 and 15.5 trace the annual differences in the natural logarithms of the variables.

The first step of the estimation procedure is to determine the degree of integration of each of the time series (in logarithmic form). For this purpose, we employed the augmented Dickey-Fuller (ADF) statistic to test the null hypothesis of a unit root, i.e. test the hypothesis of non-stationarity. As revealed in Table 15.2, all four variables are found to be non-stationary in levels (logs), but stationary in first differences.[7] The variables are therefore integrated of order one, I(1).

---

[6] All series are taken from the website of the National Economic Institute. Accessible online http://www2.stjr.is/frr/thst/rit/sogulegt/english.htm.
[7] The regression equations included a constant and a trend term. The lag-length used was determined on the basis of the Akaike information criteria.

Table 15.1 Descriptive statistics: GDP, production of marine products and capital stock in fixed prices (1990) billion kr., labour in thousands of man-years. Period is 1963-2000.

	Minimum	Maximum	Mean	Std dev
GDP	115.7	476.5	280.4	104.6
Marine products	19.1	78.8	52.3	18.4
Capital stock	401.0	1,469.7	896.9	314.8
Labour	67.4	141.7	108.6	22.3

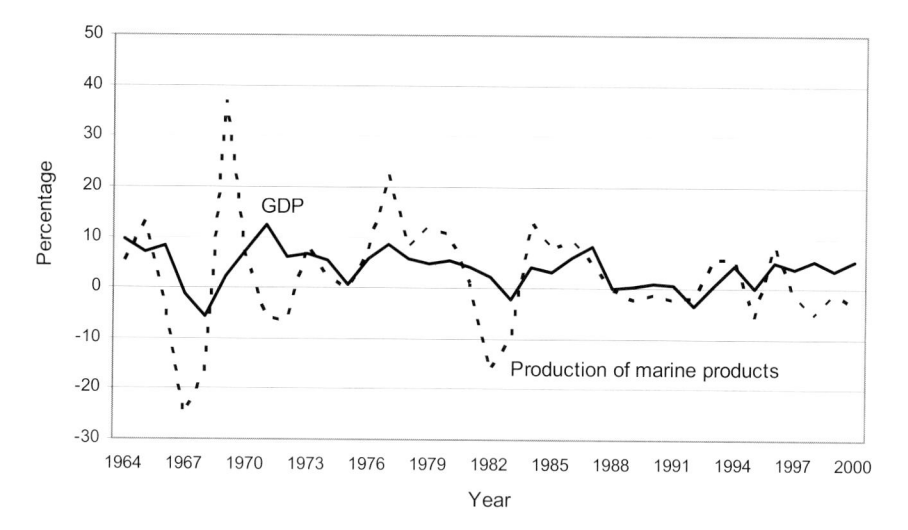

Figure 15.4 Percentage changes in GDP and production of marine products 1964-2000. (National Economic Institute, Tables 1.1, 4.5 and 4.6.)

Having established that all four variables are integrated of the same order, the next step is to check if there exists a co-integrating relationship between them. For this purpose we employ the Johansen maximum likelihood procedure as described above. To determine the lag length to be used, we first estimated a vector-autoregressive (VAR) model. Using the Akaike information criteria, a lag length of 2 was deemed appropriate. Consequently, the Johansen test was applied using one-period lags and including an intercept in the co-integrating relation.

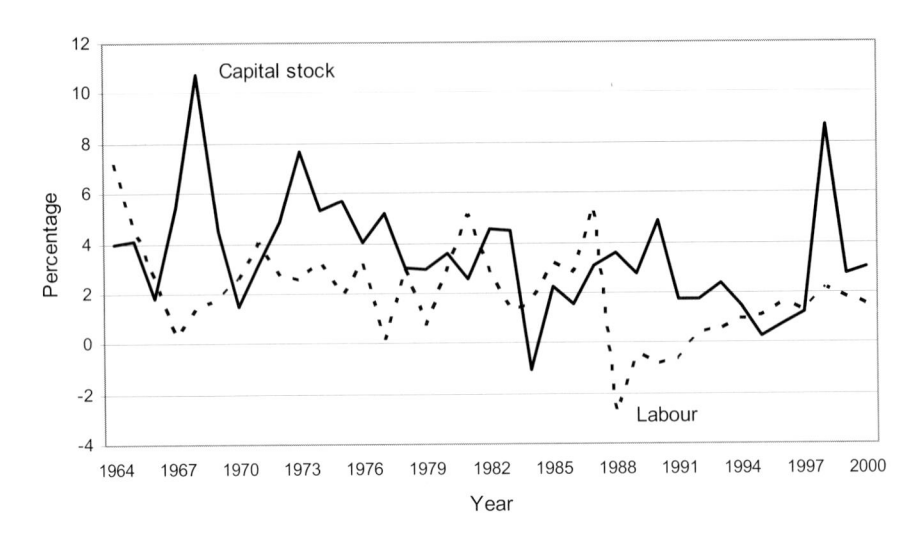

Figure 15.5 Percentage changes in the capital stock and labour employed by the fishing sectors 1964-2000. (National Economic Institute, Tables 2.4 and 5.2.)

Table 15.2 Results from augmented Dickey-Fuller tests for stationarity 1963-2000.

	Value	Lags
Levels		
GDP	-2.023	1
Marine production	-1.670	0
Capital	-2.725	0
Labour	-1.458	0
First differences		
GDP	-4.812*	0
Marine production	-7.225*	1
Capital	-5.441*	0
Labour	-4.863*	0

* denotes 1% level of significance

The Johansen's trace tests suggest the existence of one co-integration relationship at the conventional 5% level of statistical significance (see, Table 15.3). The co-integrating equation, expressed in terms of the one-period lagged error term μ_{t-1} is (standard errors in parentheses):

$$\mu_{t-1} = y_{t-1} - 0.3054 f_{t-1} - 0.4541 k_{t-1} - 0.7550 l_{t-1} + 5.9021 \qquad (15.13)$$
$$\phantom{\mu_{t-1} = y_{t-1} - 0.}(0.0646)\quad(0.1145)\quad(0.2266)\quad(0.9786)$$

Table 15.3 Results from Johansen's trace test for multiple co-integration vectors 1963-2000.

Eigenvalue	Likelihood ratio	5% critical value	1% critical value	Hypothesised number of co-integrating equations
0.528	53.75	53.12	60.16	None*
0.317	26.72	34.91	41.07	At most 1
0.219	13.01	19.96	24.60	At most 2
0.108	4.10	9.24	12.97	At most 3

*denotes rejection of the hypothesis at 5% significance level

All four parameters are significant at the 5% level or better. Rearranging gives the following long-run production function:

$$y_{t-1} = -5.9021 + 0.3054 f_{t-1} + 0.4541 k_{t-1} + 0.7550 l_{t-1} + \mu_{t-1}, \quad (15.14)$$

which represent the same long-run relationship as hypothesised in equation (15.10). Since all variables are in logs, the parameters may be regarded as elasticities. The output elasticity of the fisheries is thus estimated as 0.31, i.e. increasing the size of the fisheries production by 1% will in the long-run lead to a 0.31% increase in GDP. The output elasticity of capital is higher, or 0.45, and that of labour higher still or 0.76. The parameters associated with k t-1 and l t-1 sum to 1,21 indicating long-run increasing returns to scale, but the sum rises to 1,51 when the marine sector parameter is included.

As established in section 15.3, the output elasticity of the fisheries corresponds to the output elasticity of the economic base presented in equation (15.7). Thus we have:

$$0.31 = \alpha_1 = E(y, f) = \frac{dy}{df} \frac{f}{y}, \quad (15.15)$$

where, as before, f represents the output of the fisheries sector. Assuming that the share of fisheries in GDP amounts to 0.1, it easy to see that the economic base multiplier, dy/df, amounts in this case to 3.1.

To test the robustness of the above analysis, we employed the same procedure to check whether other economic output measures could play the same macro-economic role as fisheries production. Two extensions of this kind were carried out. First, as discussed earlier most of the marine products produced in Iceland are exported. Since marine products have made up most of Iceland's exports, it can be argued that we have so far really been examining the role exports – and not the marine sector – play in the generation of Iceland's GDP. It would therefore be interesting to check whether the value of all merchandise exports or the value of total exports (goods and services) exhibit a similar or stronger long-run relationship to that of maritime production. Second, we examined whether other similarly sized production sectors will be found to play a comparable role in generating GDP as the fishing industry. For this purpose we have combined the construction and transport sectors, whose aggregate output is similar to that of the fisheries sector.

Turning to the first extension, we used the Johansen maximum likelihood procedure to test for the existence of co-integrating relationships between GDP capital, labour and merchandise exports. Following the same procedure as before, we found that instead of one co-integrating relationship there appeared to exist three (see Table 15.4). Moreover, since some of the coefficients consistently had the wrong sign or were statistically insignificant, none of the long-term relationships estimated could be interpreted as a long-run production function.

Table 15.4 Results from Johansen's trace test for multiple co- integration vectors using export of goods as proxy for marine production 1963-2000.

Eigenvalue	Likelihood ratio	5% critical value	1% critical value	Hypothesised number of co-integrating equations
0.878	161.86	53.12	60.16	None[**]
0.787	84.08	34.91	41.07	At most 1[**]
0.437	25.80	19.96	24.60	At most 2[**]
0.139	5.52	9.24	12.97	At most 3

[**] denotes rejection of the hypothesis at 1% significance level

Replacing merchandise exports with total exports produced similar results. As shown in Table 15.5, we find evidence of three co-integrating equations instead of one, and, due to wrong signs of coefficients or statistical insignificance, none of those can be interpreted as a long-run production function.

We conclude that overall exports do not seem to play the same role in generating the Icelandic GDP as does the fisheries sector. They do not appear to constitute economic base industries in the same economically meaningful way as the fishing industry.

Table 15.5 Results from Johansen's trace test for multiple co-integration vectors using export of goods and services as proxy for marine production 1963-2000.

Eigenvalue	Likelihood ratio	5% critical value	1% critical value	Hypothesised number of co-integrating equations
0.891	157.44	53.12	60.16	None[**]
0.709	75.39	34.91	41.07	At most 1[**]
0.478	29.67	19.96	24.60	At most 2[**]
0.141	5.64	9.24	12.97	At most 3

[**] denotes rejection of the hypothesis at 1% significance level

Finally, we used the Johansen methodology to test if similar results could be obtained using a different production sector. Our choice fell up on the construction and transport sectors, which together employ a similar share of the work force, and contribute a similar fraction of the GDP according to the national accounts as the fishing industry. Here, a VAR(4) was found to be appropriate, and applying the Johansen test with three lags yielded the results shown in Table 15.6.

Table 15.6 Results from Johansen's trace test for multiple co-integration vectors; transport and construction sectors 1963-2000

Eigenvalue	Likelihood ratio	5% critical value	1% critical value	Hypothesised number of co-integrating equations
0.901	126.67	53.12	60.16	None**
0.832	71.29	34.91	41.07	At most 1**
0.591	28.47	19.96	24.60	At most 2**
0.253	7.00	9.24	12.97	At most 3

** denotes rejection of the hypothesis at 1% significance level

The results are similar to those found in the two previous cases with exports. We find clear evidence of three co-integration equations, and thus conclude that there does not exist a single, long-run production function between GDP, output in the construction and transport sector, capital stock and labour.

Our failure to detect a long-run relationship between GDP and alternative macro economic output measures, i.e. exports and the construction/transport sectors, lends support to the hypothesis that the relationship between the fishing industry output and GDP estimated above is indeed significant and indicative of an underlying structure in the Icelandic economy. It may also be regarded as an added support for our contention that the fishing industry is a base industry in the Icelandic economy. There may of course be other base industries; in fact, there probably are, but exports and the construction/transport sectors are unlikely to represent these other base industries.

Since there exists only one long-run relationship between GDP, the fisheries, capital and labour, it is straightforward to estimate the ECM for our relationship in the form specified in equation (15.8). The model was first estimated with three lags of each variable – marine production, capital and labour – and subsequently reduced using F-tests. Estimates for the reduced equation are shown in Table 15.7.

According to the results expressed in Table 15.7, changes in fisheries output and capital affect economic growth with lags of up to 3 years. Changes in labour, on the other hand, affect economic growth only in the same year. Economic growth is found to be quite sensitive to marine production. A 1% increase in the output of marine products will increase economic growth by 0.14% in the same year. This corresponds roughly to the weight the national accounts give to the fisheries sector. Expansion in the fisheries sector will further increase economic growth in the subsequent time periods by a total of 0.2%. The total short-run effect over 4 years thus amounts to 0.34%. The parameter associated with the error correction term, μ_{t-1}, is -0.21, indicating a rather slow adjustment to equilibrium – half of the disequilibrium gap is bridged in about 2.9 years. Note, however that this adjustment parameter is statistically not very well determined. The parameters associated with capital and third lag of marine production are also weakly determined.

Table 15.7 Economic growth equation, ECM-model, dependent variable is GDP_t 1963-2000.

Variable	Coefficient	Std error	t-statistic
Constant	0.0130	0.0129	1.0054
ΔMarine production$_t$	0.1376	0.0428	3.2127**
ΔMarine production$_{t-1}$	0.1013	0.0370	2.7378**
ΔMarine production$_{t-2}$	0.0669	0.0344	1.9430
ΔMarine production$_{t-3}$	0.0341	0.0459	0.7439
ΔCapital$_t$	0.2891	0.2767	1.0447
ΔCapital$_{t-1}$	0.0950	0.1859	0.5111
ΔCapital$_{t-2}$	0.0447	0.1640	0.2725
ΔCapital$_{t-3}$	0.1380	0.2313	0.5969
ΔLabour$_t$	0.8816	0.2809	3.1382**
μ_{t-1}	-0.2134	0.1817	-1.1747
R^2	0.641		
R^2 adj.	0.526		
S.E. of regression	0.025		
SSR	0.016		
Log likelihood	82.049		
DW	1.387		
Akaike	-4.297		
Schwarz	-3.893		

** and * indicate significance at the 1% and 5% level respectively

15.6 CONCLUSION

In this chapter we set about trying to clarify the role the fisheries have played in Icelandic development over the last 40 years. To this end, statistical methods were used to estimate the contribution of the fishing industry to GDP and the short-run effects the fishing industry has on GDP. To our knowledge this is the first time that these methods have been used to analyse base industries.

Using data on GDP, the production of the marine sectors, capital stock and labour, it was shown – using co-integration methods – that there exists a long-run relationship between these four variables which can be interpreted as production function. According to these results, the long-run output elasticity of the fisheries – the economic base – is 0.31, and the corresponding fisheries multiplier thus 3.1. The error correction model put forward on the basis of these results shows that there is a short-run effect of changes in production of the fishing industry on economic growth. A 1% increase in output of that sector will raise economic growth in the same period by 0.14% and 0.2% in subsequent periods. The total short-run effect thus amounts to 0.34%. The parameter of the error correction term was estimated as -0.21, indicating a rather sluggish adjustment to short-run disequilibrium.

As maritime products make up the lion's share of Icelandic exports, we repeated the exercise using two proxies for output of the fishing industry; exported goods and exported goods and services. The results obtained in both cases did not correspond to those obtained earlier. We also tested if using a different sector, in our case combining data for the construction and transport sectors, would yield identical results, but also drew a blank in this case.

The statistical analysis undertaken in this study strongly suggests that the importance of the fishing industry to the Icelandic economy is far greater than official statistics reveal. Thus it appears that the fishing industry constitutes a base industry in the Icelandic economy. It follows that removing this sector would probably have quite dramatic implications for the Icelandic GDP, although this impact would in all likelihood be far less than half a century ago, when the fishing industry was even more important than it is now.

While applied to the fishing industry in this chapter, we believe that our approach and methodology can be applied to search for base industries in general. All that is needed for this are the usual national economic statistics, GDP, labour usage and capital, as well as output or activity measures for the industries suspected to constitute base industries. With these data in hand, our methodological steps can be retraced to obtain results that may be interpreted in the same manner as we did.

Acknowledgements

Helpful suggestions by two anonymous referees are gratefully acknowledged.

References

Arnason, R. 1994. *The Icelandic Fisheries: Evolution and Management of a Fishing Industry.* Fishing News Books, Blackwell, Oxford.

Engle, R. and C. Granger. 1987. Co-integration and error correction: Representation, estimation and testing. *Econometrica* 50: 987-1006.

Frey, D.E. 1989. A structural approach to the economic base multiplier. *Land Economics* 65(4): 352-58.

Dickey, D.A. and W.A. Fuller. 1981. Likelihood ratio statistics for autoregressive time series with a unit root. *Econometrica* 49: 1057-72.

Gonzalo, J. 1994. Five alternative methods of estimating long-run equilibrium relationships. *Journal of Econometrics* 60: 203-33.

Granger, C.W.J. 1969. Investigating causal relations by economic models and cross-spectral methods. *Econometrica* 37: 24-36.

Hazari, B.R. 1970. Empirical identification of key sectors in the Indian economy. *The Review of Economics and Statistics* 52(3): 301-05.

Innis, H.A. 1930. *The Fur Trade in Canada: An Introduction to Canadian Economic History.* Yale University Press, New Haven.

Johansen, S. 1988. Statistical analysis of cointegration vectors. *Journal of Economic Dynamics and Control* 12: 231-54.

Johansen, S. 1995. *Likelihood-based Inference in Cointegrated Vector Autoregressive Models.* Oxford University Press.

Johansen, S. and K. Juselius. 1990. Maximum likelihood estimation and inference on cointegration with applications to the demand for money. *Oxford Bulletin of Economics and Statistics* 52: 169-210.

Jónsson, G. 1999. Hagvöxtur og Iðnvæðing: Þróun Landsframleiðslu á Íslandi 1870-1945. Þjóðhagsstofnun, Reykjavík.

Jónsson, G. and M.S. Magnússon (eds). 1997. *Icelandic Historical Statistics*. Statistics Iceland, Reykjavik.

Jónsson, S. 1984. Sjávarútvegur Íslendinga á 20. öld. Hið íslenska bókmenntafélag, Reykjavík.

Krumme, G. 1968. Werner Sombart and the economic base concept. *Land Economics* 44(1): 112-16.

Lenzen, M. 2003. Environmentally important paths, linkages and key sectors in the Australian economy. *Structural Change and Economic Dynamics* 14: 1-34.

Leontief, W. 1936. Quantitative input and output relations in the economic system of the United States. *Review of Economic Statistics* 18, 39-59.

McGilvray, J. 1977. Linkages, key-sectors and development strategy. In W. Leontief (ed.) *Structure, System and Economic Policy*. Cambridge University Press: 49-56.

National Economic Institute. 2000. *Historical Statistics*. Accessible online http://www2.stjr.is/frr/thst/rit/sogulegt/english.htm.

Nordal J. and V. Kristinsson. 1987. *Iceland 1986*. Central Bank of Iceland, Reykjavik.

North, D. 1955. Location theory and regional economic growth. *Journal of Political Economy* 63(3): 243-58.

Schaffer, W.A. 1999. Regional impact models. In *Web Book of Regional Science*. Regional Research University, West Virginia University. Accessible online http://www.rri.wvu.edu.WebBook/Schaffer/chap02.html.

Schulz, S. 1977. Approaches to identifying key-sectors empirically by means of input–output analysis. *Journal of Development Studies* 14: 77–96.

Sims, G. 1972. Money, income and causality. *American Economic Review* 62: 540-52.

Snævarr, S. 1993. *Haglýsing Íslands*. Heimskringla. Háskólaforlag Máls og Mennningar, Reykjavík.

Tiebout, C.M. 1956a. Exports and regional economic growth. *The Journal of Political Economy* 64(2): 160-64.

Tiebout, C.M. 1956b. The urban economic base reconsidered. *Land Economics* 31(1): 95-9.

Tiebout, C.M. 1962. *The Community Economic Base Study*. Community for Economic Development, New York.

Watkins, M.H. 1963. A Staple Theory of economic growth. *The Canadian Journal of Economics and Political Science* 29(2): 141-58.

16

Factor Use and Productivity Change in a Rights-Based Fishery

Basil M.H. Sharp
Chris J. Batstone

16.1 INTRODUCTION

Although the economic arguments for rights-based fishing are well known, few countries have fully embraced this approach to fisheries management. New Zealand's quota management system (QMS) was introduced in 1986 and has been an exemplar of fisheries management. As an early leader in rights-based fishing, it is of interest to note that legislative reforms were largely based on economic principles. Regardless of the power of economic reasoning, political will is essential if theory is to be translated into operational policy. In the 1980s the New Zealand government had the mandate necessary to implement sweeping reforms. Fisheries management was among the first policies to be reformed (Sharp, 1997).

Beginning with the status quo, the instruments of reform take time to produce the outcomes expected. Without empirical evidence we can only speculate as to the underpinnings of the lags, such as the uncertainty as to how the QMS operates, the time taken to figure out appropriate capital-labour ratios, changes in vulnerable biomass levels, and so on. However, the fact remains that in an output controlled fishery with tradable rights, firms must necessarily focus on the cost of harvest. If firms are price takers then the focus will be greater. There is little empirical evidence of technical change in a rights-based fishery. As Jin *et al.* (2002) have noted, technical change is an important factor affecting productivity growth. In the New England groundfish fishery innovative technologies, such as fish finders and electronic navigation, contributed to an annual increase in total-factor productivity of 4.4%. Fox *et al.* (2003) point to regulatory constraints adversely affecting productivity.

In this chapter we look for evidence of technical change in New Zealand's rock lobster fishery over a nine-year period. In the second section, we provide an overview of technical change. An economic model provides testable hypotheses. The third section describes the state of New Zealand's rock lobster fishery prior to the introduction of tradable rights. Stocks were depleted and profit margins were low; in short, too many vessels were chasing too few fish. In 1990 rock lobster was introduced into the QMS and we show the turn-around in vessel numbers and catch per vessel, given a relatively constant total allowable commercial catch. Data show firms substituting capital for labour. Technical change, as measured by the rate at which costs change over time, is estimated using a translog cost function that includes input prices, output, and biomass. Within-year observations are partitioned into four groups on the basis of firm output. This enables us to control for firm size and estimate the rate of technical change according to firm size. Biomass is included as a state-of-nature variable. The econometric results presented in the fourth section show the rate of technical progress improving over the nine year period. The chapter concludes with a general discussion.

16.2 TECHNICAL CHANGE

The standard economic model sees a firm with a given production relation – for example, describing how labour and capital are combined to harvest fish – responding to changes in the relative factor prices. If we assume two factors labour (L) and capital (K), with input prices w and k respectively, then the price ratio is $-w/k$ and output q is produced at minimum cost using the combination of labour and capital (L^*, K^*) shown in Figure 16.1. From this model of firm behaviour we derive the firm's cost function and input demand functions that provide the theoretical foundations for econometric modelling.

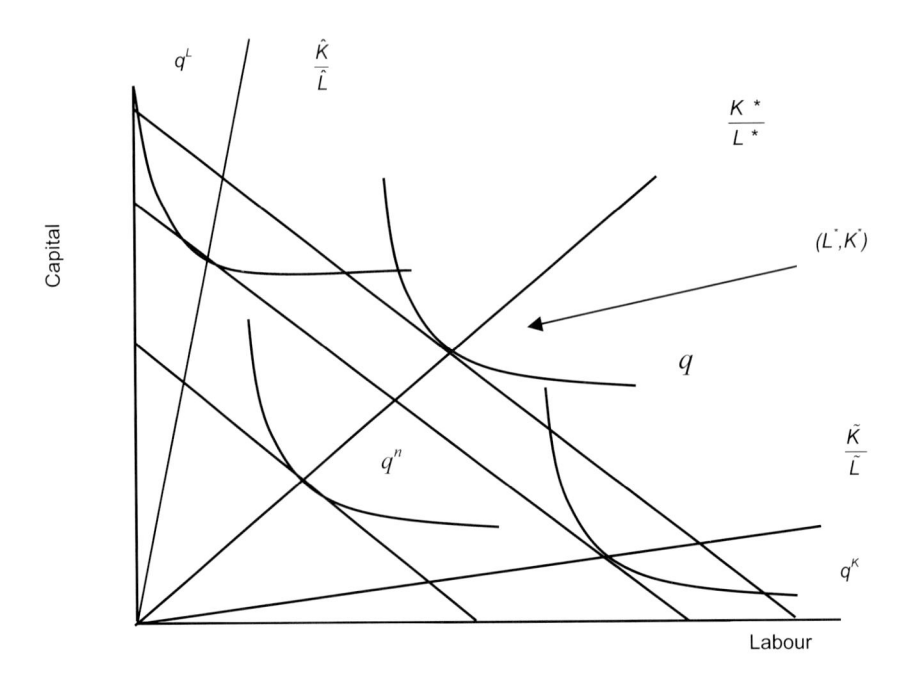

Figure 16.1 Effect of technical change on capital-labour ratio, q^n.

Technical change alters the mapping of inputs into output. For example, an investment in mechanised pot-lifting equipment might replace one labour unit. Investing in a global positioning system (GPS) enables the firm to position its vessel more accurately over productive sites that, in turn, increases the unit contributions of capital and labour. It is useful to broaden out the notion of technical change to include innovations that can be implemented at points along the chain of supply to consumers. Process innovations are defined as any adopted improvement in technique that reduces average costs at given input prices. The standard approach to technical change focuses on process innovation. In contrast, product innovations alter the output itself. The distinction between process and product innovation may blur in practice. For example, a new filleting process might lower the average cost of recovered product and change the nature of the product itself. In this chapter we focus technical change as it relates to process innovation in the harvesting sector.

The most straightforward illustration of technical change is illustrated by shifting the isoquant q in towards the origin. Three different categories of technical change are shown in Figure 16.1; cost neutral (q^N), labour saving (q^L) and capital saving (q^K). We discuss each in turn. Cost neutrality means that cost minimising input ratios K^*/L^* are independent of the technology. In terms of Figure 16.1, given $-(w/k)$, technical change that results in $q \to q^N$ is cost neutral because it preserves the marginal rate of technical substitution along the ray K^*/L^*; $q \to q^L$ is labour saving because $\hat{K}/\hat{L} > K^*/L^*$ and $q \to q^K$ is capital saving because $\tilde{K}/\tilde{L} < K^*/L^*$.

In a rights-based fishery the firm is free to adjust harvest by trading quota in the market. For example, returning to Figure 16.1, the firm producing at q^N could purchase $q - q^N$ tonnes of quota, shifting the isoquant out to q. The ability to make within season adjustments to quota introduces considerable flexibility by enabling the firm to better utilise scarce factors of production. Firms with larger vessels could, for example, lower average harvesting costs by buying quota; firms with smaller vessels might choose to sell a share of their seasonal quota entitlement.

16.3 ECONOMIC MODEL

In fisheries, the production function is complex because of the need explicitly to account for the state of the biomass (Clark, 1990). In this chapter we use the translog cost function for estimating technical change (Stevenson, 1980):

$$\log c(w,q,b,t) = \alpha_0 + \sum_{k=2}^{4} \lambda_k D_k + \sum_{i=1}^{2} \alpha_i \ln w_i + \gamma \ln q + \delta t + (\frac{1}{2})\sum_{i=1}^{2}\sum_{j=1}^{2} \alpha_{ij} \ln w_i \ln w_j$$
$$+ (\frac{1}{2})\gamma^*(\ln q)^2 + (\frac{1}{2})\delta^* t^2 + \sum_{i=1}^{2} \phi_i t \ln w_i + \sum_{i}^{2} \psi_i \ln w_i \ln q + \theta t \ln q \qquad (16.1)$$
$$+ \eta t \ln b + \sum_{i=1}^{2} \eta_i t \ln w_i \ln b + \tau t \ln b \ln q + (\frac{1}{2})\eta^* t(\ln b)^2$$

where D_k is a binary variable representing firm size, k represents 2, 3, 4; w_1 represents labour wage, w_2 represents capital price measured by annual depreciation; q represents harvest; t represents time, t represents 1, 2, ..., b represents biomass, measured in tonnes. Symmetry conditions and linear homogeneity in prices are assumed. Invoking Shephard's Lemma we obtain input cost shares which together with equation (16.1) provide the basis for estimation:

$$S_i = \alpha_i + \sum_j \alpha_{ij} \ln w_j + \phi_i t + \psi_i \ln q + \eta_i t \ln b . \qquad (16.2)$$

Following Ohta (1974) the rate of technical progress is reflected in cost reduction:

$$\dot{T} = -\frac{\partial \ln C(q,w,b,t)}{\partial t} = -(\delta + \delta^* t + \sum_i \phi_i \ln w_i + \theta \ln q + \ln b + \sum_i \ln w_i \ln b + \ln b \ln q + \ln b^2) . \qquad (16.3)$$

The growth in technical change can be broken down into (i) pure technical change $(\delta + \delta^* t)$, (ii) non-neutral technical change $(\sum \phi_i \ln w_i)$, and (iii) scale augmenting technical change $(\theta \ln q)$. The contribution of biomass to changes in cost over time is given by $\ln b + \sum \ln w_i \ln b + \ln b \ln q + (\ln b)^2$.

If there is no technological change effect then:

$$H_0^1 : \delta = \delta^* = \phi_W = \phi_K = \theta = 0 . \qquad (16.4)$$

Technological change is neutral at a non-constant exponential rate of $\delta + \delta^* + \theta t \ln q$ if:

$$H_0^2 : \phi_W = \phi_K = 0 \qquad (16.5)$$

Elasticity of scale is given by:

$$\varepsilon = \frac{\partial \ln c}{\partial \ln q} = (\gamma + \gamma^* \ln q + \sum_i \psi_i \ln w_i + \theta t + t \ln b) . \qquad (16.6)$$

$\varepsilon < 1$ indicates diseconomies of scale, $\varepsilon > 1$ economies of scale, and $\varepsilon = 1$ constant returns to scale. An indication of minimum efficient firm size (MES) is given by $\partial \varepsilon / \partial t = 0$; if $\theta > 0$ then MES can be achieved at a lower level of output, if $\theta < 0$ then MES can be achieved at a higher level of output. Thus, the final hypothesis we test is that there is no change in the MES, that is:

$$H_0^3 : \theta = 0 \qquad (16.7)$$

16.4   THE ROCK LOBSTER FISHERY

During the late 1940s profitable export markets for rock lobster developed which fuelled entry into the fishery. Figure 16.2 shows the number of vessels increasing through 1955, tapering off slightly in the early 1960s and then increasing again in the 1970s. Average harvest per vessel peaked at around 15 tonnes in the mid-1950s and generally fell off through 1980 to around 4 tonnes. This pattern of development is characteristic of newly exploited fisheries where the rapid increase in landings is sustained by harvesting larger sized age classes. With unlimited entry production will progressively decrease as smaller animals are harvested. During the early stages of development entry was limited by restrictive licensing which was removed in 1963. In 1963, 950 vessels harvested an average of 4.8 tonnes. At the time a moratorium on the issue of new licenses was introduced in the early 1980s, 1,125 vessels landed 4,534 tonnes, an average of 4.0 tonnes per vessel. The moratorium was removed prior to rock lobster being introduced into the QMS.

Prior to the moratorium, a survey of registered fishers in 1979 provides the following insights into the economic conditions of commercial rock lobster fishing: (i) 38 per cent of the respondents reported a harvest of less than one tonne, 42 per cent between one and six tonnes; (ii) most vessels were fished single handed, the average crew size was 1.6; (iii) the break-even income and quantity increased significantly 1976-1978 for most vessel classes (Annala, 1983). Against this backdrop of declining average harvest and lower profit, 78% of the respondents indicated that they would not support a competitive total allowable catch (TAC) and 64% were against a catch quota per vessel. It is of interest to note that the survey was undertaken some eight years before rights-based fishing was first introduced; the option of tradable rights was not firmly on the horizon in 1978.

16.5 QUOTA MANAGEMENT SYSTEM

Institutional change is one of the more powerful instruments that government can use to provide the conditions necessary for economic growth. Significant changes to the institutional foundations of New Zealand's fisheries management system were introduced in the mid-1980s (Sharp, 1997). In 1986 various forms of input control were replaced by the QMS. The QMS has become something of a beacon for many countries faced with the twin ills of excess capacity and stock depletion. Structurally, the QMS is quite simple. One key instrument is the TAC which is set after the Minister of Fisheries considers advice from fisheries scientists and stakeholder groups, including commercial, recreational and traditional users. Once the TAC has been determined, a total allowable commercial catch (TACC) is set for each quota management area (QMA), taking into account non-commercial fishing interests (TNCC) and any other relevant environmental social, cultural or economic factors. Thus the TAC = TACC + TNCC.

Individual transferable quotas (ITQ) are the other key instrument. When first implemented in 1986, ITQ rights were specified by weight, transferable, divisible, transformable and issued in perpetuity. Reforms introduced in 1996 saw harvesting rights redefined as a percentage share of the TACC for a quota species within each QMA. On the first day of the fishing year each quota share spawns an annual right known as an annual catch entitlement (ACE) to catch a specified quantity. While ITQ shares are a fixed percentage, the ACE generated each season will vary according to the TACC.

In many respects New Zealand's QMS aligns reasonably well with the conditions necessary for efficiency as laid out by Arnason (1990). There are however, two major exceptions. First, when setting the annual TAC the Minister is required to move stocks towards maximum sustained yield (MSY). From the viewpoint of economic theory it is well known that MSY is not consistent with maximum economic yield (Clark, 1990). Therefore, it is highly unlikely to deliver full economic efficiency even if positive rents exist in the fishery. Second, differentiated harvesting rights will also work to reduce efficiency. Quota prices, which are only available for commercial rights, enable demand estimation along the lines of Batstone and Sharp (2003). However, no specific rights attach to allowances made for recreational harvest; their harvest is managed by daily bag limits. Thus, managers must resort to non-market valuation methods if they want to compare relative values at the margin. With these limitations in mind, competitive quota markets should ensure that relatively more efficient firms get to harvest fish.

Rock lobsters were introduced into the QMS in 1990. The fishing year runs 1st April through 31st March and fishers must hold ITQ to at least 3 tonnes. Firms can freely trade harvesting rights, limited in the case of rock lobster by aggregation limits of 10% holdings in any one of the 10 quota management areas.

Vulnerable biomass provides fisheries managers with an estimate of the biomass available to the fishery at the beginning of each fishing season (Breen et al., 2002). Estimates of vulnerable biomass in the rock lobster fisheries take into account minimum legal size, the restriction on berried females, size-selectivity and seasonal vulnerabilities. Median biomass estimates are available for seven of the 10 rock lobster fisheries management areas. Table 16.1 shows the average median vulnerable biomass – for those quota management areas with biomass estimates – increasing over the period. On the other hand, the TACC available to industry (aggregated over all quota management areas) has generally declined, as the TACC in 2000 was 78% of that in 1992. Industry response to a declining TACC is seen in the percentage of harvesting rights exercised increasing from 85% to 96% over the same period. Catch per unit effort (CPUE) over the period has increased from 0.60 kg per pot lift in 1992 to 1.14 kg per pot lift in 2000. Sustained increases in CPUE support independent evidence that the biological state of the fishery has improved quite markedly over the period. Increases in the CPUE could have also come about from technical change. As noted earlier, we estimate the gains – in terms of lower costs – from innovation, controlling for improvements in the biomass.

Table 16.1 Biomass, total allowable commercial catch and commercial harvest 1992-2000.

Year	Vulnerable biomass (t)	TACC (t)	Catch (t)	Caught (%)	CPUE (kg/pot lift)
1992	431	3,616	3,066	85	0.60
1993	452	3,265	2,644	82	0.54
1994	569	2,913	2,755	95	0.66
1995	660	2,913	2,622	90	0.76
1996	756	2,913	2,536	87	0.88
1997	925	2,954	2,645	90	0.96
1998	1039	2,865	2,553	89	1.07
1999	1046	2,927	2,718	93	1.17
2000	994	2,849	2,748	96	1.14

Source: Clement and Associates (2001), Breen et al. (2002), Annala et al. (2001)

The post-QMS trend in vessel numbers and average landings stands in stark contrast to the pre-QMS trend shown in Figure 16.2. In the early 1980s 1125 vessels landed an average of 4.0 tonnes per vessel. Figure 16.3 shows the number of registered fishing vessels declining over the 1992-2000 period and the catch per vessel increasing. In 1992, two years after rock lobster were introduced into the QMS, 526 vessels harvested an average 5.8 tonnes; in 2000, 341 vessels harvested an average 8.12 tonnes.

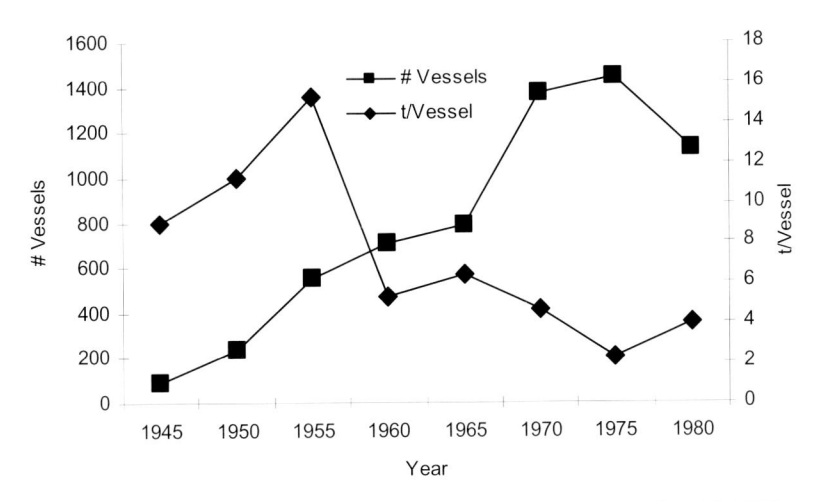

Figure 16.2 Number of registered vessels and harvest per vessel 1945-1980.

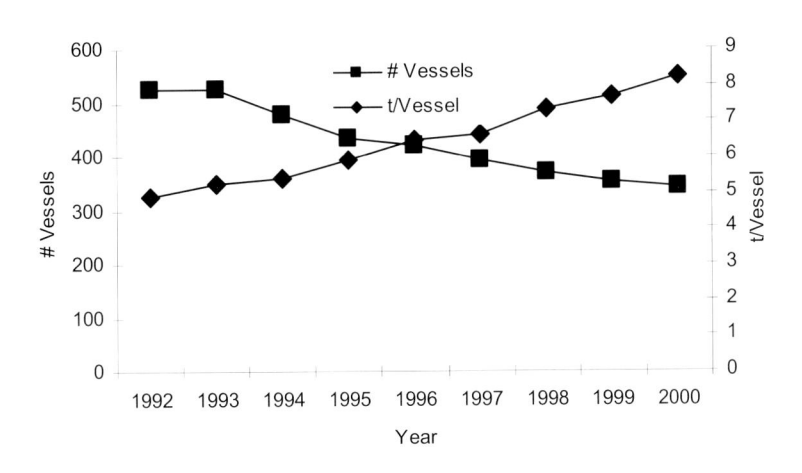

Figure 16.3 Number of registered vessels and harvest per vessel 1992-2000.

Aside from biological sustainability, technical change and innovation will have great influence on the capacity of industry to sustain economic growth. Bearing in mind the economic and regulatory environment that fishing firms operate within, it would seem obvious that the avenues to higher profit in this particular industry are limited *inter alia* to technologies that economise on scarce harvesting rights, organisational innovations, value-adding processes, and stock enhancement.

The New Zealand rock lobster industry faces harvest limits, receives no subsidies from government and exports more than 90% of the harvest. Bearing in mind that each firm is a price taker, the scope for profit enhancement is contingent on higher world prices and lowering harvesting costs. Given that both the TACC and world prices are exogenous, there should be a strong incentive for firms to augment profit by focussing on factors influencing cost. In addition, changes in the biological state of the stock could have a significant impact on profit (Jin *et al.*, 2002; Fox *et al.*, 2003). If the TACC is set at a level less than maximum sustainable yield, and is held constant over a period of time, then growth in stock biomass should work to lower harvesting costs.

Organisational innovations, including firm level innovations and industrial groupings – such as the formation of a quota owner association – fall within the category of disembodied technological change because they require no specific capital. In contrast, embodied technological change is linked to specific inputs, such as seabed mapping, colour sounders, improved vessel design, and so on. For example, improvements in GPS technology have greatly increased the accuracy of pinpointing productive sites, economising on time and increasing the effectiveness of trapping that, in turn, increases harvest per labour unit. Improved vessel design can facilitate greater substitutability between labour and other factors of production. Unfortunately, we are unable to observe disembodied technical change. In this particular industry scale also has aggregate implications for long-run growth and for the structure of industry. Larger vessels can increase the scale level at which decreasing returns set in. If we can identify increasing returns to scale then economic growth within this output-constrained industry remains a possibility.

16.6 DATA

The Annual Enterprise Survey (AES) provides financial information by industry groups. The AES forms the basis of national income accounting variables such as value added, gross output and gross fixed capital formation. To be included in the population enterprises must have annual revenue of at least $30,000. Observations on firms within the rock lobster fishing industry are available over nine years, from 1992 through 2000. It should be noted that these are firm level observations, not vessel. The data are confidential. Input prices are in NZ$ 1992.

Unfortunately, data on individual firms over this period are not contiguous so conventional panel data models are not appropriate. In order to control for scale effects we used a set of four dummy variables based on whether firm output fell within four equally spaced percentiles in any one year. This also provided a basis for testing the hypothesis attributed to Schumpeter (1934) that larger firms innovate more than smaller firms.

Table 16.2 shows average harvest per enterprise increasing over the period; output per unit of capital decreased. The data show output per unit of labour has increased by over 60%. The observation that fishing firms are economising on labour relative to capital is further highlighted by the trend in cost shares. In 1992, on average, labour costs accounted for 26% of total cost, in 2000 labour costs had fallen to 16% of total cost.

Table 16.2 Summary statistics for New Zealand rock lobster industry (NZ$).

Year	Cases (N)	Quantity (tonnes)	Labour units[a]	Capital per labour unit ($000)	Output per capital ($000)	Output per labour unit
1992	55	6.84 (4.36)	2.96 (1.95)	65.11 (61.20)	0.2260 (1.248)	2.53 (1.49)
1993	72	6.74 (5.09)	2.96 (1.81)	65.41 (57.05)	0.0676 (0.0796)	2.36 (1.34)
1994	73	6.48 (3.72)	2.89 (1.58)	79.73 (91.37)	0.0691 (0.0833)	2.35 (1.15)
1995	51	8.09 (6.06)	3.22 (1.75)	72.91 (72.08)	0.0589 (0.0540)	2.57 (1.41)
1996	58	10.35 (20.01)	3.68 (2.92)	76.73 (78.74)	0.1091 (0.3522)	2.76 (2.68)
1997	47	7.64 (5.13)	3.08 (1.75)	71.11 (68.50)	0.0970 (0.2030)	2.67 (1.63)
1998	90	10.30 (19.05)	2.50 (1.47)	132.88 (178.88)	0.0805 (0.1929)	4.11 (2.49)
1999	96	8.11 (11.84)	2.70 (2.25)	125.29 (168.26)	0.0544 (0.0594)	3.39 (2.81)
2000	101	9.09 (15.43)	2.47 (1.92)	145.35 (195.64)	0.0853 (0.1776)	4.05 (3.59)

[a] Average number of workers per enterprise per year.
Source: Annual Enterprise Survey (Statistics New Zealand). Capital is measured as $10,000 per unit.

16.7   ECONOMETRIC RESULTS

Equations (16.1) and (16.2) were estimated using a generalised least squares (GLS) procedure based on Zellner's method (1962). The system was corrected for heteroscedasticity and autocorrelation. Table 16.3 summarises the econometric results. Coefficients that attach to the binary variables are positive, significant, and increase with firm size. Parameters that attach to capital and output accord with expectations and are significant. The statistical significance of technical change is measured by a likelihood ratio test with L_R and L_U as the maximum likelihood values of the restricted and unrestricted model respectively. Each hypothesis tested *viz.* H_0^1 (no technical change effect), H_0^2 (technical change is neutral at a non-constant exponential rate), and H_0^3 (no change in the minimum efficient size firm) was rejected at the 1% level.

Table 16.3 Parameter estimates of standard time trend model.

Parameter	Estimate	t-statistic
α_0	9.2222	5.135***
λ_2	0.0734	1.716*
λ_3	0.0996	1.960**
λ_4	0.1087	1.725*
α_L	-0.0014	-0.411
α_K	1.0014	274.423***
γ	-1.3516	-4.804***
δ	4.6482	1.037
α_{WK}	-0.0573	-18.176***
α_{WW}	0.0876	20.920***
α_{KK}	-0.0379	-5.051***
γ^*	0.1716	7.608***
δ^*	-0.0182	-3.396***
ϕ_W	-0.0427	-0.942
ϕ_K	0.0427	0.942
ψ_W	-0.0014	-0.411
ψ_K	0.0014	0.411
θ	0.0077	0.044**
η	-1.3336	-1.090
η_L	0.0047	0.754
η_K	-0.0047	-0.754
τ	-0.0038	-0.154
η^*	0.2031	1.134
Log-L	488.091	
R^2	0.85	
Wald χ^2_{11}		294.1159***
D-W	1.9043	
σ for GLS	0.005	

Note: significance levels indicated by * = 10%, ** = 5%, and *** = 1%.

Table 16.4 shows an estimated annual rate of technical progress improving from 12.90% in 1992 to 11.28% in 2000. Bearing in mind that the results are based on firm level data, we note that prior to 1990 too many boats were chasing too few fish. Over the period 1980 through 2000 the number of vessels declined by 70%. General uncertainty about the operation of the QMS would have been minimal because the introduction of rock lobster lagged 4 years behind other high valued species. However, uncertainty over the likely path of the TACC could have worked against more rapid innovation during the early 1990s. Technological progress gains further momentum in the mid-1990s when we see a marked increase in the capital–labour ratio and output per unit of labour. The average rate of technical change for the period was 2.68%.

Calculated scale elasticities are also reported in Table 16.4 and suggest that that rock lobster fishing firms are characterised by diseconomies of scale. This finding is consistent with the sign of the parameter ($\theta > 0$) that indicates minimum efficient size

Table 16.4 Percentage annual rate of technical progress, scale and elasticity of cost.

Year	$\dot{T}$	Scale	ε
1992	-12.09	0.684	1.514
1993	-6.11	0.664	1.559
1994	-2.20	0.650	1.584
1995	1.06	0.665	1.582
1996	3.05	0.643	1.638
1997	4.59	0.622	1.678
1998	6.65	0.621	1.670
1999	8.43	0.605	1.736
2000	11.28	0.644	1.636
Average	2.68	0.641	1.630

can be achieved at lower levels of output. One possible explanation is that firm-level capital has increased in an environment characterised by a declining TACC. By using dummy variables we were able to calculate technical change according to firm size. The average rate of technical change increased as firm size increased and then dropped off (Table 16.5). Firms in category 3 had the highest average rate of technical change (3.6%); smaller firms in category 1 had the lowest rate of technical change (1.93%).

Table 16.5 Firm size and technical progress.

Firm size (output within)	Number	$\dot{T}$
[Minimum, 25th percentile)	117	1.93
[25th percentile, median)	125	2.32
[median, 75th percentile)	158	3.66
[75th percentile, maximum]	243	2.60

16.8   DISCUSSION AND CONCLUSIONS

Prior to the early 1980s commercial rock lobster fishing in New Zealand was characterised by low profitability and unsustainable harvest levels. Too many boats were chasing too few fish. A moratorium limiting further entry into the fishery was implemented in the early 1980s. Legislation introduced in 1983 provided the framework for rights-based fishing and beginning in 1986 commercial species were gradually introduced into the QMS. Rock lobster was introduced into the QMS in 1990 constraining commercial harvest to a TACC directed towards MSY and, importantly, enabling fishers to trade in the market for quota. Rights to harvest are a necessary factor of production that fishers combine with labour, capital and other inputs to harvest rock lobster. Because individual fishers can only legally harvest up to their quota, they face an on-going incentive more effectively to utilise their scarce harvesting rights.

The opportunity for fishers to increase profits in a rights-based fishery where the total allowable harvest is constrained and firms face world prices is limited. Over the years 1992-2000 the rock lobster TACC actually declined, the percentage of the TACC caught increased, as did the CPUE. Independent estimates of the vulnerable biomass generally increased over this period. Unfortunately we are unable to get vulnerable biomass estimates for every quota management area and had to settle for an average based on the median estimates available.

Summary statistics for the rock lobster fishery show average landings, and landings per labour unit increasing over time. Overall, the rate of technical progress has shown steady improvement, switching from being regressive, but decreasingly so, early in the sample period to positive. The rate of technical progress seems to sit reasonably well alongside the trend toward increasing output per unit of labour. Comparing the average rate of 2.8% per annum with results from other industries and fisheries is of interest. There are a number of studies reporting estimates of technical change in other industries. To illustrate, Ball and Chambers (1982) found the rate of technical change in the US meat industry to be negative, around -3%; for the alcoholic beverage industry, Xia and Buccola (2003) report annual cost reductions arising from technical change of between around 2% over the 1958-96 period. Banks *et al.* (2001) provide results on the impact of technical progress on fishing effort within the context of the Multi-Annual Guidance Programmes. For example, for the Danish cod trawl fishery in the Baltic Sea they estimate that the contribution of technical progress was 1.8% per annum over the 1987-99 period. Similar rates of technical progress were estimated for the Gulf of Lions trawl fishery. They note that the nature of the fisheries management system is a major factor determining the impact of technical progress. We are not aware of empirical work on technical change in rights-based fisheries.

New Zealand's QMS is based on ITQ working within the constraints of a TACC. Over the sample period the TACC for rock lobster has declined and the median vulnerable biomass has generally increased. The percentage of the TACC harvested has increased to 96%. Within this environment, and recognising that producers face world prices, the evidence indicating cost reductions at an average rate of 2.68% is impressive. One obvious policy issue relates to the TACC. When setting the TACC the Ministry of Fisheries relies almost exclusively on the results of stock assessment research. Presumably there is a tradeoff between biomass improvements and the TACC. Would a different TACC trajectory yield levels of vulnerable biomass that would make a stronger contribution to cost reductions and therefore profit? In other words, greater economic gains might be achieved by placing greater reliance on economic evidence at the time when the annual TACC is set.

Acknowledgements

John Annala, Paul Breen and Andrew Jeffs provided assistance and advice on vulnerable biomass. This research is supported by the New Zealand Foundation for Research Science and Technology. We also appreciate the assistance of Statistics New Zealand who provided the data and access to their secure data laboratory. The results of this study are the work of the authors, not Statistics New Zealand. The continued support of the New Zealand Rock Lobster Council and the Ministry of Fisheries is gratefully acknowledged. We are grateful for comments received from two referees and, naturally, assume responsibility for any remaining errors.

References

Annala, J.H. 1983. *New Zealand Rock Lobsters: Biology and Fishery*. Fisheries Research Division, Occasional Publication No. 42, Ministry of Agriculture and Fisheries, Wellington.

Annala, J.H., K.J. Sullivan and C.J. O'Brien. 2001. Report from the Mid-Year Fishery Assessment Plenary, November 2001. *Stock Assessments and Yield Estimates*. Ministry of Fisheries, Wellington.

Arnason, R. 1990. Minimum information management in fisheries. *Canadian Journal of Economics* 23: 630-53.

Ball, V.E. and R.G. Chambers. 1988. An economic analysis of technology in the meat products industry. *American Journal of Agricultural Economics* 64: 699-709.

Banks, R., S. Cunningham, W.P. Davidse, E. Lindebo, A. Reed, E. Sourisseau and J.W. de Wilde. 2001. *The Impact of Technological progress on Fishing Effort*. Report, European Commission, Brussels.

Batstone, C.J. and B.M.H. Sharp. 2003. Minimum information management systems and ITQ fisheries management. *Journal of Environmental Economics and Management* 46: 492-504.

Breen, P.A., S.W. Kim, P.J. Starr and N. Bentley. 2002. Assessment of the red rock lobsters (*Jasus edwardsii*) in area CRA 3 in 2001. *New Zealand Fisheries Assessment Report 2002/27*.

Clark, C.W. 1990. *Mathematical Bioeconomics*. John Wiley & Sons, New York.

Clement and Associates. 2001. *New Zealand Commercial Fisheries: The Atlas of Area Codes and TACCs*, 2001/2002, Clement and Associates, Tauranga.

Fox, K.J., R.Q. Grafton, J. Kirkley and D. Squires. (2003). Property rights in a fishery: Regulatory change and firm performance. *Journal of Environmental Economics and Management* 46: 156-77.

Jin, D., E. Thunberg, H. Kite-Powell and K. Blake. 2002. Total factor productivity change in the New England groundfish fishery: 1964-1993. *Journal of Environmental Economics and Management* 44: 540-56.

Ohta, M. 1974. A note on duality between production and cost functions: rate of returns to scale and rate of technical progress. *Economic Studies Quarterly* 25: 63-5.

Schumpeter, J.A. 1934. *Theory of Economic Development*. Harvard University Press, Cambridge, MA.

Sharp, B.M.H. 1997. From regulated access to transferable harvesting rights: lessons from New Zealand. *Marine Policy* 21(6): 501-17.

Stevenson, R. 1980. Measuring technological bias. *American Economic Review* 70 (1): 162-73.

Xia, Y. and S. Buccola. 2003. Factor use and productivity change in the alcoholic beverage industries. *Southern Economic Journal* 70(1): 93-109.

Zellner, A. 1962. An efficient method for estimating seemingly unrelated regressions and tests for aggregation bias. *Journal of the American Statistical Association* 57: 585-612.

17

Scientific Uncertainty and Fisheries Management

William E. Schrank
Giulio Pontecorvo

17.1 THE PROBLEM

After more than a quarter century of intensive fishery management, longer in many cases, many commercial marine fisheries are depleted or even commercially extinct. Debate continues as to why this is so and what should be done about it. To face this question is not to assume that all fishery management has failed, or that all marine fisheries are in danger. But the constant concern with overfishing, as reflected in reports of intergovernmental agencies (FAO, 1992; FAO, 2004), national agencies (USNMFS, 1999; Marine Research Institute 2002, 2003), and non-governmental organisations (Pew Oceans Commission, 2003), suggests at the very least that there is worldwide concern over the state of ocean fisheries and concomitantly, fisheries management. It has reached the stage where some scientists foresee the end of fisheries for wild ocean fish, to be replaced not only by aquaculture, but by deep sea ranching as well (Marra, 2005).

Any fishery management technique that sets total allowable catches (TACs) relies on fisheries science, specifically assessments of the strength of the fish stock, to determine catch limits. Other techniques, e.g. short seasons and gear and licence limitations are usually based on scientific advice. If they are so stringent and effective that effort is sufficiently limited as to present no danger to the stock, they are effectively similar to the small core fishery referred to below. Our concerns here are: (1) to what extent is the problem the result of the state of fisheries science; and (2) what can be done about it.

One point of view is that the quality of fishery science must generate sufficiently improved forecasts to permit the fishery managers to resist pressure from the industry. For instance, one argument is that a critical link missing from the science of fisheries is an understanding of the natural variation in fish stocks, in particular long-term cyclical swings (Steele and Hoagland, 2003). An alternative point of view is that fish stocks have been overfished for economic and cultural reasons that are independent of the quality of the science underlying fishery management. Therefore, new methods of fishery management, less dependent on scientific results, are necessary (Zeller and Russ, 2004).

All science is an evolutionary process. Fisheries science has not yet evolved to the point where it can produce measurements with the necessary accuracy and precision to permit fishery management to proceed with the confidence of all participants in the industry. Our hypothesis is that, given the structure and behaviour of the industry, and the inadequate science, current fisheries management techniques must become less reliant on scientific precision. Ours is a continuation of past discussions. More than 20 years ago, for instance, M. Sissenwine, a senior fisheries biologist with the United States government, focused on the uncertain environment of fishery scientists and managers, concluding that 'fishery managers must apply regulatory methods which are more robust with respect to current population size estimates' (Sissenwine, 1984: 27). More recently, Rosenberg (2003) has argued against managing to the margins. The problem arises regardless of whether the margin is defined as maximum sustainable yield (in the United States and widely throughout the world), or the more conservative $F_{0.1}$ (as in Canada), or 25% of the biomass (as was the case in Iceland).

Scientific advice can be based on either of two poles: accuracy and precision. If fisheries management depends on TACs, and TACs depend on stock assessments, then it is clear that inaccurate assessments that are biased upwards, for whatever reason, will lead to excessive TACs and overfishing. As will be seen below, this has been a factor in the northern cod failure, the problems with Icelandic cod, and with New Zealand's orange roughy. The second pole is that of precision. Imprecise stock assessments form a weak basis for setting TACs and can lead to pressures being applied to fishery managers that result in the permitting of excessive catches. It is this second pole that is the focus of this paper. Potential inaccuracies in fishery science simply reinforce the point of this paper, that weak science is a fundamental cause of the fishery crisis.[1] Scientific estimates are subject to unexplained residual errors. These errors arise from two sources. First, there is the innate stochastic nature of fish biology and the ocean environment. This white noise might have relatively low variance, in which case it presents no major problem. If the variance of the white noise is large enough, there might never be adequate precision in the scientific estimates and the problems of a fishery management dependent on science may never be solved. At this point in time it is impossible to judge the severity of the problems generated by white noise because one cannot isolate this component of the residual. Second, there is that component of the residual that arises from a lack of knowledge. Over time, science evolves and the variance arising from the second component is reduced. Currently there are enormous gaps in scientific knowledge, including the long cycles central to the argument of Steele and Hoagland (2003) as well as shorter-term associations among the environmental and biological factors affecting fish stocks. The two sources of error require different analyses, but one effect common to both is that they place doubt in the minds of the public on the results of scientific study and thus on the managerial decisions based on them.

When fishery managers attempt to reduce fishing effort substantially in response to scientific advice, economic forces often lead participants in the fishery to resist, at least in part because of the immobility of capital and labour in the industry.[2] Were there great confidence in the advice of fishery scientists, the resistance would be less. Perhaps even more important, governments, non-government organisations, and individuals who do not share the financial interests of those involved in the fishery

---

[1] This is not to ignore the serious effects on fisheries of illegal, unreported and unregulated fishing. Even if these problems were to be resolved, however, the difficulties generated by problems with science would remain.
[2] The classic paper on the non-malleability of fleet capital is Clark, Clarke and Munro (1979).

would have less sympathy for the industry's arguments; there would be less tendency to succumb to them. Despite the long history of concern over uncertainty in fishery science,[3] there has been little focus on what a critical role it plays. Why, for instance, given scientific evidence of declining stocks, does overfishing continue? [4]

The reason is that, although fishery science has made many important discoveries about the population dynamics of fish stocks, it has not, as yet, evolved to the point where it can make convincing forecasts of the sustainable catch. The uncertainty implicit in the absence of such forecasts often leads the fishing industry, with its competitive structure, short planning horizons (high discount rates)[5] and often weak financial position, to respond by resisting any catch limitation imposed by fishery managers in response to scientific advice. Given the uncertainty inherent in such advice, there is often an outcry from fishermen that a substantial cutback is unnecessary, that the recommendation is based on flawed science. Thus, pressure is applied to fisheries managers, and this pressure is often effective, as when scientific advice in 2000, and again in 2002, to close the North Sea cod fishery was overridden by European Union fishery managers who compromised by setting a total allowable catch equal to half the previous year's level (BBC News, December 15, 2000; New York Times, November 7, 2002). Not surprisingly, by 2004, the condition of the North Sea cod stock had not improved (ICES, 2004).

Why is it that economic and political pressure is so often effective? Oceanic environments are highly complex structures, with substantial stochastic components. Fishery science, in general, has tended to focus on single species in isolation. Usually such critical additional factors as climatic effects and predator–prey relations are used to explain past errors, rather than being incorporated into the basic analysis used for forecasting. The estimates generated by scientists have a large variance.[6] When scientists estimate that biomass is falling and recommend that fishing should be at least partially curtailed, this recommendation is enmeshed in uncertainty. In essence, they are saying that unless fishing effort is curtailed, there is a high probability that the stock will continue to diminish, although that probability is seldom quantified. What a fisherman sees is that he is being asked to cut back on fishing which, in turn, would lead to a reduction in his earnings. If he does not cut back, then the fishery may fail. Caught between an environmental maybe and a negative economic certainty, the fisherman (or fishing company) will use whatever economic and political tools are

---

[3] The classic paper on uncertainty in fisheries management is that of Sissenwine (1984), referred to earlier. Aspects of uncertainty are briefly touched upon in fishery management textbooks (see Hannesson, 1993 and Clark, 1985).

4 Overfishing is one of those amorphous words that means that too many fish are being caught, but how many is too many is rarely defined. Since, with the opening of a virgin fishery, the biomass declines, such a decline in itself cannot constitute the definition. By overfishing, we mean that the level of fishing, if continued, will lead to the commercial extinction of the stock. Implicit is the unsustainability of the stock at the current rate of exploitation. A conventional definition is fishing effort where catch exceeds the maximum sustainable yield (Clark, 1976, 28-9). Since maximum sustainable yield (MSY) is a static concept in an environment that is constantly in flux, we doubt the effectiveness of MSY in defining overfishing.

[5] We ignore the uncertainty problems associated with determining appropriate discount rates (Weitzman, 2001; Clark, 1985; Clark, 1976).

[6] For examples of the standard deviations associated with fish population estimates, see the cod abundance tables in Lilly *et al.*, 2003.

available to pressure the fishery managers to ignore, or mitigate, the application of the scientific recommendation. All too often the fishermen are successful.

Thus we are led to the conclusion that a major reason for the failure of past fishery management regimes, and the likely failure of more recent ones, is the uncertainty in the forecasts generated by fishery science. Perhaps with unlimited money thrown into science over a long time period the uncertainty could be narrowed sufficiently to negate the economic and political pressure.[7] Alternatively, there may be technological breakthroughs in the estimation and forecasting of fish populations. But until either of these unlikely events occurs, fishery management should either be abandoned as a waste of effort and money, or techniques of fishery management be adopted which are less dependent on uncertain scientific estimates for their foundation. This, in essence, is the conclusion reached by Zeller and Russ (2004), Sissenwine (1984) and Rosenberg (2003). Any system dependent on TACs, which are themselves sensitive to current stock assessments, in itself cannot be a solution to the problem.

It should also be obvious that the problems discussed here are less critical for lightly fished stocks. Nonetheless, the issue remains important because the recognition of potentially increased catches can quickly lead to the overexpansion of fishing capacity. Then the problems discussed here kick in.

17.2 THE NORTHERN COD EXAMPLE

Perhaps the most spectacular failure in scientific fishery management was the collapse of Newfoundland's northern cod stock in 1992 (DFO, 1992).[8] After more than a dozen years of fishing moratoria, the fishery remains closed today and shows no signs of recovery (DFO, 2004). From 1850 to 1930 the northern cod catch gradually rose from less than 200,000 tonnes to about 350,000 tonnes.[9] Then the catch levelled off until the late 1950s when, with the arrival of European distant water fleets, the catch started to rise, by 1960 exceeding 400,000 metric tons. Foreign catches continued to rise dramatically until, in 1968, the total northern cod catch reached a peak of 810,000 metric tons, of which only 119,000 metric tons were caught by Canadian fishermen. With some minor bumps, the catch fell consistently for the next decade, until it hit bottom at 139,000 metric tons in 1978. It is difficult to argue that the cause of this decline was not the overfishing of the 1960s. In the early 1970s quota systems intended to restrict catch were introduced and in 1977 fisheries jurisdiction was extended to 200 miles from the Canadian coast (Lear and Parsons 1993: 66). Although the stated goal of the extension of jurisdiction was the protection and growth of the stock (Canadian Minister of External Affairs quoted in the New York Times of June 5, 1976: 5), what actually happened were tremendous increases in Canadian

---

[7] Economic pressure as the term is used here refers to pressures on the components of the fishing industry which stimulate the industry to apply political pressure to the fishery managers to increase TACs or to make other adjustments which are economically beneficial, at least in the short term, to the industry.

[8] The press release announcing the moratorium on commercial fishing of northern cod stated that the recent 'devastating' decline in the stock was 'due primarily to ecological factors' and that the moratorium would last for less than two years.

[9] These figures are based on customs data for exports, converted to round weight (Source: Harris, 1990: 23).

capital and labour in the industry (Schrank, 1995: 291), increases which virtually guaranteed renewed overfishing.[10] There was a mild recovery of the catch during the 1980s, reaching 269,000 metric tons in 1988, then accelerating declines until the stock collapsed in 1992.[11] But with extended jurisdiction effective at the start of 1977, Canada managed nearly the entire range of the northern cod. Shortly thereafter, a conservationist fishing mortality reference point of $F_{0.1}$ was adopted that was intended to limit catches to a level that would sustain the stock indefinitely.[12] The management techniques failed. Was the cause of the failure inherent in the process of fishery management (in that it could not prevent overfishing), or could it be attributed to environmental change? It has been suggested that the critical question facing fishery science is how to distinguish between these two potential causes of stock depletion (Steele and Hoagland, 2003). After all, water temperatures cooled during the 1980s and early 1990s (Drinkwater, 2002: 116). Possibly in response, there may have been shifts in the fish population, from north to south where reproduction and growth were restricted (DeYoung and Rose, 1993).[13] Could that have caused the collapse? Some would argue that similar changes in water temperature had occurred in the past, with less serious effects on the fish stocks, thereby 'proving' that the problem is not climatic but overfishing (Hutchings and Myers, 1994). That argument is contradicted by another which states that when historical climatic conditions were similarly negative for the fish, the stock had not recently suffered the kind of overfishing that had occurred in the 1960s (Drinkwater, 2002: 119). Perhaps additional environmental factors, such as salinity changes or alterations in the number of predators (e.g. seals) or prey (e.g. capelin), also played a role (Conover *et al.*, 1995). A likely hypothesis, given what we know (or think we know), is that the basic cause was overfishing, not only during the 1960s, but later as well. The coup de grâce, however, was probably delivered by climatic changes. This time, the stock had been sufficiently weakened by overfishing that when climatic conditions changed again, in a favourable direction, the stock was unable to respond. It is too early to predict that the stock is commercially extinct in the long term, but the possibility is very real.

That the weight of economic and political pressure in getting to this point depended in large part on scientific uncertainty can be demonstrated by describing some of the events that led to the 1992 moratorium. In response, at least in part, to complaints by inshore fishermen that offshore trawlers were destroying the northern cod stocks, an expert committee of distinguished international fishery biologists was convened to review local fishery science. The committee concluded that recent stock assessments

---

[10] The increases in capacity were largely stimulated by an unbridled optimism over the future of the fishery, combined with a Canadian firm's expansion to increase its role in the world cod trade coupled with the expansionary response of local firms which envisioned decreases in their market shares if they did not also increase capacity.

[11] Sources of catch figures: for 1959-1991 (Bishop and Shelton, 1997: 60), for 1992-1995 (FRCC, 1996: 29), for 1996-2001 (DFO, 2002: 3) and for 2002-2003 (DFO, 2004: 2).

[12] The Canadian government in 1980 adopted the '$F_{0.1}$' criterion which had been developed earlier at the FAO as a more conservationist fishing mortality target than the prevailing criterion of maximum sustainable yield (Gulland and Boerema, 1973). There was no magic in the new criterion; if followed, the theory demonstrated that fisheries policy would be safer in protecting the fish than were the maximum sustainable yield criterion to be used.

[13] Lilly *et al.* (2003) cite examples from the literature which deny that any such migration occurred. The issue remains unresolved.

had overestimated the size of the stock (Alverson, 1987).[14] In response, the Canadian scientists in their assessment for 1989 stated that a reduction in the TAC of more than half was necessary to restore the stated fishing mortality target, $F_{0.1}$, of federal policy (CAFSAC, 1989: 50). The government was now faced with a dilemma. The inshore fishery had actually seen a slight recovery in the previous couple of years. The offshore catch was falling, slightly, but until very recently the catch per unit effort (CPUE) for the offshore had remained more or less constant, implying no loss of productivity (Tsoa, 1996: 48-49). Conventional fisheries science, assuming an even, rather than the possibly more realistic 'patchy', distribution of fish, considered changes in the CPUE indicative of changes in the fish population (Rothschild *et al.*, 1996: 400). So, had there been a drastic fall in the fish population, and if so, had it happened suddenly? Regardless of the speed of decline, the old question remains relevant: could it have been the result of overfishing, or was it caused by environmental changes?

The $F_{0.1}$ reference point had been adopted shortly after the extension of jurisdiction to aid in protecting the stock. What the scientists now said was that to achieve $F_{0.1}$ the TAC would have to be cut in half. But how certain were the scientists? How important was the goal of achieving fishing mortality equivalent to $F_{0.1}$? What were the costs of achieving this goal? The labour force and rural communities were captives of the fishery, with immobile capital and labour and little hope of short or medium term economic diversification. If fishing effort were reduced by half, boats would be tied up, plants would close or operate on short hours, many thousands of fishermen, fish plant workers, their families and others would be unemployed.[15] The government would have to sustain these people for an undefinable time. Any government would seek ways to avoid both this social welfare expense and, in the case of a federal state like Canada, the political pressure the federal government would experience from the wealthier provinces that, in effect, would be providing the subsidies. At the same time, the government would not want to adopt policies that would seriously disrupt the socio-economic fabric of Newfoundland society. Rather than invoke such a drastic measure, the federal Department of Fisheries and Oceans established a new committee to determine what was to be done. This committee concluded in mid-1989 that nothing drastic or threatening had occurred to the northern cod stock to date, that the implicit scientific recommendation of a reduction in the catch by half was too drastic given its obvious social and economic repercussions, and that, as a reasonable compromise, a considerably smaller cut in TAC for 1990 was acceptable (Harris, 1989). With slight modification this recommendation was accepted. Within two years the cod were gone.

In retrospect, it is obvious that the government simply postponed the day when it had to intervene financially on a large scale (Schrank, 1997).[16] In a sense, the government adopted the approach that was least politically damaging and the least economically costly in the short term, although this approach proved to be incredibly short-sighted. It is too glib to attribute the government's decision to 'politics', i.e. to

---

[14] Stock overestimation is not solely a phenomenon of fishery management 'gone wrong' as one might argue is the case in Newfoundland. Much more recently, stock-size overestimation and unintended high fishing mortalities have characterised the ITQ cod fishery of Iceland (Rosenberg *et al.*, 2002; Marine Research Institute, 2003).

[15] At the time that the northern cod moratorium was declared in 1992, the Canadian government assumed that 10,000 fish plant workers and 9,000 fishermen would be unemployed and therefore eligible for support payments as a result of the closure (DFO, 1992).

[16] For more general discussions of fisheries subsidies, see Milazzo (1998) and Schrank (2003).

pressure on the federal government from the industry and its allies on (Rosenberg, 2003).[17] The implication of this glibness is that all the government had to do was resist political pressure. It also suggests that the regulatory regime would profit by an institutional change whereby 'politics' was removed from the process. In fact, the government's action was a rational financial and social, if unwise and short-sighted, response to economic factors. Politics played a role, as it does in all decisions in a democracy, but even if political pressures were absent (and given the economic pressures on people, how could they have been), it is likely that the government would have taken the action that it did. The political and economic pressures were the consequence of the geographical distribution and economic structure of the industry.[18]

Unfortunately, fishery scientists have been able to offer neither a consensus explanation of the loss of the northern cod stock, nor a consensus explanation of the failure of the stock to recover. With the moratorium having been declared in 1992, the 1994 stock assessment offered an opportunity to summarise what had happened. But with a total reported 1993 catch of only 11,000 tonnes, the cod population was seen to be continuing to decline. This catch was too low to account for the continuing fall in the fish population. Fishery scientists were unable to offer a definitive reason for the continued decline in the stock. Reasons advanced were: (a) environmental factors; (b) incorrect 'tuning' of the stock assessment procedure; and (c) underestimation of the 1993 catch. The conclusion drawn was that 'it is not possible to determine which of these is correct' (DFO, 1994).

This situation has not improved. The 2003 stock assessment of northern cod (Lilly *et al.*, 2003) accepted that, for a decade following extended fisheries jurisdiction, the stock had been overestimated, leading to excessive TACs and excessive catches. The assessment notes that 'controversy continues regarding the time course and causation of the collapse', followed by capsule summaries of alternative theories and an extensive bibliography. A set of theories and contradictory evidence concerning a putative, and important if true, southward shift in the northern cod, mentioned above, is presented as 'illustrative of the degree of uncertainty'. Noting that uncertainty about the timing of the decline of the stock is at the root of the problem, three not-quite-contradictory hypotheses are discussed: (a) errors in survey technique may have shown more rapid population changes than actually occurred; (b) catches might have been grossly underestimated; and (c) natural mortality might have increased. Nearly a decade later, the 1994 explanations are still available, none have been eliminated.[19]

---

[17] Scientists may also exert pressure on government but their 'political' as opposed to 'scientific' weight is far less than that of the fishing industry. They have too few votes and are unlikely to garner much public support, even when they are permitted actively to work to pressure their employers (usually the government).

[18] Resistance to quota cuts is inherent in the economics of the system and is independent of the institutional system. But the manner of opposition and adaptation may vary according to cultural background. To avoid major cuts in TACs when scientifically warranted, Iceland adopted a system whereby inter-annual changes were restricted to 30,000 metric tons. This limit was clearly implemented to stabilise the system by avoiding too rapid expansion when stock size increased and moderating business dislocation when stock size decreased. In the Icelandic context, Rosenberg specifically warned against limiting downward adjustments when they are warranted (Rosenberg *et al.*, 2002: 20, Marine Research Institute, 2003).

[19] For developments in the Newfoundland fishery since the moratorium, see Schrank (2005).

Once again, the northern cod case is an extreme example, but problems of estimation, of under-reported catches, and of insufficient knowledge regarding environmental processes are nearly universal. At different times and places, fishery managers of various stocks may be more or less affected by these problems.

17.3 FISHERY MANAGEMENT

At the first substantive meetings of the Third United Nations Conference on the Law of the Sea in 1974 (UN, 1983: 191), agreement was quickly reached to allow coastal nations to control fisheries to 200 nautical miles from shore (Parsons, 1993: 238). By 1977, the 200-mile limit for fisheries jurisdiction had been declared by many coastal states which played major roles in the world's commercial fisheries (*Fish and Other Marine Life*, New York Times Index 1976: A Book of Record: 537-540). With extended control, there was a great expansion in the application of fishery management techniques. TACs were the predominant form of control, although restrictive licensing and other input and output controls were also put in place. The idea was that TACs would be set at levels that would permit just enough fishing to sustain the fish stock. Fishery science played a key role in that estimates had to be made of the size of the existing stock, the rate of reproduction and growth, and the rate of natural mortality, in order to determine the allowable level of catches.

It was within this sophisticated and scientific framework of fisheries management that stocks like the northern cod collapsed. Declines of fish stocks and catches of major species were common (see for instance, Stamatopoulos, 1993), and a general warning on the precarious position of commercial oceanic fisheries was issued in 1992 by the Food and Agriculture Organization of the United Nations (FAO, 1992). Their warning was heard, if not always heeded, and the issue of the maintenance of fish stocks became part of the world's political agenda (WTO, 1999). International negotiations started concerning 'responsible fishing' (FAO, 1995) and the removal of governmental subsidies, which were seen as a stimulus to excessive fishing (Munro and Sumaila, 2002). New management techniques were adopted, or perhaps old and formerly informal techniques were updated: community management, marine protected reserves, and ITQs, individual transferable quotas assigned to a single vessel or enterprise, rather than the older, general, TACs. Accurate science was still crucial.

Reasons for the failure of fisheries management are many: (a) the Law of the Sea Convention did not fully assign jurisdiction over stocks: continental shelves extended beyond the 200-mile limit, and highly migratory species were excluded, therefore excessive fishing continued beyond the limit, on the high seas, and in areas just outside the 200-mile limit of 'neighbouring' coastal states (as in the case of the doughnut hole in the Bering Sea, surrounded by areas under the jurisdiction of the United States and Russia); (b) excessive fishing by local fishermen permitted in response to political pressure exerted by fishermen and their allies in response to economic pressures; (c) inadequate science; (d) inadequate enforcement of regulations; and (e) conflicting goals among players in the industry, from individual fishermen through to government.[20]

---

[20]Scientists are concerned with the conservation of the stock, fishermen are interested in making a living, fish companies are interested in keeping their capital employed and making a profit,

17.4 CONCLUSIONS

So what is to be done? Neither community control nor ITQs in themselves can solve the problem, since both depend on questionable stock assessments.[21] Marine protected areas, if they occupy a sufficiently large portion of the stock's habitat and the fish are not excessively mobile, might work for some species (Hannesson, 1998). Exhortations to behave 'responsibly' are not going to be effective in the face of scientific uncertainty, potential profits, and the need for fishermen to make a living. We have elsewhere suggested that, regardless of the management regime, TACs should be reduced dramatically to such an extent that even when environmental conditions are seriously deleterious to the survival of the stock, the probability that the permitted level of fishing effort would result in the commercial extinction of the stock would be very low (Pontecorvo and Schrank, 2001). Recall that in past times, the northern cod survived negative environmental conditions. It was only when the negative environmental conditions occurred after the stock was seriously weakened by overfishing, that disaster occurred. Under this new regime, the domestic fishery would be cut substantially, requiring buyouts, and would not be permitted to grow when the stock grew. This 'small core' fishery would rarely, if ever, require the reductions in catch that create political problems.[22] Presumably political pressure from fishing communities will be less than at present because there will not be thousands of fishermen needing potentially excessive catches to make a living. There would be a limited number of fishermen catching a sustainable quantity of fish. The size of the core need not be determined with great certainty because of the large safety margin built into the system. When there were an excess of fish in the stock, rather than let them die of old age, temporary rights to catch the excess could be auctioned off with the receipts being divided between the government and the small core fishermen. This depends on the existence of risk takers, presumably distant water fleets of foreign nations. With local fishermen gaining some benefit from the auctioning process, they presumably would not be so anxious to clamour for a share of the increased stock than they would be otherwise. If there were excess fish to be auctioned over several years, then that situation might be considered normal, and economic benefits might revert to

---

governments are interested in social stability and remaining in office. Fishery managers must balance these goals, subject to the constraint that they are employees of government.

[21] ITQs are susceptible to the problems of scientific uncertainty and error as are all systems that depend on TACs. The classic case of a fishery failure under an ITQ system was that of New Zealand's orange roughy into the early 1990s, when excessive fishing was permitted because of errors in the understanding of the population dynamics of the species. Industry challenged the government on its implementation of sharply reduced quotas on the grounds of incorrect science. However, the reductions were implemented and, with increased scientific understanding, the stock has apparently now recovered (New Zealand, 1997; Roberts, 1991; Seafood Industry Council, 2003). In this example, the problem was simply inaccurate science, not scientific uncertainty. Political pressure probably played no role.

[22] Doubleday (1993) considers the reliability of scientific advice and its effect on fishery management. Among the possibilities he discusses for minimising the potential damage to the fishery from inadequately precise scientific advice is the maintenance of a low exploitation rate which would not be overly sensitive to scientific estimates of fish populations (380-82).

the domestic fishery, perhaps in the construction of fish processing facilities. This must be avoided since, with excess capacity, the owners and employees would argue in the face of scientific evidence that cutbacks were not necessary, that the scientific results were not reliable. We would be back at the situation that exists today. The system must be so designed that stocks deemed safe to catch in excess of the small core TAC must be caught and processed in a manner that yields no economic benefits to the domestic community except for the revenues from the auction, and neither community nor government can be allowed to become dependent on those revenues. The design of such a system would not be easy, but it is conceptually possible. An additional benefit of the small core regime is that the excess stock could be conservatively estimated so that there would be no attempt to set the level of fish harvesting by optimisation with respect to any single or group of reference points. Marginalist calculations would, and should in this case, be abandoned (Rosenberg, 2003). Larger safety factors could be built into estimates of allowable catches than are possible under current management regimes. The small core proposal would encounter serious opposition on economic, political and social grounds (Polacheck, 2002), and many details would have to be worked out. But the alternatives may mark the end of the commercial marine fisheries of wild stocks. Until, and unless, scientific uncertainty can be greatly reduced, a small core fishery, or an alternative fishery management technique not heavily reliant on scientific estimation, seems necessary.

All fields of human endeavour face various degrees of uncertainty. Yet it is rare to find an industry such as fisheries where such a fundamental element of the industry as the current supply of the raw material is so enmeshed in uncertainty.[23] Fisheries are different. The assignment of full property rights to fish might meliorate but cannot avoid the problems caused by such uncertainty. In addition, the difficulties that arise from excessive time discounting leading to political, social and economic pressures are all exacerbated by the fundamental problem of uncertainty in supply. What can be done is to develop techniques of fishery management less sensitive to this uncertainty.

References

Alverson, D.L. 1987. *A Study of the Trends of Cod Stocks Off Newfoundland and Factors Influencing Their Abundance and Availability to the Inshore Fishery*. A Report to the Honourable Tom Sidden, Minister of Fisheries, Canada. Task Group on Newfoundland Inshore Fisheries.

Bishop, C.A. and P.A. Shelton. 1997. *A Narrative of NAFO Divs. 2J3KL Cod Assessments from Extension of Jurisdiction to Moratorium*. Canadian Technical Report of Fisheries and Aquatic Sciences #2199. Fisheries and Oceans Canada.

CAFSAC. 1989. Advice for 1989 on the management of cod in Divisions 2J3KL. In *Canadian Atlantic Fisheries Scientific Advisory Committee Annual Report* (Including Advisory Documents), XII. Ottawa: Fisheries and Oceans Canada: 45-58.

Clark, C.W. 1976. *Mathematical Bioeconomics: The Optimal Management of Renewable Resources*. New York: John Wiley and Sons.

---

[23] Most industries face uncertain demand. Fisheries are no different in this respect from other industries. Similarly, most industries face some uncertainty in supply, e.g. that caused by labour or transportation disruptions. What distinguishes fisheries is a more basic uncertainty in supply.

Clark, C.W. 1985. *Bioeconomic Modelling and Fisheries Management*. New York: Wiley.

Clark, C.W., F.H. Clarke and G.R.Munro. 1979. The optimal exploitation of renewable resource stocks: Problems of irreversible investment. *Econometrica* XLVII: 25-47.

Conover, R.J., S. Wilson, G.C.H. Harding and W.P. Vass. 1995. Climate, copepods and cod: Some thoughts on the long-range prospects for a sustainable northern cod fishery. *Climate Research* V: 69-82.

DeYoung, B. and G.A. Rose. 1993. On recruitment and distribution of Atlantic cod (*Gadus morhua*) off Newfoundland. *Canadian Journal of Fisheries and Aquatic Science* L: 2729-41.

DFO. 1992. Crosbie Announces First Steps in Northern Cod (2J3KL) Recovery Plan, News Release of Fisheries and Oceans Canada NR-HQ-92-58E, July 2, 1992. Dartmouth, NS: Department of Fisheries and Oceans (DFO).

DFO. 1994. Report on the Status of Groundfish Stocks in the Canadian Northwest Atlantic. Dartmouth, NS: DFO.

DFO. 2002. Northern (2J+3KL) Cod Stock Status Update. DFO Science Stock Status Report A2-01. Dartmouth, NS: DFO.

DFO. 2004. Northern (2J+3KL) Cod Stock Status Update. Canadian Science Advisory Secretariat Stock Status Report 2004/011. Dartmouth, NS: DFO.

Doubleday, W.G. 1993. Reliability of scientific advice on fishery management measures. In L.S. Parsons and W.H. Lear (eds) *Perspectives on Canadian Marine Fisheries Management: Canadian Bulletin of Fisheries and Aquatic Science #226*. Ottawa: National Research Council of Canada: 369-83.

Drinkwater, K.F. 2002. A review of the role of climate variability in the decline of northern cod. *American Fisheries Society Symposium* XXXII: 113-30.

FAO. 1992. *Marine Fisheries and the Law of the Sea: A Decade of Change*. Rome: Food and Agriculture Organization of the United Nations (FAO).

FAO. 1995. *Code of Conduct for Responsible Fisheries*. Rome: FAO.

FAO. 2004. *The State of World Fisheries and Aquaculture*. Rome: FAO.

Fish and Other Marine Life. New York Times Index 1976: A Book of Record: 537-40.

FRCC. 1996. *Building the Bridge: 1997 Conservation Requirements for Atlantic Groundfish*. Report to the Minister of Fisheries and Oceans (FRCC.96.R.2). Ottawa: Minister of Public Works and Government Services Canada.

Gulland, J.A. and L.K. Boerema. 1973. Scientific advice on catch levels. *Fishery Bulletin* LXXI: 325-35.

Hannesson, R. 1993. *Bioeconomic Analysis of Fisheries*. New York: Halsted Press.

Hannesson, R. 1998. Marine reserves: What would they accomplish? *Marine Resource Economics* XIII: 159-70.

Harris, L. 1989. Independent Review of the State of the Northern Cod Stock, Prepared for the Honourable Thomas E. Siddon, Minister of Fisheries. Ottawa: Minister of Supply and Services.

Harris, L. 1990. Independent Review of the State of the Northern Cod Stock Final Report, Prepared for the Honourable Thomas E. Siddon, Minister of Fisheries. Ottawa: Minister of Supply and Services.

Hutchings, J.A. and R.A. Myers. 1994. What can be learned from the collapse of a renewable resource: Atlantic cod, *Gadus morhua*, of Newfoundland and Labrador. *Canadian Journal of Fisheries and Aquatic Science* LI: 2126-46.

ICES. 2004. North Sea (Subarea IV) (North Sea), Skagerrak and Kattegat (Division IIIa) and the Eastern Channel (Division VIId). Report of the ICES Advisory Committee on Fishery Management and Advisory Committee on Ecosystems, 2004: Vol 1: 313-41. Copenhagen: International Council for the Exploration of the Sea (ICES).

Lear, W.H. and L.S. Parsons. 1993. History and management of the fishery for northern cod in NAFO Divisions 2J, 3K and 3L. In L.S. Parsons and W.H. Lear (eds) *Perspectives on Canadian Marine Fisheries Management: Canadian Bulletin of Fisheries and Aquatic Science #226*. Ottawa: National Research Council of Canada: 55-90.

Lilly, G.R., P.A. Shelton, P.A. Brattey, N.G. Cadigan, B.P. Healey, E.F. Murphy, D.E. Stansbury and N. Chen. 2003. An Assessment of the Cod Stock in NAFO Divisions 2J+3KL in February 2003. Canadian Science Advisory Secretariat Research Document 2003/023. Ottawa: DFO.

Marine Research Institute. 2002. State of Marine Stocks at Iceland 2001/2002 – Prospects for 2002/2003 – English Summary. Accessible online www.hafro.is/Astand/2002/engl-sum.htm (accessed December 12, 2002)

Marine Research Institute, 2003. English Summary of the State of Marine Stocks in Icelandic Waters 2003/2004 – Prospects for the Quota Year 2004/2005. Accessible online www.hafro.is/Astand/2003/engl-sum-03.pdf

Marra, J. 2005. Commentary: When will we tame the ocean. *Nature* CDXXXVI July 14.

Milazzo, M. 1998. *Subsidies in World Fisheries: A Re-examination*. Washington D.C.: The World Bank.

Ministry for the Environment, New Zealand. 1997. The state of our fish. Chapter 9 in *The State of New Zealand's Environment 1997*. Accessible online www.mfe.govt.nz/publications/ser/ser1997/html/

Munro, G. and U.R. Sumaila. 2002. The impact of subsidies upon fisheries management and sustainability: The case of the north Atlantic. *Fish and Fisheries* III: 233-90.

Parsons, L.S. 1993. Management of marine fisheries in Canada *Canadian Bulletin of Fisheries and Aquatic Science #225*. Ottawa: National Research Council of Canada.

Pew Oceans Commission. 2003. *America's Living Oceans: Charting a Course for Sea Change*. Washington D.C.: Pew Charitable Trusts.

Polacheck, T. 2002. Will 'small core' fisheries solve the fishery management dilemma? *Marine Policy* XXVI: 369-71.

Pontecorvo, G. and W.E. Schrank. 2001. A small core fishery: A new approach to fishery management. *Marine Policy* XXV: 43-8.

Roberts, P.R. 1991. New Zealand fishery: Background, management issues, and groundfish data. In W.E. Schrank and N. Roy (eds) *Econometric Modeling of the World Trade in Groundfish*. Dordrecht: Kluwer Academic Publishers: 487-505.

Rosenberg, A.A., G. Kirkwood, M. Mangel, S. Hill and G. Parkes. 2002. *Investigating the Accuracy and Robustness of the Icelandic Cod Assessment and Catch Control Rules*. Tampa, FL: MRAG Americas Inc.

Rosenberg, A.A. 2003. Managing at the margins: The overexploitation of fisheries. *Frontiers in Ecology and the Environment* I: 102-06.

Rothschild, B.J., S.G. Smith and H. Li. 1996. The application of time series analysis to fisheries population assessment and modeling. In V.F. Gallucci, S.B. Saila, D.J. Gustafson and B.J. Rothschild (eds) *Stock Assessment: Quantitative Methods and Applications for Small-Scale Fisheries*. Boca Raton, FL: Lewis Publishers: 354-402.

Schrank, W.E. 1995. Extended fisheries jurisdiction: Origins of the current crisis in Atlantic Canada's fisheries. *Marine Policy* XIX: 285-99.

Schrank, W.E. 1997. The Newfoundland fishery: Past, present and future. In S. Burns (ed.) *Subsidies and Depletion of World Fisheries: Case Studies*. Washington D.C.: World Wildlife Fund: 35-70.

Schrank, W.E. 2003. Introducing Fisheries Subsidies. Fisheries Technical Paper #437. Rome: FAO.

Schrank, W.E. 2005. The Newfoundland fishery: Ten years after the moratorium. *Marine Policy* XXIX: 407-20.

Seafood Industry Council. 2003. Orange Roughy: A New Zealand Success Story. (Press Release by Ministry of Fisheries 01 Dec 2003) News Centre Press Release. New Zealand.

Sissenwine, M.P. 1984. The uncertain environment of fishery scientists and managers. *Marine Resource Economics* I: 1-30.

Stamatopoulos, C. 1993. *Trends in Catches and Landings: Atlantic Fisheries, 1970-1991*. Rome: FAO.

Steele, J. and P. Hoagland. 2003. Are fisheries 'sustainable'? *Fisheries Research* LXIV: 1-3.

Tsoa, E. 1996. The collapse of the northern cod fishery: Predator-prey and other considerations. In D.V. Gordon and G.R. Munro (eds) *Fisheries and Uncertainty: A Precautionary Approach to Resource Management*. Calgary: University of Calgary Press: 45-59.

UN. 1983. The Law of the Sea: United Nations Convention on the Law of the Sea with Index and Final Act of the Third United Nations Conference on the Law of the Sea. New York: United Nations.

USNMFS. 1999. Our Living Oceans: Report on the Status of US Living Marine Resources, 1999. Washington: United States National Marine Fisheries Service.

Weitzman, M.L. 2001. Gamma discounting. *American Economic Review* XCI: 260-71.

WTO. 1999. Benefits of Eliminating Trade Distorting and Environmentally Damaging Subsidies in the Fisheries Sector: Annex I - Promote Sustainable Development by Eliminating Trade Distorting and Environmentally Damaging Fisheries Subsidies, WT/CTE/W/121 (Submission by five member states). Geneva.

Zeller, D. and G.R. Russ. 2004. Letter to the Editor. *Fisheries Research* LVII: 241-45.

18

Spatial-Temporal Stock Assessment Analysis with Application to the Scotia-Fundy Herring Fishery

Daniel E. Lane

18.1 INTRODUCTION

It is notoriously difficult to provide regular, updated, and improved estimates of fish stock abundance (e.g. numbers or total biomass of fish of a particular species or stock in a particular region). Problems arise due to the high stochastic variability of stock sizes, stocks' dynamic migration behaviour, and the natural fluctuating marine environment in which fish live (Hilborn and Walters, 1992). Nevertheless, management strategies for sustainable fisheries exploitation and efficient socioeconomic performance of the commercial fisheries sector such as those developed by Gordon Munro in Munro (1979), Clark *et al.* (1979), and Munro and Clark (2003) depend on reliable estimates of stock status. In recent years, Canadian government and fisheries and oceans mandates have been experiencing more restrictive budgets on science activities, superseding issues (e.g. broader oceans management mandates), and the high cost of maintaining regimes, ships, and highly trained personnel to monitor the regular fish stock assessment process. Consequently, the net contribution of fisheries stock assessments – especially in stocks that have declined – are being challenged (DFO, 2004a). For these reasons, in this period at the beginning of the twenty-first century, the capacity to carry out fisheries stock assessments and subsequent analyses in Canada are in decline. In response, this chapter presents an alternative, inclusive, and cost-effective framework for spatial and temporal analysis of in-season fishing activity, illustrated for the case of the Scotia-Fundy commercial herring fishery (North Atlantic Fisheries Organization (NAFO) Divisions 4VWX).

The objective of the chapter is to present a spatial-temporal model for the intraseasonal analysis of fisheries populations in support of in-season commercial management decisions in localised areas. The underlying 'core process' of the model is presented that describes in-season migration dynamics of population substocks, and estimates stock status for each spatial area and in each period of the season. It is shown how repeated measures from catch and other observations during the season are used to update the partially observed core process (i.e. observed with error). The model is applied to actual data from the 4VWX herring fishery and the results discussed. Implications of the model for in-season decision making as well as in longer term exploitation strategies are also discussed. The direct participation of commercial fishermen in the development and application of the model offer an economically efficient alternative to the current government-led stock assessment methods that are suffering from budget cuts and a disappearing institutional knowledge base.

18.2 BACKGROUND

Atlantic herring fisheries have been important to the economies of coastal nations for centuries (Kurlansky, 2003). For the case of the Scotia-Fundy herring population, the largest herring stock in the north-west Atlantic, there are several identifiable and discrete spawning grounds that form complex stock dynamics in NAFO Divisions 4VWX (Stephenson, 1990; Sinclair, 1988; Sinclair and Iles, 1985). Inadvertently, problems from adopting groundfish-style aggregated management arose in the early 1980s, e.g. the decline of Trinity Bay herring, that caused herring management difficulties. Recently, this complexity has been addressed in participative management decision-making through more explicit commercial involvement in the decision-making process (Power *et al.*, 2004; Stephenson *et al.*, 1998; Lane and Stephenson, 1995).

In herring, as in all fish stocks, longitudinal stock assessment practices are important. For example, at a conference in the 1960s in Fredericton, New Brunswick, Alfred Needler, Deputy Minister of Fisheries for Canada, declared that there were 'inexhaustible herring resources' in north-western Atlantic Canada. Subsequently, the Norwegian spring spawning herring in the north-eastern Atlantic collapsed giving rise to a system of national quotas in Canada and the north-eastern Atlantic herring stocks in the early 1970s. Following the demise of Norwegian spring spawning herring, the north-western Atlantic herring of Georges Bank was over-exploited and the herring stock there also collapsed. Then, in 1975, faced with real concern for herring in Canadian waters, the Minister of Fisheries for Canada, Roméo LeBlanc, announced the 'Bay of Fundy' Project that outlawed fishing for herring for reduction into fishmeal, and a separation of herring harvesters and processors. This led to improved prices for herring as food. Subsequently, the Department of Fisheries in Canada restructured the domestic herring management rules and determined what herring 'belonged' to which fishermen. Canadian catches rose along with the price of herring on world markets. At the same time, there were further concerns of applying groundfish-style aggregated stock assessment results to disaggregated groups of spawning herring (Stephenson *et al.*, 1993).

In the 1990s, herring prices fell causing difficulties in returns to the homogeneous domestic herring fleet that also coincided with stock resource issues and indications that herring stocks were actually in decline. During this period, traditional stock assessments were applied to herring using the available groundfish methods of virtual population analysis (VPA) (Rivard, 1988). In reaction to the difficulties with the groundfish-style VPA applied to the pelagic herring stock, scientists and fishermen developed a participative in-season protocol that applied to herring fishing on a regular in-season basis. Currently, the 'survey-assess-fish' in-season protocol is ensconced in the management of 4VWX herring (Power *et al.*, 2004).

This brief review of herring outlines the need for ongoing stock monitoring and assessment in this and all other commercial stocks. However, factors in recent times are conspiring against stock assessment continuation. Recent reports by Canada's Fisheries Resource Conservation Council (FRCC) reiterates the need to support scientific efforts to continue stock assessment amid noted shifting of federal funds away from this activity (FRCC, 2001, 2002, 2003). The frustration of support for existing expertise in scientific assessments parallels the switch to higher valued and more available invertebrate stocks (lobster, crab and shrimp) in north-western Atlantic waters during declines and even moratoria of traditional commercial groundfish stocks. At the same time, while these more abundant invertebrate resources do not extend science assessment efforts, demographics are such that existing groundfish capabilities and expertise in scientific assessments is rapidly disappearing from government ministries.

Science reviews are quick to adopt important ecological and oceans management issues in favour of improved analysis of biodiversity versus species analyses, as is prevalent in many American studies (NOAA, 1998) and in the revision of mandates of federal fisheries ministries such as the Canadian Department of Fisheries and Oceans (DFO, 2004a). At the same time, funding restraints, internal mandate reviews and apparent uncertainty in 'science priorities' have permeated recent discussions among Canadian fisheries administrators. The end result has been emphasis on improved oceans monitoring and away from continued stock assessment in support of commercial fisheries when arguably stock assessment requirements (when stocks are lower) are even more required than ever (DFO, 2002).

In 2003, a DFO conference was sponsored among groups interested in the 'future of aquatic science' entitled *Aquatic Science 2020: Workshop Report* (DFO 2004b). Curiously, representatives of the fishing industry were not present at this meeting, and accordingly, commercial fisheries were relegated to a receding role in the future of Canada's oceans. The 2020 report suggested that commercial fisheries would be diminished and that stock assessment and management activities could be replaced by the wider oceans management issues, e.g., oceans monitoring, and science in support of marine protected areas.

A recent determining event in the stock assessment discussion occurred serendipitously when the main DFO research survey vessel, the aptly named, *CCGV Alfred Needler,* caught fire in August 2003 before a stock assessment trip planned for renewing scientific stock information in the Gulf of St. Lawrence. It followed that usual stock assessment and population analysis for groundfish stock status was not produced in 2003 for the Gulf of St. Lawrence stocks in conformity with past longitudinal numerical analyses (DFO, 2004c). The initial indication was that the vessel would not be prepared for further stock assessment work except perhaps later in a broader 'oceans' concept. While this apparently was later overturned, the impact on the future of traditional stock assessments was very clear. In fact, in October 2003, after the fire, stock status reports for the commercial cod stock in the Scotia-Fundy area that had been surveyed by the Needler before the fire, were being prepared. It was decided that the traditional population analysis would not be done in that period because 'the Needler fire...will create great uncertainty in any comparisons of subsequent survey results'. (DFO, 2003a: 7; Clark and Hinze, 2003). Curiously, the Scotia-Fundy stock status report used data from the same period to carry out a standard VPA in which TACs were recommended to increase (DFO, 2003b; Hurley *et al.*, 2003).

Concern for the declining importance of stock assessments, and in order to demonstrate the existence of a more inclusive, cost-effective alternative option for estimating stock abundance and catch limits, has led to the development of a spatial-temporal model presented below. The intra-seasonal analysis illustrated below is in support of managing the commercial exploitation of herring in the Scotia-Fundy area.

18.3 SPATIAL-TEMPORAL ANALYSIS MODEL

Consider the in-season spatial-temporal behaviour of a fish-stock complex comprised of discrete subpopulations that regularly mix over defined marine grid locations as a single species. Define subpopulations as a dynamic system of 'spawning groups' that have characteristic spawning behaviour (in space and time). Distinct 'spawning groups' are characterised by specific seasonal migration patterns. Define also the state of the dynamic system by the geographical location ('zone') and assessed stock status (age-aggregated adult spawning stock biomass weight) of each of the spawning groups during each period of the season.

For the case of 4VWX herring, spawning groups are designated by annual routes that proceed from overwintering areas to summer feeding and pre-spawning areas, to late summer and fall spawning grounds, and finally the return to overwintering (Stephenson *et al.*, 1993). Spawning group routes and their stock status in the current period provide information needed for estimating the adult biomass–stock status of each group (and the whole stock) in subsequent periods. The period-to-period transition is modelled as a first-order Markov chain (non-stationary) where the states are represented by the discrete set of state levels for each spawning group (Lane *et al.*, 2004; Lane, 1989). The Markov chain is characterised by a set of probability transition matrices for each period that links the state of the current period (i.e. adult stock biomass level in zone) with the state in the next period (zone and adult biomass level according to the defined stock routes and birth–death processes). In this manner, the status of the stock is described by a probability distribution over the states for each successive time period beginning with a starting point. The state probability distributions are updated using Bayes' statistics after each period by repeated observations about the stock from catch and other ecosystem observations on the assumption that this information has some pre-specified degree of reliability with respect to contributing to the actual population level.

The partially observable Markov chain methodology adopted for application to the 4VWX Scotia-Fundy herring stock assessment (Lane *et al.*, 2004; Lane, 1989) is defined below:

Let $k \in \{1,2,...52\}$ denote the kth week of the year, and let $z \in Z$ denote a geographical zone in the fish habitat area with $|Z| = Nz$, the number of zones in the fishery. Let X_k be a Nz-dimensional random variable whose components represent the population levels of the stock in the various zones at the end of the kth week, so the time series $\{X_k\}$ describes the dynamics of the 'core process', the actual changing state levels of the fish stock throughout the year.

Define $p_{ijk} = \Pr(X_k = j | X_{k-1} = i)$, the probability of moving from one state of adult biomass level of value i at the end of the $(k-1)$th week to a state biomass level vector of j at the end of the kth week, where i and j are vectors of length N_X, representing pairwise discrete levels of stock biomass across all zones. Then the probability transition matrices, P_k, describe the non-stationary (i.e. weekly varying) process of stock spatial–temporal biomass dynamics from zone to zone and week to week throughout the season.

The core process of the actual state dynamics is only partially observable, i.e. commercial or survey catches or other fishing observations are considered samples of the adult biomass observed with error. Let the N_z-dimensional vector random variable, Y_k, define the 'imperfect' observation process, i.e. the series of actual catch statistics by zone and week. The components of these vectors have values that represent discrete levels of observations of adult stock biomass status in a zone during a particular week. Moreover, the observation information is assumed to be standardised, i.e. catch per unit effort does not vary with stock size on pelagic herring stocks and fishing effort is assumed to be relatively constant in each period; at the same time, no fishing effort implies no catch and consequently no additional information as to stock size. However, a catch of no fish when there has been fishing effort expended, may provide considerable information about actual stock status. Let:

$$q_{jlk}(z) = \Pr(Y_k = l | X_k = j), z \in Z, \qquad (18.1)$$

define the state-to-observation function that describes observations as a function of the known state of the core process, the actual adult biomass level, $X_k = j$. Assuming standardised observations at a particular fishing zone, the state-to-observation or 'reliability' function (18.1) above may be expressed more parsimoniously as a function of the observation, l and the actual state, j only, and independent of the week, k in which the observations occur. Simply we write: $q_{jlk} = q_{jl}$ for all k. The resulting matrices $Q(z)$ are the 'signal' or 'reliability' matrices specific to zone, z. They describe the probability distribution of all possible observations for given levels of the actual state in a zone and are provided as input data to the model.

The sufficient statistic (a quantity of smaller dimension than the information vector that contains all the information necessary for control purposes) for the state representation (Bertsekas, 1976) π, is defined as:

$$\pi_j(k) = \Pr(X_k = j | I_k), j \in N, \qquad (18.2)$$

where $\pi_j(k)$ is the state variable of the system defining the interaction between the partially observed actual adult biomass size, X_k and the available information about it, I_k, where $I_k = (Y_k, ..., Y_2, Y_1)$ comprises the catch and other ecosystem observations to date. Using Bayes' formula for the revision of probabilities, $\pi_j(k+1)$, the updated state population probability distribution is defined by the transfer function T_k as:

$$T_k(\pi \mid I)_j = \pi_j(k+I) = \frac{q_{jlk} \sum_i p_{ijk} \pi_i(k)}{\sum_{i,j} q_{jlk} p_{ijk} \pi_i(k)} \qquad (18.3)$$

Let $\pi(0)$ (= π (52) of the previous year) describe the initial probability distribution of the stock biomass at the beginning of the year. The $\pi_j(0)$, probabilities of initial biomass levels, are assumed to be known through historical estimates or from research survey data and are provided as starting input data. Equation (18.3) uses Bayes' Theorem to infer information about the true level of the stock biomass through the sufficient statistic (2). Since $\pi(k)$ is the probability distribution across the biomass levels, this can also be computed in each period by zone as an expectation and then summed over all zones to obtain total aggregate estimates of adult biomass. Assessed estimates are updated for each period k based on the core transitions, P_k, and the information I_k.

18.4 MODEL DEVELOPMENT AND APPLICATION

The 4VWX commercial herring fishery takes place along the Southern and Western Scotian Shelf and into the Bay of Fundy in Atlantic Canada (Figure 18.1). Figure 18.1 divides the whole management area into 15 separate fishing zones for catch-reporting purposes.

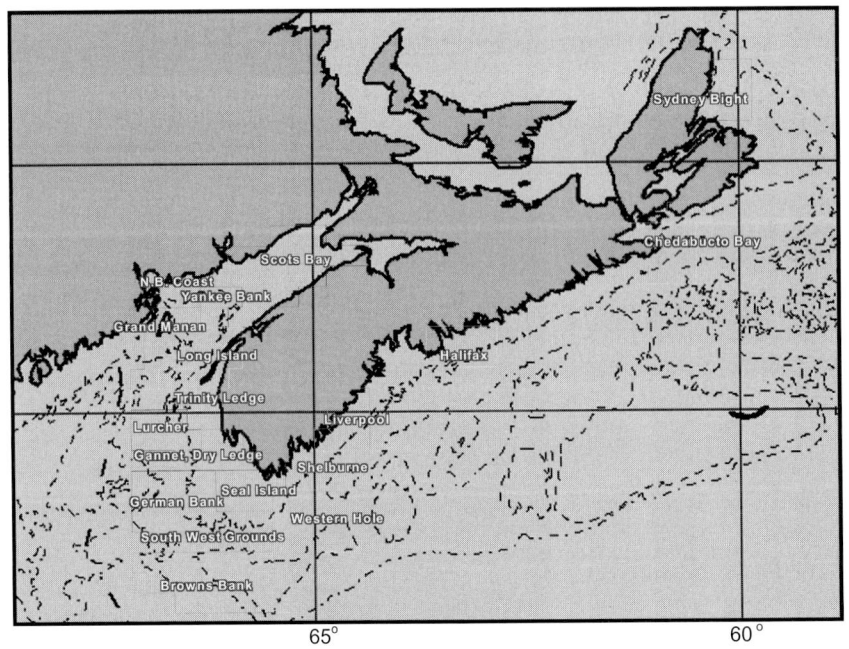

Figure 18.1 Map of the 4VWX commercial herring fishery grounds including designated geographical location or fishing zones in the Bay of Fundy and along the Scotian Shelf.

Estimates of the total adult herring biomass and the proportional biomass of the individual spawning groups are estimated from historical stock assessment estimates (Power *et al.*, 2004). Table 18.1 defines the six major herring spawning groups in 4VWX, the estimated annual ranges of the proportion of each subgroup strength within the total population, and the specific spawning zone and peak spawning periods that characterize the 4VWX herring stock complex.

Table 18.1   4VWX herring spawning groups including relative proportions of the total adult spawning biomass, and timing of the spawning period.

Group	Substock name	Minimum proportion	Maximum proportion	Expected proportion	Spawning zone no	Spawning period
1	Scots Bay/ Chedabucto	5.25%	18.75%	12%	8	July 20-Aug 15
2	Scots Bay/ Grand Manan	1.75%	6.25%	3%	8	July 20-Aug 15
3	Trinity	10%	30%	20%	3	Aug 15-Sept 15
4	German Bank	20%	60%	40%	7	Sept 1-Oct 15
5	Seal Island	10%	30%	20%	6	Sept 1-Oct 15
6	Lurcher	2%	10%	5%	4	Aug 15-Sept 15

While each year there are inter-annual differences in run-timing and locations, the expected behaviour of the herring substocks are generally consistent from year to year (Stephenson *et al.*, 1993). For example, Table 18.2 describes the migration pattern for the Scots Bay spawning substock.

Table 18.2   Migration route of the Scots Bay–Chedabucto substock spawning group and identification of the substock group major activity.

Weeks	Zone No.	Zone name	Substock activity
1-6	9	Chedabucto Bay	Overwintering
7-17	12	Diffuse	Widespread, unavailable for harvest
18-27	2	Long Island	Pre-spawning, feeding aggregation
28	13	Yankee Bank	Feeding
29	10	N. B. Coastal	Feeding
30-32	8	Scots Bay	Spawning
33-48	12	Diffuse	Widespread, unavailable for harvest
49-52	9	Chedabucto Bay	Overwintering

The probability transition matrices, P_k, describe the probabilistic stock adult biomass size dynamics from zone to zone and week to week. The matrices together describe the 'chain' of migration events for each substock. The matrices are determined by simulating the relative size and passage of each substock into the spatial grid of the fishing zones defined by the substock's expected migration routes. The population generation model simulates adult biomass in each zone and week over many simulated trials. These values are discrete and defined as states in the ranges, e.g. high, medium, and low adult biomass states define the core process. The probability transition matrices are calculated by recording the incidence of change of adult biomass from one discrete state to another over each week and zone for the simulated trials.

18.5 POPULATION GENERATION MODEL

The population generation model begins by generating an initial total adult stock biomass for the aggregate of all substocks and then distributes this total into substock components according to the relative proportions data from Table 18.1. Total adult biomass of the 4VWX herring stock is estimated to vary between 100 and 500 thousand tonnes with an annual average near 300 thousand tonnes as used in the herring model (Power *et al.*, 2004). Estimated adult biomass for the spawning groups are simulated and tracked along their respective migration patterns over time and the total stock aggregates are calculated by keeping track of the presence of all simulated substock biomass in each zone and week (Lei, 2004).

Modelled herring groups are each assumed to move through five 'waves' or run sequences measured on successive intervals of 1 week. A 'main wave' is accompanied by an 'early' and 'late' wave, i.e. the early wave follows the same route as the main wave but is always one week ahead of the main wave; the late wave is always one week behind its main wave, etc. 'Very early' (preceding the main wave by two weeks) and 'very late' (lagging the main wave by two weeks) run timings round out the migrating herring options, as observed from data. Run timings designation also requires model specification of relative biomass size in each wave. These have been identified in the model as a selection from various beta distributions and may include: (1) uniform-waves with the same weighting; (2) normal form wave designated location of the symmetrical mode; (3) truncated uniform-reduced set of equally weighted waves; (4) skewed left; and (5) skewed right as user-defined non-symmetrical distributions. Run-time wave distributions are shown in Figure 18.2.

The herring population generation model is designed in three component parts: (1) input data graphical interface; (2) population model computational engine; and (3) output analysis graphical interface. Input data on stock size, migratory routes and run timing population model simulation trials are executed, and summary data are tracked and recorded. The output analysis graphical interface illustrates the amount of adult herring biomass in a particular zone each week of the year and includes output reports for: (1) biomass week–zone analysis, (2) biomass analysis by spawning group, and (3) total spawning group biomass analysis. For example, Figure 18.3 shows the arrival and departure of summer feeding herring between weeks 19 and 29, then the presence of spawning herring during weeks 35 to 41 on German Bank (Zone 7), a major spawning group in the mouth of the Bay of Fundy. The population generation simulation model reflects the actual variability of the adult biomasss substocks dynamics replicating the rich spatial-temporal variability of the actual 4VWX herring population structure (Deng, 2000).

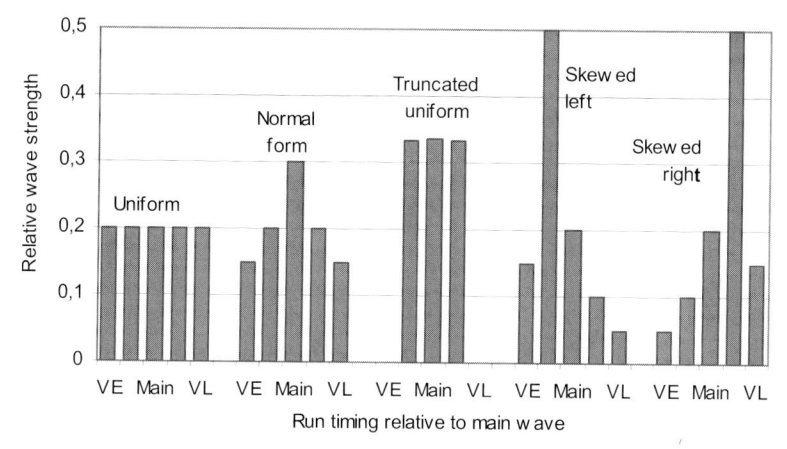

Figure 18.2 Spawning group run timing migration distribution patterns. For each of the five distribution patterns (uniform, normal, truncated uniform, skewed left and skewed right), the relative wave strength assigns five different temporal 'waves' to define each pattern: the 'very early' (VE) wave (2 weeks in advance of the main wave); 'very late' (VL) wave (2 weeks in advance of the main wave); the 'early' (E) wave (1 week in advance of the main wave), 'late' (L) wave (1 week in advance of the main wave).

18.6 STATE DEFINITION

Estimates from the population generation simulation model are differentiated into adult state biomass range levels as noted in Table 18.3. For 10 levels of state population size in each of 15 geographical zones, then the cardinality of $\{X_k\}$, the total weekly state possibilities set, is $|\{X_k\}| = (10)^{15}$. Enumeration of all possible states of the system by week is prohibitive. In order to deal with this problem of the curse of 'dimensionality', the state representation must be greatly simplified. Firstly, we note that there are six (6) identified spawning substocks of herring in the 4VWX stock complex. Secondly, we note from analysis that substocks occur rarely in the same zone for the same week (e.g. during the overwintering or the prespawning periods), so that many zones are empty of appreciable amounts of like-migrating herring in any given week. Thirdly, many zones have herring for only a few weeks of the year and these never attain high population levels or attract significant, consistent fishing effort. Moreover, spawning groups move relatively independently of each other and their absolute sizes are also independent. Finally, herring generally move into a zone from at most two other adjacent zones during a single weekly period. For these reasons, we adopt the simplifying assumption that the weekly state dynamics of each zone be considered independently in this model. At any given point in time, the actual state of the system may therefore be described by a single state level specified independently for each of the 15 zones.

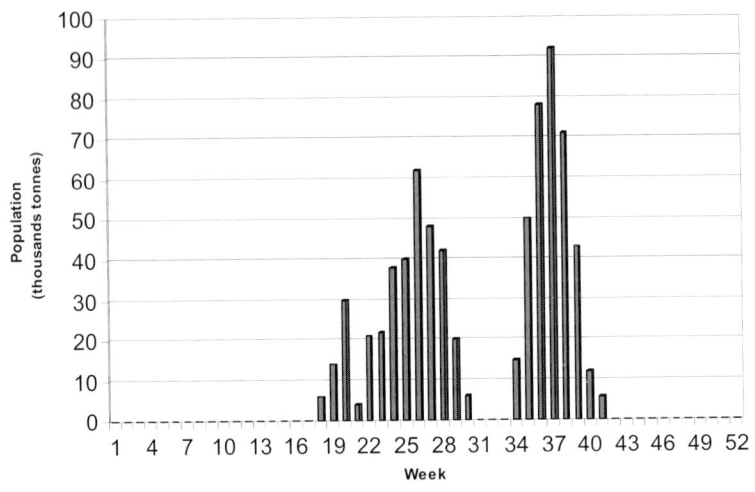

Figure 18.3 Population generation model biomass week–zone analysis for a single simulated population of the weekly incidence of herring on German Bank, the Zone 7 spawning group. The figure illustrates the population generation model Excel-based graphic results and also includes estimates for (1) biomass week-zone analysis, (2) biomass analysis by spawning group, and (3) total spawning group biomass analysis (Deng, 2000).

Table 18.3 Herring discrete population state biomass level definitions (000 t).

Population state level	1	2	3	4	5
Fuzzy definition	Zero	Extremely low	Very low	Low	Medium low
Cell range definition	0	1-25	25.1-50	51-75	75.1-100
Cell midpoint	0	12.5	37.5	67.5	87.5

Population state level	6	7	8	9	10
Fuzzy definition	Medium	Medium high	High	Very high	Extremely high
Cell range definition	101-150	151-200	201-250	251-300	300+
Cell midpoint	125	175	225	275	325

18.7 STATE-TO-STATE TRANSITIONS INCIDENCES

The 52 week-over-week (non-stationary) probability transition matrices, P_k, are estimated from the incidences of state-to-state change over all the trials of the population simulation model. For a given stock size and period, the single movement of herring from one zone at a given biomass state level into another adjacent zone is recorded as an incidence of the transition of the stock in the original zone to a decreased state level in the zone over that period. Similarly, the adjacent state to which the herring moved records an increase in its state level over that same period, *ceteris paribus*. When all population generation trials are recorded along with state levels in each zone over each period, the total incidence matrix of the count of the movement from each state level to all others for each zone in each week is determined. These transitions, for a given transition period (one week to another), are counted over $n=500$ population simulation trials. The probability transition matrices represent expected transitions between states for each of the 15 zones independently. The 10x10 states zonal transition matrices are normalised to form 15-10x10 zone-independent matrices that constitute the probability transition matrices for moving among assessed states within a single zone over each weekly period of the year (Storey, 1997).

Table 18.4 Zone 9 (overwintering area in Chedabucto Bay) biomass probability transition matrix for week 6 (mid-February) to week 7 at the beginning of the spring migration. Probability distributions by row represent the expected probability that the row state condition ('from state') makes the transition during week 6 to the column state condition ('to state') in week 7.

$a \backslash b$	1	2	3	4	5	6	7	8	9	10
1	1.0	0.0	0.0	0.0	0.0	0.0	0.0	0.0	0.0	0.0
2	0.0	1.0	0.0	0.0	0.0	0.0	0.0	0.0	0.0	0.0
3	0.0	1.0	0.0	0.0	0.0	0.0	0.0	0.0	0.0	0.0
4	0.0	0.478	0.522	0.0	0.0	0.0	0.0	0.0	0.0	0.0
5	0.0	0.19	0.772	0.038	0.0	0.0	0.0	0.0	0.0	0.0
6	0.0	0.039	0.454	0.493	0.014	0.0	0.0	0.0	0.0	0.0
7	0.0	0.0	0.18	0.532	0.261	0.027	1.0	0.0	0.0	0.0
8	0.0	0.0	0.0	0.5	0.462	0.038	0.0	1.0	0.0	0.0
9	0.0	0.0	0.0	0.143	0.714	0.143	0.0	0.0	1.0	0.0
10	0.0	0.0	0.0	0.0	0.0	0.0	0.0	0.0	0.0	1.0

[a] From state levels (week 6)
[b] To state levels (week 7)

For illustration, consider that during the sixth week of the calendar year (mid-February), herring are moving from their overwintering zone into a diffuse state (designated as non-geographical area, Zone 12) en route to their respective spawning grounds. The probability transition matrix for week 6 to week 7 from the primary overwintering area in Chedabucto Bay, designated as Zone 9 is provided in Table 18.4. The lower diagonal non-zero values in Table 18.4 denote a transition period where the herring are vacating Zone 9 as they begin the movement from the overwintering site en route to the spawning grounds in the Bay of Fundy.

18.8 STATE-TO-OBSERVATION SIGNAL MATRIX

DFO requires the reporting of herring catch data on a daily basis by longitude and latitude location. Confidential purse seine catch data are obtained from vessel logbook reports. Observed catch data are restricted to harvests by purse seiners (80+% of the herring harvested annually in the 4VWX fishery is caught by purse seiners) for the period 1991 to 1998 (Liu, 2000). Varying weather conditions, precise locations of fish, and fish behaviour all affect catch levels so that even with the same available stock of harvestable herring and the same fishing effort, there will be variability in the actual catch results. Catch reporting is not without error so that reported catches may differ from the actual catch and fishing mortality. In the model, actual catch observations in each zone are differentiated into catch classes. Each zone is assigned to a catch class based on the magnitude of the historical catches in that zone. Table 18.5 describes the differentiated catch levels assigned to Class 1 zones.

Table 18.5 Catch levels for class 1. Class 1 includes herring fishing zones 6, 7, and 8 except for the spawning periods in zones 6 and 7 when they are classified as class 5 (see Table 18.6) instead.

Discrete catch observation level	Range of catches Minimum(t)	Maximum(t)
1	0	0
2	1	500
3	501	1000
4	1001	1500
5	1501	2000
6	2001	2500
7	2501	3000
8	3001	3500
9	3501	4000
10	4001	4500
11	4501	5000
12	5001	5500

Each class has its own state-to-observation reliability or signal matrix derived from its empirical catch histories and catch levels. The state-to-observation signal matrix for Zone 9 is recorded in Table 18.6. This matrix represents the probability of a discrete catch level l, $l=1,...12$ occurring (Table 18.5), for a given population state level j, $j=1,...10$ (Table 18.3). These probability values apply to catches in Zone 9 for every weekly period, k.

Table 18.6 Zone 9 (overwintering area in Chedabucto Bay) state-to-observation signal (reliability) matrix. Probability distributions by row represent the probabilities that the discrete catch observation levels (columns 1 through 12) for any week of standard fishing effort in this zone row depend on the actual population state level ('state level' row) present there. For a given population level (row), individual cells are the probabilities that the particular catch level (column) will occur.

	Discrete catch observation level											
	[1	2	3	4	5	6	7	8	9	10	11	12]
1	0.90	0.01	0.01	0.01	0.01	0.01	0.01	0.01	0.01	0.01	0.01	0.00
2	0.53	0.10	0.13	0.10	0.06	0.03	0.01	0.01	0.01	0.01	0.01	0.00
3	0.20	0.10	0.24	0.20	0.10	0.10	0.01	0.01	0.01	0.01	0.01	0.01
4	0.15	0.10	0.20	0.22	0.15	0.11	0.02	0.01	0.01	0.01	0.01	0.01
5	0.10	0.10	0.18	0.20	0.15	0.15	0.07	0.01	0.01	0.01	0.01	0.01
6	0.10	0.10	0.10	0.20	0.20	0.14	0.05	0.04	0.04	0.01	0.01	0.01
7	0.10	0.10	0.10	0.12	0.18	0.20	0.07	0.05	0.04	0.02	0.01	0.01
8	0.05	0.05	0.06	0.10	0.12	0.15	0.10	0.11	0.10	0.07	0.06	0.03
9	0.05	0.05	0.05	0.05	0.10	0.10	0.15	0.10	0.10	0.10	0.10	0.05
10	0.05	0.03	0.03	0.05	0.10	0.10	0.10	0.12	0.12	0.10	0.10	0.10

(rows labeled: $\langle State\ level \rangle$)

18.9 INITIALISATION AND MODEL PROCESS

To start the process, an initial 'prior' probability distribution for the state of the population in each zone is required. The initial probability distribution is a vector of 10 state elements for each of the 15 zones, a 150 element vector. The first week of the year (the first week of January) is during the overwintering period for the stock. The 4VWX herring stock is assumed to overwinter in only two zones of its habitat range: Chedabucto Bay – Zone 9 and around Grand Manan Island – Zone 1 (Figure 18.1). Catches on the 4VWX stock occur during the early part of the calendar year on overwintering stock primarily around Chedabucto Bay (Zone 9). In mid-February, the fishery becomes inactive until prespawning aggregations reappear in the Bay of Fundy and the first spawning activity occurs at Scots Bay in summer (July–August). From that point onward, the fishery remains active on prespawning and spawning aggregations of herring generally proceeding north to south at the designated spawning grounds until the end of the spawning period (and end of the management year) near the end of October. Finally, catches of herring on the overwintering grounds take place at the end of the calendar year.

Model estimation occurs from the beginning to the end of the calendar year. Updating is done based on actual catch data. In periods where no catches typically occur (e.g. the substantial spring period from February to June, and the early winter period from mid-November to mid-December), the model uses the underlying core process transitions (probability transition matrices, P_k) to update the zonal state level probabilities for estimated assessed stock.

18.10 RESULTS

The intra-seasonal model described here is used: (1) as an independent spawning stock biomass estimation procedure; (2) as an in-season management operational decision support system; and (3) as a stock projection system for policy evaluation in strategic management. These applications are discussed below.

18.10.1 Spawning stock biomass estimation procedure

The Bayesian updating framework with repeated measures from weekly catch data over the course of the season generates in-season expected adult biomass values. The model provides these weekly updates on herring adult stock biomass in the form of probability distributions for each zone and state abundance level. The estimates of age-aggregated, spatially-defined spawning stock biomass depend on the starting values, the core transition parameters, and the defined level of detail of the differentiated state levels. A series of calibration analyses described below has been carried out for the illustrative herring data.

Model sensitivity examines how well the model predicts total stock size. In a series of 52 week (annual) simulations, the initial stock size was set and weekly catches by zone were generated randomly according to the probability distributions contained in the signal matrix Q for the assumed known actual stock size in each zone and period. The spatial-temporal state updating equation (18.3) was used to calculate expected total stock size in successive periods beginning with the start of year prior with weekly updating through randomly generated catches. Results from 10 one-year (52 week) trials using different random number streams (for generating different sets of weekly catch data in the fishery) were used to compute weekly zonal expected population estimates and standard deviation measures for simulated total stock size. Model estimates were compared to the simulated actual population size for low, medium, and high assumed cases. Table 18.7 summarises the results of the 10 independent annual trials.

The results indicate that the model is sensitive to changes in the assumed adult biomass. At the historical range of low (138 thousand tonnes) and high (330 thousand tonnes) spawning stock biomass, the model assessed estimates drift back to the mid-range population levels of about 250 thousand tonnes. This is the effect of lengthy periods of the 52 week calendar year without any catches and updating resulting in model revisions through the core transitions. Since the in-season core transitions are meant to define 'regular' herring population state levels dynamics, population values tend to return to these levels when there are no observations to indicate otherwise. Improved results for estimating stock size in the model occur when the model is applied solely to the fishing and sampling period and avoiding the calendar that includes periods of no fishing.

Table 18.7 Results of random trials for three simulated populations (low, medium and high) and corresponding model estimates of adult spawning stock biomass (in thousands of tonnes) compared to the actual values.

Simulated population input	Average actual simulated	Actual standard deviation	Initial model estimate	Model average estimate	Model standard deviation
1. LOW (N=10 trials)					
All weeks of the year*	138	1	138	251	5
First 4, last 8 weeks of the year	139	2	138	228	19
2. MEDIUM (N=10 trials)					
All weeks of the year*	237	1	238	265	5
First 4, last 8 weeks of the year	239	3	238	242	21
3. HIGH (N=10 trials)					
All weeks of the year*	333	2	338	272	5
First 4, last 8 weeks of the year	337	5	338	251	21

* Denotes model updating occurred for all 52 weekly periods during which fishing took place; no purse-seine fishing occurs historically in the 4VWX during late winter and spring, weeks 6-22.

Reliable model results are obtained for model updates when spawning stock size is estimated weekly for the years 1992-1998 using modified estimates of spawning stock biomass based on the annual total VPA estimates. Table 18.8 shows the results of the aggregate year-over-year estimates by spawning group from the model analysis using generated catches and compared to the annual VPA spawning-stock estimates. These results also show a tendency for the model to be 'sticky' near the overall average spawning-stock biomass of 300 thousand tonnes compared with the annual VPA estimates. The VPA spawning stock biomass estimates range from 128 to 433 thousand tonnes whereas the model estimates range from 228 to 337 thousand tonnes for 1992-1998. It is noted however, that significant observations and updating in German Bank result in estimates that better reflect the variability (ranging from 74 to 121 thousand tonnes) similar to the variability of the VPA estimates of Table 18.8.

Table 18.8 Expected value estimates of predicted model adult spawning stock biomass (SSB) after adjusting annually for the VPA estimate (thousand tonnes) 1992-1998 (Power *et al.*, 2004).

Spawning group	1992	1993	1994	1995	1996	1997	1998
Scots Bay	54.2	54.9	54.7	52.6	55.8	53.6	55.7
Trinity	77.5	77.4	78.2	74.1	77.8	77.7	75.9
Lurcher	24.5	24.5	24.5	24.7	24.6	24.8	24.5
Seal Island	59.9	57.7	58.7	56.7	53.8	56.6	60.6
German Bank	109.9	100.5	74.2	74.0	104.5	111.6	120.6
Total estimated model SSB	326.1	315.0	290.2	282.1	316.5	324.4	337.3
VPA estimate of SSB	318.6	241.0	132.9	127.5	306.5	357.0	432.8

18.10.2 In-season management

The spatial–temporal model serves as a valuable tool for in-season weekly management decisions arising from the 'survey-assess-fish' protocol. As such, weekly updating information provided by the model is consistent with the in-season management approach adopted for this stock. Weekly observations from the ongoing fishery at spawning zones provide observations for updating expected stock status at each zone. The model applies this opportunity to define what is 'expected' and to give an operational basis for definition of 'normal' versus 'abnormal' events as the fishery progresses thereby enabling a more immediate in-season management response to planned exploitation (Stephenson *et al.*, 1998).

Weekly spawning group estimates are based on constructing the occurrence of herring in the zone of interest during the designated spawning periods. For example, on German Bank (Zone 7) spawning occurs each year during weeks 35 through 41 (early September to mid October). The update of weekly abundance of German Bank spawners for 1998 are provided in Figure 18.4. The operational graphic illustrates the appearance of appreciable herring throughout weeks 19 through 31 well before the noted German Bank spawning period. It is important to note for herring management purposes that these German Bank herring appear in feeding aggregations and include herring from other spawning zones, e.g. Scots Bay. These herring may also be captured on German Bank at this earlier time period as fish for the bait or shore-based processing for food-fillet markets.

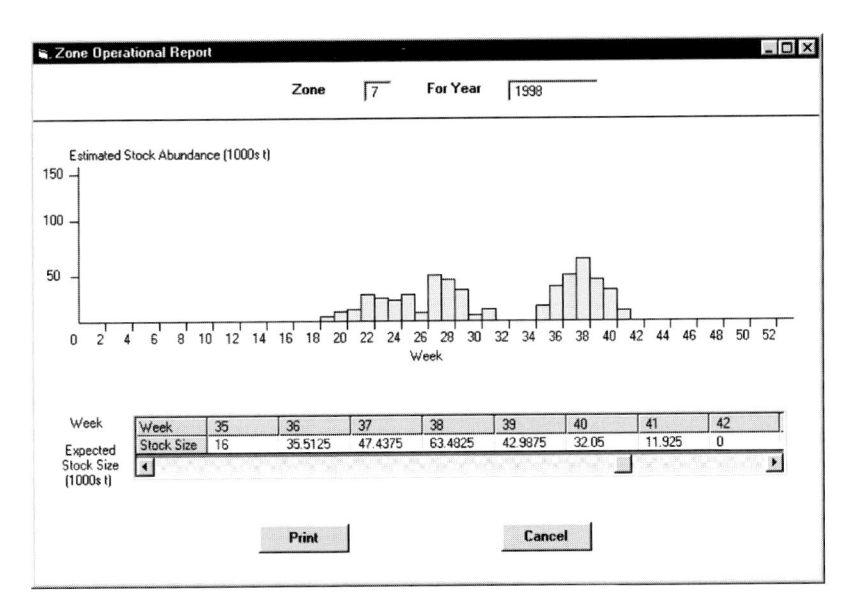

Figure 18.4 Model operational report results for German Bank in 1998. The figure illustrates the operational report for weekly spawning group adult biomass expected values estimated from the updated probabilities for the biomass state level in each week for German Bank. The significant spawning activity is isolated as occurring in weeks 35 through 41; evidence of German Bank prespawning and feeding of significant spawning groups also are evidence in weeks 19 through 31 (Lui, 2000).

18.10.3 Strategic Management

The model serves as a valuable tool for strategic planning and longer-term protection of individual spawning groups by exploring alternative annual herring TAC fishing policies. The model explores the impact of projected catch levels by adopting observed patterns of historical catch observations by the fleet. This simulates fishing seasons and the impact on expected stock assessment estimates in the different spawning groups. For a given projection year, modelled strategic catches are determined for each spawning group at alternate TAC strategies, e.g. 100 thousand tonnes, and 75 thousand tonnes, and assuming a rescaled pattern of actual catches as occurred spatially and temporally throughout a year, e.g. 1998. Following actual results, the model predicts catches for the spawning groups, e.g. Lurcher, Trinity and Seal Island. Alternatively, 'normal' estimated abundance for the same projection year based on the same alternate TAC strategies of 100 thousand tonnes and 75 thousand tonnes, would demonstrate the ability of the fleet to catch weekly amounts corresponding to historical stock availability. If at the lower TAC strategy projection, effort can be considered 'normal', then the estimated abundance is determined to be lower overall as well. As before, comparing this information to actual catches allows managers to anticipate the status of the stock on the repeated, updated catch measures during the spatial–temporal operation of the ongoing season.

18.11 DISCUSSION

The spatial–temporal model presented here is meant to reaffirm the need for continued fish stock assessment despite the potential shift in DFO priories in recent times. Opportunities to include all decision-makers to meet the expressed DFO desire of improved management sharing and stewardship (DFO, 2004a, 2002) and stream-lined costing via participatory repeated measures provides a viable alternative to aggregated VPA and groundfish-style estimates while encouraging the direct use of what has been referred to as 'anecdotal' or 'ancillary' information from fishermen regarding stock spatial-temporal information. The results also show that incorporating stock life cycle (substock migration information) in the spatial–temporal model improves on the VPA estimation and therefore, the resulting fish policy management. The approach represents the need to capture the fishery as a more complete spatially and temporally defined system that encompasses joint human and biological activity. It is recognised that disenfranchising or ignoring any part of the fisheries system will imperil the potential benefits of real shared management and required stewardship of the fish.

The model evaluates stock impacts of alternative fisheries management policy decisions throughout its analysis. As well, commonly applied limitations on catching times and catching locations can be analysed directly including determining measurable effects due to 'area closures' or 'small fish protocol' regulations on fish abundance and stock exploitation.

Modelling operational and strategic management issues in 4VWX herring permits fishery observations from multiple sources for interpreting valuable repeated spatial–temporal in-season stock information. These observation data include procedures to link the model explicitly to the considerable 'ancillary' and 'anecdotal' input from fishermen's observations about the ecosystem as well as their recorded catch data. Such information includes the incidence of major herring predators (e.g. whales) that would provide positive evidence of herring. Qualitative measures could also be used to adjust the model 'observation level' to reflect all combined, regular information about the stock in each zone and period. As well, the spatial-temporal model presented here directly incorporates biologically known information on stock life cycle and migration patterns that to date have not been included directly in stock assessment analyses (Stephenson and Lane, 1994).

The development of the cost-effective model adds to the sparse field of spatial–temporal modelling for marine resources assessment and management on both Canada's east and west coasts. The decision-aid software illustrated here for applications in the Scotia-Fundy herring fishery represents a contribution toward assisting decision makers in the in-season co-operative management of this fishery. Moreover, the model requires data on fish biology and life cycle dynamics. These data are the precise expertise of fisheries scientists and as such are more directly pertinent to their knowledge base than the second-hand statistical requirements of traditional VPA numerical methods that do not accommodate aspects of fish biology. The general framework of the spatial–temporal modelling exercise can be adapted to other marine resources, including taking into account biological knowledge about the seasonal dynamics of groundfish stocks, locations of known spawning groups, and historical in-season exploitation patterns on the fishing grounds. The current analysis suggests that a feasible stock estimation procedure incorporating aspects of spatial–temporal fish biology is less costly and more important than the cost of annual research vessel surveys and the extensive aging of samples and poorly understood numerical aggregate population analyses associated with traditional VPA. Finally, the intuitive nature of the proposed spatial–temporal model rebalances management decision making by attributing knowledge of fish science to stock estimation, and by implicating fishermen directly in the observation process leading to increased co-management and enhanced stewardship of the resource.

Acknowledgements

Herring purse seiner catch data from logbooks are compiled by the Department of Fisheries and Oceans Canada at the St. Andrews Biological Station. The use of these data is acknowledged with thanks. The author acknowledges with much appreciation the assistance and collegiality of the St. Andrews Biological Station Staff: Rob Stephenson, Gary Melvin, Mike Power and Jack Fife. This chapter was supported in part by Natural Sciences and Engineering Research Council (NSERC) of Canada Operating Grant OGP0122822, and from the Social Sciences and Humanities Research Council (SSHRC) of Canada strategic grant project on *Fisheries Uncertainty and the Precautionary Approach*.

I am grateful for the opportunity to celebrate the work and achievements of Professor Gordon Munro who I can proudly say had and continues to have a profound impact on my own academic and applied understanding of the betterment of valuable marine resources that began during my tutelage with him as a graduate student at the University of British Columbia.

References

Bertsekas, D.P. 1976. *Dynamic Programming and Stochastic Control*. Academic Press, New York.
Clark, C.W., F.H. Clarke and G.R. Munro. 1979. The optimal management of renewable resource stocks: Problems of irreversible investment. *Econometrica* 47: 25-47.
Clark, D. and J. Hinze. 2003. Assessment of cod in Division 4X in 2003. Canadian Science Advisory Secretariat, *CSAS Res. Doc.* 2003/115.
Deng, X. 2000. *A comparative analysis of fish stock assessments methods: Spatial-temporal versus VPA*. MSc thesis in Systems Science, University of Ottawa.
DFO. 2002. *Canadian Waters*. Ocean Management, Department of Fisheries and Oceans (DFO), Ottawa. Accessible online http://www.dfo-mpo.gc.ca/canwaters-euaxcan/oceans/index_e.asp
DFO. 2003a. Cod on the southern Scotian Shelf and Bay of Fundy (Division 4X/5Y) in 2003. *DFO Canadian Science Advisory Secretariat Stock Status Report*. 2003/050. DFO, Ottawa.

DFO. 2003b. Haddock in the southern Scotian Shelf and Bay of Fundy (Division 4X/5Y). *DFO Canadian Science Advisory Secretariat Stock Status Report.* 2003/051. DFO, Ottawa.

DFO. 2004a. *Fisheries and Oceans Sustainable Development Strategy.* Progress Report on 2001-2003 Strategy. Communications Branch, DFO, Ottawa.

DFO. 2004b. *Aquatic Science 2020: Workshop Report. Montreal Quebec, May 6-8, 2003. DFO/2004-155.* Accessible online http://www.dfo-mpo.gc.ca/science/aquatic_2020/aquaticscience2020_e.pdf.

DFO. 2004c. Cod on the southern Gulf of St. Lawrence. *DFO Science Status Report.* 2004/003. DFO, Ottawa.

FRCC. 2001. *2001 Conservation Requirements for Scotian Shelf and Bay of Fundy Groundfish Stocks and Redfish Stocks.* Fisheries Resource Conservation Council (FRCC), Ottawa.

FRCC. 2002. *2002/2003 Conservation Requirements for Groundfish Stocks on the Scotian Shelf and in the Bay of Fundy (4VWX), in Sub-Areas 0, 2+3 and Redfish Stocks.* FRCC, Ottawa.

FRCC. 2003. *2003/2004 Conservation Requirements for Groundfish Stocks on the Scotian Shelf and in the Bay of Fundy (4VWX5Z), in Sub-Areas 0,2+3 and Redfish Stocks.* FRCC, Ottawa.

Hilborn, R. and C.J. Walters. 1992. *Quantitative Fisheries Stock Assessment: Choice, Dynamics & Uncertainty.* Routledge, Chapman & Hall, New York.

Hurley, P., G.A.P. Black, J.E. Sumner, R.K. Mohn and P.A. Comeau. 2003. Assessment of the Status of Division 4X/5Y Haddock in 2003. *Canadian Science Advisory Secretariat, CSAS Res. Doc.* 2003/104. DFO, Ottawa.

Kurlansky, M. 2003. *Salt: World History.* Penguin Books.

Munro, G.R. 1979. The optimal management of transboundary renewable resources. *Canadian Journal of Economics* 12(3): 355-76.

Munro, G.R. and C.W. Clark. 2003. Fishing capacity and resource management objective. In S. Pascoe (ed.) *Technical Consultation on the Measurement of Fishing Capacity.* Food and Agriculture Organization of the United Nations (FAO), Rome.

Lane, D. 1989. A partially observable model of decision making by fishermen. *Operations Research* 37(2): 240-54.

Lane, D.E. and R.L. Stephenson. 1995. Fisheries management science: The framework to link biological, economic, and social objectives in fisheries management. *Aquatic Living Resources* 8: 215-21.

Lane, D., R. Stephenson, S. Storey, Y. Liang, Q. Lui, X. Deng. 2004. *Spatial–Temporal Analysis of the Scotia-Fundy Herring Fishery.* Working Paper. School of Management, University of Ottawa.

Lei, L. 2004. *A system simulation model of stock estimation methods: spatial–temporal and VPA.* MSc thesis in Systems Science, University of Ottawa.

Lui, Q. 2000. *Assessing the dynamic status of fish resources.* MSc thesis in Systems Science, University of Ottawa.

NOAA. 1998. *Ecosystem-based Fishery Management: A report to Congress by the Ecosystem Principles Advisory Panel.* National Oceanic and Atmospheric Administration (NOAA), Washington.

Power, M.J., R.L. Stephenson, K.J. Clark, F.J. Fife, G.D. Melvin and L.M. Annis. 2004. *Evaluation of 4VWX herring.* CAFSAC Research Document 2004/030. Canadian Atlantic Fisheries Scientific Advisory Committee (CAFSAC), Ottawa.

Rivard, D. 1988. Collected Papers on Stock Assessment Methods. CAFSAC Research Document 88/61. CAFSAC, Ottawa.

Sinclair, M. 1988. *Marine Populations: An essay on population regulation and speciation.* Washington Sea Grant Program, University of Washington Press, Seattle.

Sinclair, M. and T.D. Iles. 1985. Atlantic herring (*Clupea harengus*) distributions in the Gulf of Maine-Scotian Shelf area in relation to oceanographic features. *Canadian Journal of Fisheries and Aquatic Sciences* 42: 880-87.

Stephenson, R.L. 1990. Stock discreteness in Atlantic herring: A review of arguments for and against. In V. Wespestad, J. Collie, E. Collie (ed.) Proceedings of the International Herring Symposium, Anchorage, Alaska, October 23-25, 1990. (9th Lowell Wakefield Fisheries Symp.). University of Alaska, Fairbanks.

Stephenson, R.L. and D.E. Lane. 1994. Fisheries science in fisheries management: A plea for conceptual change. *Canadian Journal of Fisheries and Aquatic Sciences* 52: 2051-56.

Stephenson, R.L., D.E. Lane, D. Aldous and R. Nowak. 1993. Management of the 4WX Atlantic Herring (*Clupea harengus*) fishery: An evaluation of recent events. *Canadian Journal of Fisheries and Aquatic Sciences* 50: 2742-57.

Stephenson, R.L., D.E. Lane, D. Aldous and R. Nowak. 1998. A framework for risk analysis in fisheries decision-making. *ICES Journal of Marine Science* 55: 1-13.

Storey, S. 1997. *Spatial-temporal fish stock assessment.* MSc thesis in Systems Science, University of Ottawa.

Authors

S. Agnarson
University of Iceland

R. Arnason
University of Iceland

C. Batstone
Auckland University of Technology

T. Bjørndal
University of Portsmouth

A. Charles
Saint Mary's University

C.W. Clark
University of British Columbia

G. Cripe
Spokane Falls Technical College

M. Doyle
Iowa State University

D. Dupont
Brock University

P.V. Golubtsov
Moscow State University

D.V. Gordon
University of Calgary

R.Q. Grafton
The Australian National University

R. Hannesson
Norwegian School of Economics and Business Administration

D.D. Huppert
University of Washington

V. Kaitala
University of Helsinki,

S. Kobayashi
Nihon University

L.G. Kronbak
University of Southern Denmark

D.E. Lane
University of Ottawa

M. Lindroos
University of Helsinki

R. McKelvey
University of Montana

K.A. Miller
National Center for Atmospheric Research, Colorado

H.W. Nelson
University of British Columbia

G. Pontecorvo
Columbia University

W.E. Schrank
Memorial University of Newfoundland

A. Scott
University of British Columbia

B.M.H. Sharp
The University of Auckland

R. Singh
Iowa State University

U. Rashid Sumaila
University of British Columbia

B. Turris
Pacific Fisheries Management Incorporated

Q. Weninger
Iowa State University

Index

ad valorem royalty, 80, 82, 83, 84
Alaskan Pacific halibut fishery, 142, 152, 153
Anferova, E., 76, 85
anticipation, 87, 217, 218
Arnason R., 8, 9, 12, 14, 30, 31, 32, 58, 85, 86, 239
auctions, 26, 74, 75, 76, 77, 78, 80, 81, 83, 84, 85, 86

Baltic Sea cod, 193, 195
bargaining solution, 165, 169, 177, 196
Bering Sea, 22, 277
Bjørndal, T., 1, 2, 4, 105, 194, 205, 206, 219
blue whiting, 197, 205
buy-backs, 116, 117

Canada-United States Pacific Salmon Treaty, 9, 11, 220
capacity
  excess, 5, 87, 109, 113, 116, 134, 261, 279
  fishing, 109, 112, 113, 114, 131, 134, 136, 198, 273
capital
  adjustment costs, 137, 141, 145, 152, 153
  stuffing, 109, 116
capital theory, 1, 58, 117, 134
characteristic function, 191, 194, 195
Chile, 5, 76, 84, 86, 221
coalitions, 184, 185, 186, 187, 189, 190, 191, 192, 195
  grand, 186, 188, 189, 190, 191, 193
  restrictions, 192
cod, 17, 61, 63, 77, 88, 89, 90, 91, 92, 94, 96, 98, 99, 100, 101, 103, 193, 195, 196, 210, 244, 268, 271, 272, 273, 274, 275, 276, 277, 278, 279, 280, 281, 282, 285, 301
common knowledge, 164
common property, 28, 46, 47, 49, 58, 60, 62, 71, 118, 135, 159, 183, 194, 231, 236
community management, 277

competitive equilibrium, 202
conservation, 3, 18, 30, 60, 75, 119, 138, 195, 204, 205, 212, 220, 277
co-operative solutions, 190, 200, 205
costly capital adjustment, 138, 139, 153, 154
  adjustment of capital, 137
  adjustment of fishing, 137

differential games, 222, 232
dimensions of property rights, 33, 41, 53, 54, 55
diminishing marginal productivity, 137
diminishing returns, 138
Doubleday, W.G., 278, 280
dynamic
  optimisation, 120, 123, 124, 128, 129, 131, 137
  programming, 52, 127, 142

efficiency, 32, 33, 37, 42, 51, 54, 55, 56, 60, 72, 83, 188, 261
elasticity
  demand, 143
  intensity, 94, 96, 98, 99
enforcement, 25, 58, 66, 74, 76, 77, 89, 189, 194, 195, 277
Engesæter, S., 196, 197, 206
equity, 60, 75, 128, 212, 213, 214
escapement, 137, 138, 139, 140, 141, 145, 146, 147, 148, 149, 150, 151, 152, 155, 161, 166, 169, 199, 202
  constant, 137, 149, 150
  policy, 137, 146, 149, 150, 152
Estonia, 76, 84
European Union, 196, 272
exclusive economic zone, 3, 4, 62, 185, 196, 197, 206, 207, 215
exclusivity, 23, 24, 25, 26, 27, 28, 29, 34, 37, 41, 44, 45, 46, 48, 54, 55
extinction, 135, 200, 203, 206, 213, 272, 278

$F_{0.1}$, 271, 274, 275
Fisheries and Oceans Canada, 72, 279, 280, 301

fisheries management, 1, 2, 3, 4, 5, 55, 56, 58, 60, 61, 71, 86, 133, 137, 138, 139, 142, 149, 153, 182, 184, 185, 188, 189, 194, 219, 257, 261, 262, 268, 269, 270, 271, 272, 277, 281, 300, 302, 303
fishing effort, 25, 62, 68, 87, 98, 99, 110, 124, 125, 134, 202, 268, 271, 272, 275, 278, 287, 291, 294, 295
fishing quota, 25, 26, 27, 74, 242
flexible functional form
   quadratic, 93, 96, 105, 222, 224, 232
   translog, 258, 259
Food and Agriculture Organization, 4, 5, 9, 12, 13, 58, 72, 85, 86, 135, 138, 139, 154, 207, 217, 219, 220, 221, 270, 274, 277, 280, 282, 302
full co-operation, 186, 187, 188, 189, 190, 192, 193

game theory, 3, 4, 128, 184, 185, 190, 195, 196, 221
geoduck, 77, 78, 79, 82
Gordon, D.V., 6, 11, 105, 206, 282
Gordon, H. S., 58, 118
Gorte, R.W., 76, 78, 85
Grafton, R.Q., 72, 85, 105
growth
   compensatory, 161, 166, 169
   depensatory, 161, 166, 169
   function, 38, 39, 50, 140, 145, 153, 222, 223, 224, 225, 232
   stock, 137, 138, 139, 140, 145, 148, 150, 153, 199

Hamilton-Jacobi-Bellman equation, 233
Hamre, J., 197, 206
Hannesson, R., 14, 85, 86, 182, 206
high seas
   fisheries, 4, 183, 185, 194, 195, 205
   management, 4
Hoagland, P., 270, 271, 274, 282

IFQs, 74, 75, 76, 77, 80, 81, 82, 83, 84
   auction, 75, 80, 81, 82
incentive compatibility, 196, 197, 204

Individual Transferable Quotas, 72, 73, 261
   ITQ, 26, 27, 29, 34, 55, 58, 59, 60, 62, 64, 65, 66, 67, 68, 69, 70, 71, 85, 86, 87, 88, 89, 90, 91, 104, 105, 243, 261, 262, 268, 269, 275, 277, 278
information
   asymmetric, 177
   knowledge, 178
   private, 165, 169, 176, 178
   transparent, 169, 176, 178
international fisheries agreements, 184

Jan Mayen, 197

Kaitala, V., 3, 10, 11, 12, 135, 184, 185, 187, 188, 189, 190, 191, 193, 194, 195, 206, 222, 236

Law of the Sea, 3, 4, 6, 7, 8, 9, 30, 185, 195, 207, 221, 277, 280, 282
limited duration property rights, 51
Lindroos, M., 12, 194, 195, 205, 206
lump sum fee, 74

mackerel, 17, 192, 195, 197, 205
marine protected reserves, 277
Markov perfect equilibrium, 222, 223, 224, 225, 226, 229, 231, 232, 235
Matthiasson, T., 75, 86
measurement precision, 164, 165, 169
Milgrom, P., 81, 83, 86
model
   deterministic, 124, 198
   split-stream, 160, 173, 181, 182, 216
   stochastic, 126, 131, 133, 134, 137, 138, 140, 149, 153, 154, 155, 159, 160, 161, 166, 173, 181, 182, 183, 271, 272, 283
Morgan, G.R., 75, 86
Munro, G.R., 1, 29, 110, 119, 132, 154, 185, 196, 235, 283, 301

Nash
   bargaining solution, 165, 169
   equilibrium, 166, 186, 192, 226, 227, 228, 229, 232

new member problem, 188
New Zealand, 25, 26, 55, 56, 59, 85, 257, 261, 264, 265, 267, 268, 269, 271, 278, 281, 282
Newfoundland, 6, 21, 273, 275, 276, 279, 280, 282
non-co-operation, 186, 190, 223, 232
non-co-operative solution, 187, 198, 200, 201, 202, 203, 205
non-malleable capital, 2, 124, 128, 132
North Sea, 196, 197, 205, 206, 210, 272, 281
North Sea herring, 196, 197, 205
north-east Atlantic, 21, 22, 195, 197, 205, 219
Norway, 5, 13, 55, 56, 187, 191, 196, 197, 207, 214, 215, 219, 220, 221
Norwegian spring spawning herring, 185, 187, 192, 197, 205, 210, 214, 216, 219, 284

optimal dynamic response, 150, 151, 152
orange roughy, 271, 278
organisation, 58, 182, 188, 189, 218
overcapacity, 109, 112, 113, 114, 115
overfishing, 18, 21, 22, 109, 112, 113, 210, 215, 270, 271, 272, 273, 274, 275, 278

Pacific halibut fishery, 30, 141, 142, 152, 153, 154
partial co-operation, 186
Peña-Torres, J., 77, 86
permanence, 37, 48, 50, 55
plaice, 196
precautionary approach, 63, 219
productivity, 23, 26, 53, 111, 129, 130, 137, 159, 209, 212, 257, 269, 275
profit charge, 80, 81
profit function, 37, 38, 39, 44, 51, 53, 54, 88, 92, 93, 95, 96, 97, 105, 106, 122
property rights, 18, 24, 25, 26, 32, 33, 34, 35, 37, 40, 41, 42, 44, 45, 49, 50, 51, 52, 53, 54, 55, 56, 58, 105, 279

property rights map, 35
property rights quality, 37, 55, 56

Q-measure of property rights, 37
quality of quota, 56
quota rental charge, 80, 83, 84
quota value, 70, 71, 85
    shadow value, 39, 40, 44, 46, 48, 87, 88, 92, 94, 95, 96, 99, 103, 104

regional fisheries management, 184, 185, 188, 189, 194
renewable natural resources, 222, 232
rent
    capture, 66, 74, 75, 80, 81, 82, 84
    resource, 2, 40, 46, 49, 60, 74, 75, 78, 80, 81, 84, 86, 137, 152, 207, 244, 245
resource investment rule, 2
risk
    management, 218, 219
    neutral, 165, 166, 181
Rosenberg, A.A., 271, 273, 275, 276, 279, 281
Russ, G.R., 270, 273, 282

saithe, 196, 210
Scott, A., 7, 30, 31, 58
security, 23, 24, 26, 27, 34, 37, 41, 42, 43, 44, 54, 55
Shapley value, 190, 191, 192, 193, 195
shared stocks, 202, 206, 217
Sissenwine, M., 86, 271, 272, 273, 282
sole owner, 1, 58, 122, 124, 183, 202
sprat, 76, 196
stability of coalitions, 186
Steele, J., 270, 271, 274, 282
stochastic stock growth, 137, 138, 153
stock assessment, 63, 67, 69, 71, 75, 268, 271, 273, 274, 276, 278, 283, 284, 285, 286, 289, 299, 300, 303
Sumaila, U. Rashid, 5, 13, 14, 118, 219, 281

307

taxation, 44, 45, 46, 55, 58, 128, 223, 231, 232
technological change, 260, 264
technology, 18, 19, 21, 22, 25, 37, 38, 41, 51, 60, 69, 75, 106, 120, 140, 214, 259, 264, 269
threat point, 169, 196, 216, 217
timber sales, 78, 79
total allowable catch, 28, 59, 62, 64, 65, 67, 68, 69, 76, 77, 81, 82, 84, 87, 89, 97, 99, 104, 109, 113, 114, 116, 130, 196, 215, 216, 261, 270, 272, 275, 279, 299
tragedy of the commons, 30, 58, 181, 190
transferability, 23, 24, 26, 28, 35, 37, 41, 54, 55, 64, 65, 71

uncertainty
  scientific, 274, 278
  stock, 154
United Nations, 4, 12, 58, 135, 184, 195, 206, 207, 221, 277, 280, 282, 302
University of British Columbia, 1, 6, 9, 10, 132, 154, 301

value function iteration, 142, 145
Vetemaa, M., 76, 85, 86
veto-coalitions, 191
Vislie, J., 196, 206

Washington State, 76, 77, 78, 79, 84
whiting, 196, 197, 205, 210

Zeller, D., 270, 273, 282
zonal attachment principle, 196, 197, 198, 202, 204, 205